E-SYSTEMS FOR THE 21ST CENTURY

Concept, Developments, and Applications

Volume 2
E-Learning, E-Maintenance, E-Portfolio,
E-System, and E-Voting

E-SYSTEMS FOR THE 21ST CENTURY

Concept, Developments, and Applications

Volume 2
E-Learning, E-Maintenance, E-Portfolio, E-System, and E-Voting

Edited by
Seifedine Kadry, PhD
American University of the Middle East, Kuwait

Abdelkhalak El Hami, PhD
INSA of Rouen, France

Apple Academic Press Inc. | Apple Academic Press Inc.
3333 Mistwell Crescent | 9 Spinnaker Way
Oakville, ON L6L 0A2 | Waretown, NJ 08758
Canada | USA

©2016 by Apple Academic Press, Inc.
Exclusive worldwide distribution by CRC Press, a member of Taylor & Francis Group
No claim to original U.S. Government works
Printed in the United States of America on acid-free paper

International Standard Book Number-13: 978-1-77188-266-8 (Hardcover)

International Standard Book Number-13: 978-1-77188-264-4 (eBook)

Library and Archives Canada Cataloguing in Publication

E-systems for the 21st century : concept, developments, and applications.

Includes bibliographical references and indexes.
Contents: Volume 2. E-Learning, e-maintenance, e-portfolio, e-system, and e-voting / edited by Seifedine Kadry, PhD, American University of the Middle East, Kuwait, Abdelkhalak El Hami, PhD, INSA of Rouen, France.

Issued in print and electronic formats.
ISBN 978-1-77188-266-8 (volume 2 : hardcover).--ISBN 978-1-77188-255-2 (set : hardcover).--ISBN 978-1-77188-264-4 (volume 2 : pdf)
1. Computer networks. 2. Computer systems. 3. Electronic systems.
I. Kadry, Seifedine, 1977-, editor II. El Hami, Abdelkhalak, editor
TK5105.5.E89 2016   004.6   C2016-901696-X   C2016-901697-8

CIP data on file with US Library of Congress

Apple Academic Press also publishes its books in a variety of electronic formats. Some content that appears in print may not be available in electronic format. For information about Apple Academic Press products, visit our website at **www.appleacademicpress.com** and the CRC Press website at **www.crcpress.com**

CONTENTS

LIST OF CONTRIBUTORS

Mourine Achieng
Cape Peninsula University of Technology, Cape Town, South Africa

Ebtisam Al-Harithi
Southampton University, London – UK

Mohammed Zuhair Al-Taie
UTM Big Data Centre and Faculty of Computing, Universiti Teknologi Malaysia (UTM), Skudai, 81310 Johor, Malaysia

Rasim Alguliyev
Director of the Institute of Information Technology of ANAS, Baku, Azerbaijan

Linus Atorf
Institute for Man-Machine Interaction, RWTH Aachen University, Germany

Ramatsetse Boitumelo
Department of Industrial Engineering, Tshwane University of Technology, South Africa

María Del Carmen Caba-Pérez
Department of Economía y Empresa Universidad de Almería, Carretera de Sacromonte s/n 04120 Almería (Spain)

Te Fu Chen
Department of Business Administration, Lunghwa University of Science and Technology, Taoyuan Taiwan

Robert Costello
Graduate School University of Hull, UK

P. Devika
Department of Communication, PSG College of Arts and Science, Coimbatore- 641014, Tamil Nadu, India

Abdelkhalak El Hami
National Institute of Applied Sciences, Rouen, FRANCE

Seyyed Muhamad Mutallebi Esfidvajani
Entrepreneurship Department, University of Tehran, Tehran, Iran

Leila Esmaeili
Computer Engineering and Information Technology Department, Amirkabir University of Technology, Tehran, Iran

Harry Fulgencio
Leiden Institute of Advanced Computer Science, Niels Bohrweg 1 Leiden, 2333CA, The Netherlands

María del Mar Gálvez-Rodríguez
University of Almeria, Spain

Seyyed Alireza Hashemi Golpayegani
Computer Engineering and Information Technology Department, Amirkabir University of Technology, Tehran, Iran

Norman Gwangwava
Department of Industrial and Manufacturing Engineering, National University of Science and Technology, Bulawayo, Zimbabwe

Maryam Haghshenas
Media Management, University of Tehran, Iran

Nurussobah Hussin
Faculty of Information Management, Universiti Teknologi MARA (UiTM) Malaysia

Seifedine Kadry
American University of the Middle East, Kuwait

Adeyeri Michael Kanisuru
Department of Industrial Engineering, Tshwane University of Technology, South Africa

Mpofu Khumbulani
Department of Industrial Engineering, Tshwane University of Technology, South Africa

Shahla Mardani
Computer Engineering and Information Technology Department, Amirkabir University of Technology, Tehran, Iran

Shahla Mardani
Computer Engineering and Information Technology Department, Amirkabir University of Technology, Tehran, Iran

Nasim Matar
Applied Science University, Amman – Jordan

N. Mathiyalagan
Department of Communication, PSG College of Arts and Science, Coimbatore- 641014, Tamil Nadu, India

Chi Man Mui
Department of Information Technology, Chinese YMCA College, Hong Kong SAR, China

Alaa Amir Najim
Mathematical Department, Science College, Basrah University, Basrah, Iraq

Safa Amir Najim
Computer Science Department, Science College, Basrah University, Basrah, Iraq

Mojtaba Nassiriyar
IT Management, University of Tehran, Iran

Makinde Olasumbo
Department of Industrial Engineering, Tshwane University of Technology, South Africa

Rachid Oumlil
Department of Management, ENCG-Agadir, Ibnou Zohr

Ashis Pani

Information Systems Area, XLRI Jamshedpur, Jharkhand, India

Malte Rast
Institute for Man-Machine Interaction, RWTH Aachen University, Germany

Jürgen Rossmann
Institute for Man-Machine Interaction, RWTH Aachen University, Germany

Ephias Ruhode
Cape Peninsula University of Technology, Cape Town, South Africa

Abouzar Sadeghzadeh
Electronics & Telecoms Engineering, University of Bradford, UK

Alejandro Sáez-Martín
University of Almeria, Spain

Wan Satirah Wan Mohd Saman
School of Information Management, Universiti Teknologi MARA, Malaysia

Michael Schluse
Institute for Man-Machine Interaction, RWTH Aachen University, Germany

Roghayeh Shahbazi
IT Management, Alzahra University, Tehran, Iran

Hamid Reza Shahriari
Computer Engineering and Information Technology Department, Amirkabir University of Technology, Tehran, Iran

Siti Mariyam Shamsuddin
UTM Big Data Centre and Faculty of computing, Universiti Teknologi Malaysia (UTM), Skudai, 81310 Johor, Malaysia

Dalibor Stanimirovic
University of Ljubljana, Faculty of Administration, Gosarjeva ulica 5, 1000 Ljubljana, Slovenia

Rakesh Tiwari
SSP India Private Limited, Gurgaon, Haryana, India

Watcharapol Wiboolyasarin
Suan Dusit Rajabhat University, Bangkok, Thailand

Farhad Yusifov
Department of Information Society, Institute of Information Technology of ANAS, Baku, Azerbaijan

LIST OF ABBREVIATIONS

CD	Concept Drift
CG	Categorizing of Groups
EPSS	Electronic Performance Support Systems
LCMS	Learning Content Management Systems
LMS	Learning Management Systems
PEOU	Perceived Ease of Use
PU	Perceived Usefulness
TAM	Technology Acceptance Model
TRA	Theory of Reasoned Action
TSPE	Trustworthy Stochastic Proximity Embedding
UTAUT	Unified Theory of Acceptance and Use of Technology

PREFACE

E-based systems and computer networks are becoming standard practice across all sectors, including health, engineering, business, education, security, and citizen interaction with local and national government. They facilitate rapid and easy dissemination of information and data to assist service providers and end-users, offering existing and newly engineered services, products, and communication channels. Recent years have witnessed rising interest in these computerized systems and procedures, which exploit different forms of electronic media to offer effective and sophisticated solutions to a wide range of real-world applications.

With contributions from researchers and practitioners from around the world, this two-volume book discusses and reports on new and important developments in the field of e-systems, covering a wide range of current issues in the design, engineering, and adoption of e-systems. *E-Systems for the 21st Century: Concept, Developments and Applications* focuses on the use of e-systems in many areas of sectors of contemporary life, including commerce and business, learning and education, health care, government and law, voting, and service businesses.

The two-volume book offers comprehensive research and case studies addressing e-system use in health, business, education, security, and citizen interaction with local and national government. Several studies address the use of social networks in providing services as well as issues in maintenance and security of e-systems as well.

This collection will be valuable to researchers at universities and other institutions working in these fields, practitioners in the research and development departments in industry, and students conducting research in the areas of e-systems. The book can be used as an advanced reference for a course taught at the undergraduate and graduate-level in business and engineering schools as well.

Seifedine Kadry, PhD
Abdelkhalak El Hami, PhD

ABOUT THE EDITORS

Seifedine Kadry, PhD, has been an associate professor with the American University of the Middle East in Kuwait since 2010. He serves as editor-in-chief of the *Research Journal of Mathematics and Statistics* and the *ARPN Journal of Systems and Software*. He worked as head of the software support and analysis unit of First National Bank, where he designed and implemented the data warehouse and business intelligence. He has published several books and is the author of more than 50 papers on applied math, computer science, and stochastic systems in peer-reviewed journals. At present his research focuses on system prognostics, stochastic systems, and probability and reliability analysis. He received a PhD in computational and applied mathematics in 2007 from the Blaise Pascal University (Clermont-II) – Clermont-Ferrand in France.

Abdelkhalak El Hami, PhD, is a full professor at the National Institute of Applied Sciences in the Rouen area of France as well as Deputy Director and Head of LOFIMS Laboratory and mechanical chair of the National Conservatory of Arts and Crafts. He is the author of many articles in international journals and books on optimization and uncertainty software and has presented at many conferences. He is also an IEEE Senior Member as well as on the editorial boards of several journals. He has published several books and is the author of more than 500 papers published in international journals and conferences. He received a doctorate in engineering sciences from the University of Franche-Comté in France (1992). He received his Habilitation diploma to supervise research (HDR) in 2000.

INTRODUCTION

E-systems are an integral part of the increasingly prevalent complex, pervasive, embedded, and ubiquitous computing solutions that have been, or are being, developed.

Recent years have witnessed the rising interest in these computerized systems and procedures, which exploit different forms of electronic media in order to offer effective and sophisticated solutions to a wide range of real-world applications. Initially, the impetus toward e-systems uptake was prompted by convenience (for the customer) and cost savings (for the merchant) of e-commerce transactions. Subsequently, a rapidly growing number of government services and information sources are now available online, providing fast, reliable, and convenient global access to e-government. Innovation and research development continue to sustain this rapidly evolving area and are increasing its scope into many more areas, for which we adopt the term 'e-systems engineering.' Such developments have typically been reported on as part of cognate fields such as information and communications technology, computer science, systems engineering, and social science and engineering. The fundamental elements of e-systems engineering, based on shared standards, web services, service-oriented architecture (SOA), and distributed data, are changing the way computer and software systems are designed, architected, delivered, and consumed, arguably meriting e-systems engineering as a separate and self-contained field of study.

The developments and innovations of e-systems are becoming prevalent in many diverse domains including business, education, security, and governance. This usually involves the use and implementations of Internet technology to reproduce inter/intra organizational procedures and frameworks that are conveniently available for the end-user, offering existing and newly engineered services, products, and communication channels.

This book is a collection of chapters that could completely cover the huge coverage that e-systems encompass. The chapters are draw from different science fields like e-commerce, e-decision, e-government, e-health, social networks, e-learning, e-maintenance, e-portfolio, e-system and e-voting.

The book is organized into two volumes of 12 chapters each. Each volume is divided into five sections. Volume I sections are: e-commerce, e-decision, e-government, e-health, and social networks. Volume II sections are: e-learning, e-maintenance, e-portfolio, e-system, and e-voting.

PART 1:

E-LEARNING

CHAPTER 1

CREATIVE CRITICISM WRITING FOR E-WORLD: DESIGN FOR BLENDED CLASSROOM IN 21ST CENTURY

WATCHARAPOL WIBOOLYASARIN

Faculty of Humanities and Social Sciences, Suan Dusit University, Thailand, Fax: (66) 22418375, E-mail: watcharapol.wib@gmail.com

CONTENTS

1.1   INTRODUCTION

Comments on the public space are expressed by the emotions and feelings that occur after the read or recognition of content stories in society. Comments through the writing on the public space, especially the social media which are non-government organizations or any persons who scrutinize the words of the language from the author before publishing, when a reader or receiver with less emotional maturity may be amenable to or conflict with content awareness. To write for expressing opinions on any subject matter should be written on the logical principle called a criticism writing which is through the thinking process sequentially, through screening and evaluating carefully to be a rational, reliable, attributed to decide what to believe or what to should do, however, unless the criticism writing is comprised of creativity, author may also write this to rebuke or attack others on the logical reasoning. So the neutral comment wisely without bias must include a creative and a criticism writing or Creative criticism writing (CCW) that is most suitable for e-learners in 21st century.

    The essential for promoting and developing creative criticism writing to learners is the use of information and communication technology as an instructional media or a tool for exceedingly facilitating learning or improving abilities. In particular, the introduction of Social media as a tool to create new ways of learning can respond to the learners' expectation in enhancing their capabilities to learn different languages, work together to build awareness, develop knowledge and access to knowledge with unlimited ways. Learners can establish learning, curiosity, self-awareness, and social transformation literacy. The features of social media can be accessible to all learners without an access charge and can take advantages of social media without any restrictions. Anyone can create contents and be an editor as his needs. Learners highly took part in this process and became learning communities, freely share contents to the peer or others without time and place conditions, learning achievements are fulfilled by the application of a virtual classroom because both the synchronous and asynchronous communication between individual to individual or individual to group. It provides high flexibility for lecturers to integrate with the creative criticism writing ability, help bridge the gap of communication between the lecturer and the learners or among the learners by conveying their message or writing to the receivers via online networks defined as 70 percent, and present knowledge contents to receivers in the classroom defined as 30 percent, according to the proportion of blended learning. It can be said that communication through social media makes a sender or source material and a receiver change their roles into a participant in the communication process and change the meaning of the communication process from transmitting information to create a social reality or called Participatory communication which the lecturer provides communication channels and various participatory models through the use of social media to encourage learners to engage in different media and communication process, gives precedence to the decision and acknowledge consequences of the participation of learners in communication, no criticizing but try to improve, modify the existing thought.

An exception can be found in the fact that only communication will not drive instructional activities to be accomplished; it requires interaction process as a key part to help strengthen the learning effectiveness to learners by using the Round table writing technique to enhance the creative criticism writing ability. The round table writing technique applies the principle to produce plenty of ideas, enable all learners to express their ideas out as a circle and then the collected ideas will be further applied in creative criticism writing. Before moving on, it should be pointed out that round table writing technique is also a brainstorming strategy of cooperative learning considered as an advanced skill that requires the critical thinking ability and abstractness to come across learners.

1.2   CREATIVE CRITICISM WRITING ABILITY: CONCEPT FOR E-WORLD

The enhancement of creative criticism writing ability derived from the studies of the theoretical concepts and related researches, including contents of participatory communication, social media, round table writing technique, and creative criticism writing approach which provided more fair details as in the following subsections (Wiboolyasarin, 2012).

1.2.1   PARTICIPATORY COMMUNICATION

Participatory communication is defined as a two-way communication process of sharing information by entering into dialogues between the parties involved in development or receivers and the data source, mediated by development communicators as renewal facilitators. Both of them can switch their roles between sender and receiver or "Participants," free, and have equal access to the means of expressing their viewpoints, feelings, and experiences. Collective action aimed at promoting their interests, solving their problems and transforming their society, is the means to an end. Participatory communication process could be divided into five steps as Figure 1.1. In this chapter, all steps of participatory communication were selected as an umbrella term for the main concept.

1.2.2   SOCIAL MEDIA

Social media is a new set of tools, new technology that allows people to more efficiently connect and build relationships with each other. It is a channel for communicating between senders and receivers using software programs via Internet network and with web 2.0 technologies, allow users to create and share contents to others in all interaction level on virtual

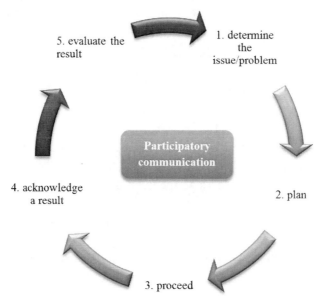

FIGURE 1.1   Process of participatory communication.

space. The entire world of social media could be classified into fifteen categories as follows; Social networking, Photo sharing, Audio, Video, Microblogging, Livecasting, Virtual worlds, Gaming, RSS and aggregators, Search, Mobile, Interpersonal. Some academicians (Mayfield, 2008: online; Sirichodok, Makaramani, Sangpum, 2012) proposed other categories of social media such as Wiki, Podcasts, and Online forum. Alongside that, we find in this chapter, social network, online forum, and wiki were selected as tools for enhancing creative criticism writing ability.

1.2.3   ROUND TABLE WRITING TECHNIQUE

Round table writing technique is a writing activity, which is a simple and effective skill-building technique to use when a high level of creativity and criticism are desired. The entire class or the smaller can participate to generate responses per one or many questions which mostly process of activities between round table writing and brainstorming are alike. Therefore, round table writing technique can be used in a brainstorming activity by organizing 6 to 10 learners per group, the leader begins the round table writing session by briefly stating the problem under consideration and should stress to the group that all idea need to be expressed. All group participants need to realize that achieving the highest possible quantity of suggestions is paramount. Participants then generate thoughts and collect information for solving the problem, a note taker writes down the states and problem of doing activity as well as all thoughts into whiteboard or paper (a pen and a paper per one group). The whole process should be less than one hour and started when the leader states questions or problem, then the first member writes a reply, says it out loud, and passes the paper to the left-handed side member, the second one does the same until the time is up, and the activity will be done. After that, choose top five ideas and decide the best idea or merge all into one. In this chapter, round table writing technique is applied with an online forum or electronic forum, instead of pen and paper as well as wiki for collaboration to create the group written works.

1.2.4 CREATIVE CRITICISM WRITING APPROACH

Creative criticism writing approach is blended critical writing and creative writing together and defined as a new high-leveled writing ability towards any subject matters derived from: (i) Studying, finding out, and compiling knowledge, (ii) Expressing any ideas, evidences, or arguments, (iii) Giving his opinions towards any subject matters, (iv) Passing on new knowledge, thoughts, experiences and imagines, and (v) Using his own various idiomatic wording.

1.3 CREATIVE CRITICISM WRITING INSTRUCTIONAL MODEL: DEVELOPMENT FOR BLENDED CLASSROOM

The creative criticism writing (CCW) instructional model for blended classroom was designed and developed based on the participatory communication with round table writing technique on social media or the theoretical concepts above. There are eight key elements, such as (i) purpose, (ii) content, (iii) instructional strategy, (iv) instructional media, (v) social media, (vi) learner, (vii) lecturer, and (viii) evaluation (Figure 1.2).

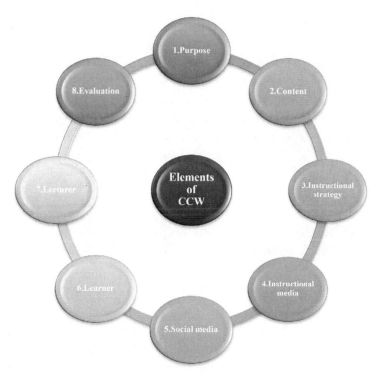

FIGURE 1.2   Elements of the blended creative criticism writing.

1.3.1   PURPOSE

Purpose is to intend targets of the instruction that when the instruction is finished, any behaviors of the learners will be changed. The defining purpose is to help the learners:

(a) Explain the principles of creative criticism writing; and

(b) Apply creative criticism writing towards any issues.

1.3.2   CONTENT

Content about the principles of creative criticism writing in the form of electronic media where learners can download from the web in order to self-studying and read in the classroom by offering concise contents that are supported with learning of learners who build new knowledge from creative criticism writing. It can be divided into definition, elements, process, plots, supporting evidences, strategies, and suggestions.

Creative criticism writing is considered as both art and science of writing, as a science that can be practiced, and when practicing repeatedly until proficiency, the written works would have been improved sequentially and good writings should use a variety of strategies to be an art, much more unique writing style especially creative criticism writing is the written works that the author should consider as a challenge, find arguments and ideas, proposals or practices further to their readers in order to present their concept to a new alternative which does not overlap with the original writings and not undermine or conflict with the other works.

1.3.2.1   Definition

Creative criticism writing is the written work with the inquiry, arrangement, and comparison of concepts, arguments, evidences and comments on any issues. They were assessed the reliability, judgment or summary with the reasons for writing and composing knowledge, experience, ideas and imagination with exotic unique idioms using the author's words, be rich and artistic authorship, demonstrate creative thinking clearly which do not require to be an academic writing report or article, but it can be a personalized written work, such as a daily journal or an essay.

1.3.2.2   Elements

No educator who has studied creative criticism writing determined its elements clearly; Wiboolyasarin (2013) has synthesized the elements of critical writing and creative writing to define into the elements of creative criticism writing as follows:

1. Understand of topic or concept of the issues.
2. Determination of plots in accordance with a topic or support the main idea of the issues.

3. Conjecture of question, controversial issues, beforehand problem from the readers, and the way to fix the gap of writing.
4. Transmission of attitude, emotion, feeling, and connotation in writing.
5. Imagination for readers to visualize in their mind and amenable attitude or idea of the author.
6. Use of the author's language by choosing words, idioms, phrases, sentences, and the placement of all above correctly as an orthography. Do not use sarcasm, sarcastic and biting others or any written works.
7. Mental and intellectual worth; not only the emotional fun, but also edify the reader's ethics. In their ethics was feeling pretty good, beneficial to themselves and the society around them.

1.3.2.3 Process

Mostly study results of creative criticism writing are brought to apply know-how in literary circles. No educator has determined the process of creative criticism writing clearly so the author must synthesize key concepts that many well-known academicians have studied the processes of critical writing (Decaroli, 1973; Ennis, 1985; Quellmalz, 1985) and creative writing (Divito, 1971; Furner, 1973; Osborn, 1963; Torrance, 1989; Wallach et al., 1962) which can be summarized as follows.

Step 1: identify the issue. Lecturer shows issue/problem for learners to perceive, recognize, and locate some causes of the problem or assigns subject or issue in each case for writing.

Step 2: gather information and consider the reliability of resources. Learners collect information from various resources to find solutions along with the reliability, accuracy, and adequacy of the information for writing.

Step 3: hypothesize. Learners handle with the information obtained from Step 2 to consider and identify relationships in deep details. Then, set a postulate to solve the problem for using as a reference guide of unclear and vague issues or information, or identify possible approaches from the available data.

Step 4: rekindling an idea/thinking. The data is also split into sections, scattering, and not winding up into ideas. It may take a while to let it go and wait for the learners to take the opportunity to judge the evidence construed in accordance with the facts or existing issues, then lead to the inference using inductive and deductive reasoning to find an alternative or an answer key.

Step 5: create imagination and write. Learners imagine and write freely without defining exactly what to write will be appropriate to solve the problem.

Step 6: found clear answers or ideas. The scattered thoughts or imaginations that learners created were linked together, vision in mind that the problem should be solved by any methods, but still cannot believe these ideas were right and appropriate.

Step 7: test, evaluate and accept. The answers or ideas in step 6 were tested or evaluated with hypothesis, proved the fact that their own ideas were accurate and fit, then accepted to implement by using the criteria in determining the reasonableness of a summary or answer key.

1.3.2.4 Plots

Creative criticism writing is a writing form for higher education. Rationally disputed is the heart of this writing, the writer need to understand in creative criticism writing and assessment criteria by asking the lecturer about the defined criteria before writing, conceive the key features of each assessment item will help determine the approach to the assessment. Priority should be placed on the good writing structure, relation with the ideas for the issue and reasoned arguments appropriately. The writing style is required official that includes the structure of the composition as follows.

1. Introduction is like the navigation or location/point on the road (write in chronological order, cause and effect) is to arrange the necessary contexts without mention of important contents placed on the first part of the writing, open act, attract attention, regardless of the issue they are writing. The art of long-winded language writing does not mention anything about the story, not repeat the conclusion or ending.

2. Main body is like the adventure along with the order of cause and effect, as defined in the introduction. The argument explained the supporting evidences and examples of reasoning is an important and longest writing consisted of knowledge, ideas, and information that the author researched and edited as a systematic procedure to expand the topics in the issue laid out in advance. In writing may depict or describe examples of a various stylistic lift or there may be several paragraphs to include the paragraph was written to be clear, the reader can understand the intention of the author as well. This is called a functional visual essence, each paragraph must only have a single main body relevantly called a unity, and the contents of each paragraph must be correlated by the forward paragraph will relate with the previous paragraph, called a semantic.

3. Conclusion is like the arrival destination, the mirror reflecting the expanded introduction to some like more similar distributions in place of arrival, cause and effect is a combination of them all and writes in order with the reasoning. Each paragraph on the story, the writer will leave with the impression, consistent with the story. The concise writing has several ways, such as leaving viewpoints impressed readers outset of the story, soliciting, encouraging the reader to follow, questioning readers to figure out the answer by themselves, and quoting proverbs, quotes, or impressive poems. It can be a short written (length should be equal to the introduction) summarized by the petition, invitation, or commendation. Avoid apology or disclaimer that the writer did not know and should not intrude into other issues.

Conclusion combines the reasoning and evidences that may be more than one conclusion, but do not have issues, concepts, and further evidences that are the argument's core into the conclusion. It should be placed in the main body.

Creative criticism writing consists of writing structures; that is, introduction, main body, and conclusion. It would have evidences, and writer's arguments emphasized on observing, hearing, connecting relations, criticizing, and concluding. The selection of writing activities should come from the level of cognitive differences, which the lecturer must firstly determine prior knowledge and experiences of the learners, then prepare the materials or activities and the advance answers to the questions asked by the learners, and assess the progress of learners on a regular basis. Classroom atmosphere should be friendly, closely with the learners to motivate them to freely think, be one's self, keep an open mind and accept advices and criticisms of the lecturer, promote direct experiences and encourage being curious, need for more knowledge on their own, and develop criticism skills as well as correctly evaluate their works and others.

Three principles of creative criticism plot:

1. Each paragraph has one single concept in the same direction with the topic.
2. Relevance of content throughout the story must have put a good plot, sequencing paragraphs organized with appropriate compilers.
3. Complete the entire contents of the key sentences in each paragraph must be clear. Sentence extended weighs reliable, accurate facts to help highlight the key completed sentences.

1.3.2.5   Supporting Evidences

Supporting evidences that the writer used are the resources of creative criticism writing by referring to the appropriate contents, harmonious supporting evidences selection should come from sources of information. It is found that each data has provided information resources for different types of providers. Researching data requires users to select the appropriate information sources to apply. For example, to find the document about the ICJ verdict, user can use the university library, public library for searching for books, journals, articles, or newspapers. Users may choose to use specific library, such as legal library, but it can be used on demand; that is, the ICJ verdict document used in litigation cases and will be available in the National Archives and the most reliable because of its original, or if one wanted to learn the local newspaper, he should use a public library in each locality. To view photos, it may be obtained from central libraries of the university. If being a historical image, it should be used one from mass media.

For the reliability of information provided information sources, press institutes, and individuals will present information or resources qualified and evaluated before releasing. An exception to this would be the Internet, there is still problems in the reliability of public information and not responsible for screening any content before presenting, but its advantage is modern information, conveniently access is available for unlimited time and place, nevertheless, on the Internet, there is a number of useful and reliable as well but users need to be knowledgeable about and choose the right search engine and have the ability to evaluate information as well.

1.3.2.6   Strategies

The important part in helping to gain more interesting, following and impressed should consider the strategies for implementation of the other writer affected writings or issues, confused or made an appeal or not. The strategies for the implementation can be done in various ways, for example.

Good writing is a written communication to meet the primary objectives, able to convey ideas and emotions put into writing to express general principles that are follows.

1. Write the correct words, no any misspelled word so that readers can understand exactly what the writer wants.
2. Use words to meet with the meanings. Some words have multiple meanings in both well-meaning and connotation. The authors should study well about word usage before writing.
3. Use words as individual levels. Some words have various levels, according to the level of individuals, including low individuals, neutral and high.
4. Compile words into euphemistic sentences correctly. The author has a better understanding of sentence structures which has the following guidelines:

    • Write the correct sentence structure.
    • Not use a foreign phrase.
    • Not wordy

5. Study many types of literatures, and then write in the correct format. It also needs to learn to write correctly and clearly.
6. Review it. The writer should read for reviewing and check for updates to the elegance of the writing as well.

1.3.2.7 Suggestions

1. Use a simple language.

Simple language is a charming language for written with simple words and expressions which are what readers' want the most because they do not waste time pondering what sentence means.

2. Do not think that others know about it.

Another blind spot of writing that is the writer often forgets to think about his offer that readers already know it. He tends to write some words, or some sentences. These makes the readers do not understand the background.

3. Write easier to read.

Write easier to read and use simple language is different. Simple language does not mean readers always understand. The language usage is simple but if the contents or stories presented are beyond the perception of the reader, they will not be worth anything.

4. Think before writing.

Before writing, one thing that must be done first is to think.

5. Write in line with a target.

Write anything, do not miss what is written to coincide with the target intended by the writer to be the first. If the writing was in a roundabout way, readers would be annoyed or bothered enough to sit and listen to people talk aimlessly.

6. Use familiar words.

Write for communication should be familiar with everyday life, it is better by choosing polite words that everyone can understand immediately, no need to open a dictionary to find the meaning.

7. Write a concise sentence.

The writer will be required to seek a broad term of meaning used to cover most in their work as much as they could, like the sentence "Mostly cloudy this morning seems so clear." If they try to use less word like "the sky is clear," it is more comprehensive and compact.

8. Leave a space

Good written works have to leave some spaces or punctuations. The writer should study more in this special case.

9. Revise before publication.

After writing, the writer must read and review the written works again. Message or any expression that is not smooth should be edited before publishing.

10. Write meaningful paragraphs.

The writer will need to set up a paragraph to direct a point and meaning conscientiously. Do not enter a new one without rules.

1.3.3 INSTRUCTIONAL STRATEGY

Instructional strategy is a how-to-teach the learners to achieve learning objectives by using deductive approach (offline) that provides learners with an understanding of the principles of creative criticism writing. Then the sample writings are given to reflect a variety of creative criticism. Simultaneously, uses the round table writing technique (online) to let learners collect various ideas to be illustrative of creative criticism writing. The details are as follows:

- Deductive method (offline) is comprised of the following steps:

    - Describe principles of creative criticism writing to the learners through any presentation program such as MS PowerPoint, Keynote, Prezi, or, etc. Learners will receive handouts or else can be downloaded as electronic files to store on their own computer.
    - Give many examples of creative criticism written works that learners will be able to apply the knowledge from them.
    - Practice creative criticism writing by distributing a writing test to each learner for applying his or her knowledge to a new situation.

- – Analyze and discuss the principles of creative criticism writing in class together. Then measure and evaluate the learner's knowledge from the group written works and writing tests.
- • Round table writing technique (online) is comprised of the following steps:
  - – Post question for each group on the online forum (one post per one group) by defining the issue for round table writing and inform web links of each forum to the class on social network.
  - – Start giving the first comment by a learner acted as a leader (one learner per one comment per one round), then next one who is on left-hand side will give the second comment via his/her computer and do this until the time runs out or have enough various ideas.
  - – Ask the opinion of the members in a group by the leader to choose the top five ideas or one of the best. If members cannot decide to select any, lecturer should assist or facilitate to.
  - – Evaluate on the chosen idea of learner overall and give feedback.
  - – Bring all the chosen ideas to create a creative criticism written work through wiki (one group per one written work).

1.3.4   INSTRUCTIONAL MEDIA

Instructional media are tools used in the instruction process to facilitate learners to learn and achieve the objectives set out efficiently by using the following instructional media.

- • MS PowerPoint used for delivering the contents of creative criticism writing.
- • Documents of the presentation include the contents of creative criticism writing for distributing in the classroom and downloading as e-files from social network.

1.3.5   SOCIAL MEDIA

Social media is the use of online networks via the Internet for round table writing and creating a creative criticism written work and a channel of communication between the learners and the lecturer as well as the learners and the learners with using any web browser to enter the online forum, wiki, and social network websites.

- Online forum (ex. http://www.proboards.com) is the use of a discussion board for round table writing to gather the ideas of learners towards the important issues to be given as the title, an outline including introduction, main body, and conclusion, as well as the supporting evidences.
- Wiki (ex. http://www.pbworks.com) is the use of a private virtual space for creative criticism writing by collecting all data from the round table writing on the online forum to create a group written work of the members in each group.
- Social network (ex. http://www.facebook.com) is the media or channel of communication between the learners and the lecturer or among the learners. It is used as a spot to post web links of online forums in each group for learners to round table write and web links of wiki for learners to create a creative criticism written works after round table writing is completed. Also, it is important to use as a way to encourage students to use online forum and wiki to achieve defined objectives.

1.3.6   LEARNER

Learner plays a major role in the study of theoretical knowledge about principles of creative criticism writing and uses social media to create a creative criticism written work with round table writing on the online forum and write together on wiki to create a group written work. The details are as follows:

- attend an orientation to make a clear understand of instructional approach, how to measure and evaluate the course, practice to use social media, and segmentation by defining roles and responsibilities of each learner in a group.
- listen to or read the detailed contents from lecture notes to understand the contents of creative criticism writing.
- test to measure capabilities of creative criticism writing before learning with the blended creative criticism writing instructional model.
- round table write about identified issues to gather the ideas of each learner. Then consider the best important ideas and consistent up to five ideas.

- select the best single idea and evaluate it to the implement as a basic data for creative criticism writing.
- combine preliminary data from round table writing to create a group creative criticism written work by the process from 6.4 to 6.6 will take some time to run over a period of three weeks, each week would be round table writing in different issues assigned by lecturer.
- present creative criticism written works, then give creative criticism comments on the written works of other groups.
- consider the other groups' comments to improve one's creative criticism written work.
- receive assessment results of group creative criticism written works with the rubrics criterion.
- test to measure capabilities of creative criticism writing after learning with the creative criticism writing instructional model and test the individual ability of creative criticism writing.

1.3.7   LECTURER

Lecturer is a facilitator in preparing contents and instructional media to provide learners with an understanding of the theoretical principles of creative criticism writing and prepare social media to help learners continue to use them effectively. The details are given below:

- hold an orientation for learners to understand instructional approach, how to measure and evaluate the course, provide social media available and train learners practically and divide them into groups by defining roles and responsibilities of each learner in groups.
- lecture contents via MS PowerPoint and let learners read the lecture notes to better understand the contents of creative criticism writing.
- hand out capability measurement pre-tests of creative criticism writing before teaching with the creative criticism writing instructional model.
- define the key issues to providing learners with round table writing. Lecturer may be involved in the process for giving the necessary advices. The process will take some time to run over a period of three weeks, each week lecturer will assign a different issue.
- listen to the presentations or read the group creative criticism written works. Then share feedbacks with each group so that learners have the data to improve better creative criticism written works.

- make additional recommendations for each group creative criticism written works improvement.
- evaluate group creative criticism written works with the rubrics criterion.
- hand out capability measurement post-tests of creative criticism writing after teaching with the creative criticism writing instructional model and the individual ability tests of creative criticism writing.

The role of the learners and the lecturer are shown in Chart 1.1.

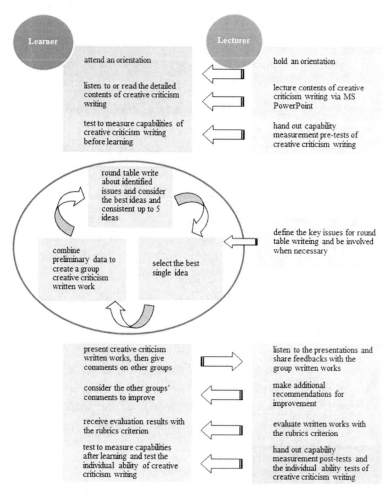

CHART 1.1 The role of the learner and the lecturer.

1.3.8   EVALUATION

Evaluation is a set of evaluation forms/tests for measuring knowledge of learners before and after learning; the individual ability test of creative criticism writing, the capability measurement pre-test and post-test of creative criticism writing, the group creative criticism written works assessment form, and the rubrics criterion for creative criticism written works and general written works detailed as follows.

The blended creative criticism writing instructional model consists of five stages (online and in class); each one takes about one week or less in the instruction process. Lecturer can merge sub-activities in each stage appropriately as detailed as follows.

Before starting the course, according to the blended creative criticism writing instructional model, lecturer should prepare the following:

1. create a group on social network such as Facebook Groups and gather learners in a class into the group.
2. create online forums and wikis for each sub-group and post all web links into Facebook Group.

Then, lecturer should do the following in classroom:

1. explain about the course, teaching and learning methods, and measurement and evaluation.
2. describe how to learn, rules for round table writing, and provide learners with documents. Then give an opportunity to ask questions.
3. divide learners into sub-groups (6–10 members) and set 30 minutes for a creative criticism writing ability pre-test (4–5 lines).
4. collect written works and score the ability of learners in each group.
5. hand out learning documents about the definition, elements, and process, plots, supporting evidences, strategies, and suggestions of creative criticism writing and lecture by presenting via MS PowerPoint. When learners are ready and understand all the data, stage 1 determine now begins online in both synchronous and asynchronous modes as follows.

Stage 1 Determine (Week 1 online in forums)

1. round table write (one person one idea per one round) to determine the title of written work along the topic that lecturer specified.

RUBRICS CRITERION FOR CREATIVE CRITICISM WRITTEN WORKS

Criterion	9–10 Excellent (A)	6–8 Good (B)	3–5 Fair (C)	0–2 Poor (F)
1. Knowledge arrangement acquired from the study (*Criticism writing*)	Respectively arrange knowledge as a paragraph hold to cause and effect principles smoothly	Arrange knowledge as a paragraph respectively hold to cause and effect principles smoothly but sometimes found 21–40% of problems	Arrange knowledge as a paragraph respectively hold to cause and effect principles a little and found 41–60% of problems	No knowledge as a paragraph with the confusion and disorganized of cause and effect principles
2. Concepts, proofs, arguments indication (*Criticism writing*)	Efficiently put forward concepts, proofs, arguments to support or refer the whole written work, these come from accurate and reliable information	Put forward concepts, proofs, arguments to support or refer some parts of written work, these come from accurate and reliable information	Put forward concepts, proofs, arguments to support or refer some parts of written work efficiently but these come from inaccurate and unreliable information	No concepts, proofs, arguments to support or refer in written work nor these come from inaccurate and unreliable information
3. Opinion expression toward any topics (*Criticism writing*)	Excellently show cause and effect in opinion expression for clear acquiescence from readers	Show cause and effect in opinion expression for acquiescence from readers at least 80% or more	Show cause and effect in opinion expression for a bit of acquiescence from readers and data may not correct	No cause and effect in opinion expression, or very little acquiescence and incorrect data
4. Novel and unique knowledge, thought, experience, and imagine transfer (*Creative writing*)	Smoothly composed of novel and unique knowledge, thought, experience, and imagine of author	Composed of quite novel knowledge, thought, experience, and imagine of author but less duplicated on other authors' works about 21–40%	Composed of knowledge, thought, experience, and imagine of author but moderately duplicated on other authors' works about 41–60%	No knowledge, thought, experience, and imagine of author and mostly duplicated on other authors' works about 61% or more

RUBRICS CRITERION FOR CREATIVE CRITICISM WRITTEN WORKS (CONTINUED)

Criterion	9–10 Excellent (A)	6–8 Good (B)	3–5 Fair (C)	0–2 Poor (F)
5. Numerous author's words and idioms usage *(Creative writing)*	– Cull varied words and as idioms of author.	– Cull words and as idioms of author consistently	– Cull words and as idioms of author to be less interesting	– No words are carefully culled as language of an author, or more extravagant than necessary
	– No words and expressions annoying others	– Appear little words and expressions annoyed others about 21–40%	– Appear moderate words and expressions annoyed others about 41–60%	– Appear words and expressions annoyed others throughout the story about 61% or more

RUBRICS FOR GENERAL WRITTEN WORKS

Criterion	4 Excellent (A)	3 Good (B)	2 Fair (C)	1 Poor (F)
1. Purpose	– Clearly present a new interesting purpose in accordance with the contents	– Present a new purpose but not in accordance with the contents	– Present an old purpose or sometimes irrelevant contents	– Not present any purpose or present unclear one
2. Title	– Properly title links to the contents, readers easily guess plots from the title	– Title links to the contents, readers reasonably guess plots from the title	– Title rather links to the contents, readers guess a bit plots from the title	– No title or the title is not linked to the contents or very weak link; reader cannot guess plots from the title
3. Introduction	– Effectively describe background of the story	– Describe background clearly enough but lack important details	– Describe background, but not link to the story	– No pattern of writing introduction, lack of clarity
4. Main body	– Properly present the author's views, clear and accurate	– Present the author's views	– Present the author's views but not clearly enough	– Not present the author's views, unclear and inaccurate
5. Conclusion	– Completely summarize concept of the story and focus on the main title linked to the plot	– Summarize key points but lack little 1–2 issues, but enough to see the perspectives	– Bring the plot to re-summarize, but lack major 3–4 issues, enough to the perspectives but not clear	– No plot summary or any core issues, may be somewhat but unclear and ineffective, invisible perspectives

RUBRICS FOR GENERAL WRITTEN WORKS (CONTINUED)

Criterion	4 Excellent (A)	3 Good (B)	2 Fair (C)	1 Poor (F)
6. Coherence	– Excellent link or support between concepts and purposes or arguments, reader understands the reasons of presentation	– Adequately link or support between concepts and purposes or arguments, reader understands the main reasons of presentation	– Link or support between concepts and purposes or arguments, reader quite understands the reasons of presentation	– Less or not link or support concepts and purposes or arguments, no reader understands the reasons of presentation or vaguely understands
7. Grammar	– No grammar error	– Only 1–2 grammar errors	– 3–4 grammar errors	– More than 4 grammars errors
8. Spelling and punctuation	– No punctuation or spelling error	– Only 1–2 punctuations or spelling errors	– About 3–4 punctuations or spelling errors	– More than 4 punctuations or spelling errors
9. APA reference and other reference formats	– Have APA references or other reference formats accorded with all items	– Have 1–2 errors in APA references or other reference formats, or accurately about 80% or more	– Have 3–4 errors in APA references or other reference formats, or accurately about 60–79%	– No APA references or other reference formats, or have more than 5 errors or accurately about 60% or less
10. Document design	– Create an interesting towards readers and perfectly design document composition	– Not interesting as it should be, but it shows the efforts in applying various techniques in designing document composition	– Not interesting in designing document composition but organize enough	– Poor and confusing to the readers because of the disorders of designing document composition

2. consider and select the top five titles.
3. choose the best single title.
4. assess the chosen title and give the title.

Stage 2 Plan (Week 2 online in forums)

1. round table write to set the outline (introduction, main body, and conclusion) and collect supporting evidences properly along the plot writing.
2. consider and select the top five outlines and score the supporting evidences.
3. choose the best single outline and sort supporting evidences from maximum to minimum scores.
4. assess the chosen outline and supporting evidences in accordance with the title considered the accuracy and clarity by lecturer.

Stage 3 Proceed (Week 2 online in forums)

1. round table write to specify the appropriate resources along with the plot writing.
2. check the reliability of the resources and consider their reasonable accuracy and clarity.
3. assess the resources based on the supporting evidences reviewed by lecturer.
4. consider and select the top five data sources.

Stage 4 Acknowledge (Week 3 online in forums and wikis)

1. round table write to collect idea organization methods, document designs, and writing strategies along the plot writing.
2. consider and select the good and suitable five idea organization methods, document designs, and writing strategies.
3. combine the best idea organization methods, document designs, and writing strategies into a written work.
4. assess the chosen idea organization methods, document designs, and writing strategies through contents in the written work.

Stage 5 Evaluation (Week 4 online in wikis and in class)

1. present the written works and share comments towards other group written works.

2. consider, revise, edit the written works to be the best.
3. assess the written works along the rubrics criterion for creative criticism written works.
4. set 30 minutes for a creative criticism writing ability post-test (4–5 lines).

Besides, the measurement and evaluation of creative criticism writing ability for undergraduate learners before and after instruction uses (a) a creative criticism written group ability evaluation form, (b) an individual creative criticism writing ability measurement form, (c) a creative criticism writing ability pre-test and post-test. The details are as follows:

1. Creative criticism written group ability evaluation form includes

   • Creative criticism written group evaluation
   • General written group evaluation

2. Individual creative criticism writing ability measurement form is tested by writing one's own creative and critical opinion towards the story (5–6 lines).
3. Creative criticism writing ability pre-test and post-test measures towards the news or story (4–5 lines).

The process of blended creative criticism writing instructional model can be presented in Figure 1.3.

The blended creative criticism writing instructional model has provided the activities that can be taught in 4–5 weeks, the first week of classes, lecturer provides the activities in step 1 determine and step 2 plan, in the second week lecturer provides the activities in step 3 proceed which allows learners to search for resources for writing, in the third and fourth weeks lecturer provides the activities in step 4 acknowledge. Learners may take one - two weeks to proceed at this step, depending on each group and in the fifth week will be an assessment of the creative criticism written works.

Educational institutions that desire to implement the blended creative criticism writing instructional model are required to prepare the tools and infrastructure needed for blended learning such as notebooks or tablets to meet the number of learners or have a computer room that can support the use of learner groups, network and high-speed wireless Internet. If the problem occurred, it was required outsource agencies or the administrator

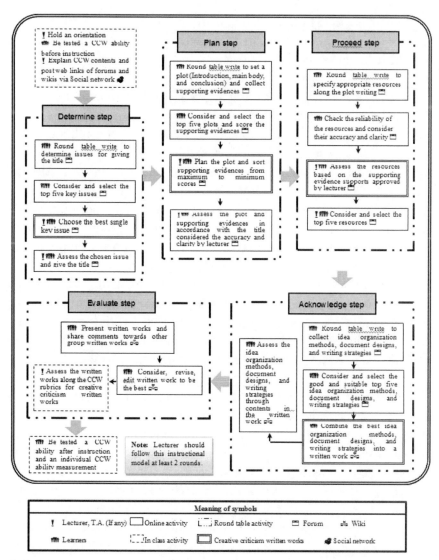

FIGURE 1.3   The blended creative criticism writing instructional model.

who can instantly fix and control problems such as, the Internet cannot be normally used, or the computer downs and cannot be fixed manually. Lecturer should develop the ability to use social media to learners before the implementation of the instructional model, in particular, the introduction to the social network, online forum, and wiki.

Moreover, this instructional model is suitable for social sciences courses or any courses that lecturer can set up a variety of issues, no conclusive answer, course content is conducive to higher-order thinking processes such as critical thinking, synthesis thinking such in Cross-cultural communication course which has studied theories related to inter-cultural communication, analyzing, interpreting, synthesizing factors affecting understanding Thai resulting from language and cultural differences as well as ways to solve problems. Lecturer should set as a topic for creative criticism writing, for examples, Thai culture or cultural issues by defining a broad topic that learners can classify as sub-topics for writing such as Chang Sip Mu (10 craft arts), Thai unique, Thai identity, beliefs in each part of Thailand, beliefs about the way of life in Thai society, linguistic and cultural change problem, Thai teen attires, cultural degeneracy in Thailand, etc. so lecturer should assign topics in accordance with the respective course. The course must be in the group of social sciences courses. Then study the course description to understand clearly before or explore course content to fit with creative criticism writing or not, so that learners begin to round table write to gather ideas on the topic that lecturer will clearly determine the one of each week in social network.

Enhancing creative criticism writing ability should be more two writing activities and take a minimum of four weeks or 28 days to develop learners to have the creative criticism writing ability in higher level. If learners used round table writing technique and create group written works more often as well as a lecturer participated in writing and gave feedbacks to the flaws in the writing, learners would learn ways to improve and develop the group creative criticism written works even further, which would affect the creative criticism writing ability to the next higher level.

The blended creative criticism writing instructional model is an activity required specific computer skills, so learners should be prepared and trained to provide them with these skills. In particular, the practice of using social media as defined by the model with fluency. Lecturer may have noticed the learner with high computer skills or technical abilities in the role as a teaching assistant or peer assistant because the communication between learner and learner is to communicate on the same level. Learners with less skill will be comfortable to seek assists from the learner together rather than seek from the lecturer. As a result, the teaching activities will be more effective and are very important conditions that should continue to prepare learners with high ability before starting the course and reinforce learners when he or she technically helps other learners.

1.4   CREATIVE CRITICISM WRITING ACTIVITY PLAN: APPLICATION FOR 21ST CENTURY LECTURER

Lecturers who implement this blended creative criticism writing instructional model should hold orientations to keep learners focus at the value and importance of creative criticism writing ability and its benefits towards self and society in order to reinforce a positive attitude on teaching the following model due to the implementation of the instructional model to effectively and efficiently achieve the purposes needs to obtain true cooperation and participation from learners within the groups and lecturer and when the learners are ready to learn, lecturer can proceed five stages as follows.

1.5   CONCLUSION

The instructional model is a blended model for integrating five stages between learning in the classroom and learning online by using social media (social network, online forum, and wiki) to be tools for writing. Online forum was used as a tool for round table writing and lecturer posted issues for each group and learners share their various ideas or brainstorming. Wiki was used as a tool for cooperative writing by all data obtained from round table writing to create a group creative criticism written work. One point of differentiation is that learners cannot start round table writing because they do not know the web links of the online forum for writing so lecturer has created a group on social network and added the learner into group. Then posted web links of online forums groups for each group. Facebook should be chosen as social network since it is a useful tool and easy to use, can be used to access information anywhere and improves the efficiency of communication between the learner and the lecturer. (Tirayakioglu, 2011). After be informing web links of the online forum, and start round table writing on social media, learners will be aware of issues that need to be written on the forum. When the ideas are much enough, learners will need to select the best idea or combine the top five

STAGE 1: Decide to choose the title along with any two topics "topic 1" for round 1 and "topic 2" for round 2 to create a creative criticism written work

Process	Method/Principle	Purpose	Learner's role	Lecturer's role	Activity	Social media	Result
Stage 1 Determine (Week 1)	1. Round table writing	To gather ideas for the title of written work.	Share their ideas to define the title of written work	1. Set the topics for round table writing for learners to gather ideas for the title of written work. 2. Discuss and share opinions with learners in each group	Round table writing and express their ideas on the topic to set the title of written work	Forum	List of titles
	2. Consider and select ideas	To consider and select the top 5 of titles according to a data selection criteria.	1. Vote only five titles per member. 2. Select the titles which members vote the most for five ones	advise and suggest learners	Vote the top five titles from the list of titles above	Forum	Top five titles
	3. Choose the best idea	To choose the best single title	Choose the best single title	Advise and suggest learners	Choose the best title from the top five titles above	Forum	Best single title

STAGE 1: Continued

Process	Method/ Principle	Purpose	Learner's role	Lecturer's role	Activity	Social media	Result
	4. Assess	1. To determine the appropriateness and clarity of idea 2. To give the title	1. Consider the accuracy and clarity of title. 2. Give the title 3. Check the possibility of finding information for writing	1. Share opinions on the title and allow learners go on writing 2. Introduce resources for writing operator	1. Check the chosen title by considering the accuracy and clarity. 2. Give the title 3. Introduce sources from various resources	Forum	Title reviewed by lecturer

STAGE 2: Plan a written work by plotting included an introduction, main body, and conclusion with supporting evidences

Process	Method/ Principle	Purpose	Learner's role	Lecturer's role	Activity	Social media	Result
Stage 2 plan (Week 2)	1. Round table writing	1. To plot the outline (introduction, main body, and conclusion). 2. To determine the supporting evidences	Share their ideas to define the outline (introduction, main body, and conclusion), and supporting evidences	1. Discuss and share opinions about the outline (introduction, main body, and conclusion) with learners in each group. 2. Suggest supporting evidences.	Round table writing and express their ideas to determine the plot (introduction, main body, and conclusion) and appropriate supporting evidences	Forum	1. List of plots (introduction, main body, and conclusion). 2. List of supporting evidences
	2. Consider and select ideas	To consider and select the top five outlines and score supporting evidences according to a data selection criteria	1. Vote only five outlines and supporting evidences per member 2. Select the plot and the supporting evidences which members vote the most for five ones	Advise and suggest learners	Vote the top five outlines and supporting evidences from the list above	Forum	1. Top five plots 2. Top five evidences

STAGE 2: Continued

Process	Method/ Principle	Purpose	Learner's role	Lecturer's role	Activity	Social media	Result
	3. Choose the best idea	1. To choose the best single outline 2. To sort the supporting evidence from scores	1. Choose the best single outline 2. Sort the supporting evidences from maximum to minimum scores	Advise and suggest learners	1. Plot the outline 2. Sort the supporting evidences from maximum to minimum scores	Forum	1. Best single outline 2. Supporting evidences sorted by scores
	4. Assess	To determine the appropriateness and clarity of outline and supporting evidences	1. Consider the accuracy and clarity of outline and supporting evidences in accordance with the title	Share opinions on the outline and supporting evidences	Check the chosen outline and supporting evidences by considering the accuracy and clarity	Forum	Outline and supporting evidence reviewed by lecturer

STAGE 3: Proceed to retrieve resources according to the outline and supporting evidences

Process	Method/Principle	Purpose	Learner's role	Lecturer's role	Activity	Social media	Result
Step 3 Proceed (Week 2)	1. Round table writing	To specify resources	Share their ideas to specify resources	Discuss and share opinions about the resources	Round table writing to specify the appropriate resources for the outline	Forum	List of resources
	2. Assess	To determine the appropriateness and clarity of resources	Consider the accuracy and clarity of resources	Express their ideas on resources	Check resources by considering the accuracy and clarity	Forum	Resources reviewed by lecturer
	3. Consider and select ideas	To consider and select the top five resources	1. Vote only five resources per member 2. Select the resources which members vote the most for five ones	Advise and suggest learners	Vote the top five resources from the list above	Forum	Top five resources
	4. Arrange the ideas	To sort top five resources from scores	Sort the resources from maximum to minimum scores	Advise and suggest learners	Sort the resources from maximum to minimum scores	Forum	Resources sorted by scores

STAGE 4: Get to know the result of collaborative writing by using the idea organization methods, document design, and writing strategies to create a written work

Process	Method/Principle	Purpose	Learner's role	Lecturer's role	Activity	Social media	Result
Step 4 Acknowledge (Week 3)	1. Round table writing	To specify the idea organization methods, document design, and writing strategies	Share their ideas to specify the idea organization methods, document design, and writing strategies	Discuss and share opinions about the idea organization methods, document design, and writing strategies	Round table writing to appropriately specify the idea organization methods, document design, and writing strategies for the outline	Forum	1. List of the idea organization methods 2. List of document design 3 List of writing strategies
	2. Consider and select ideas	To consider and select the top five idea organization methods, document design, and writing strategies	1. Vote only five idea organization methods, document design, and writing strategies per member 2. Select the idea organization methods, document design, and writing strategies which members vote the most for five ones	Advise and suggest learners	Vote the top five idea organization methods, document design, and writing strategies from the list above	Forum	Top five idea organization methods 2 Top five document design 3 Top five writing strategies

STAGE 4: Continued

Process	Method/Principle	Purpose	Learner's role	Lecturer's role	Activity	Social media	Result
	3. Combine the idea	To combine the top five idea organization methods, document design, and writing strategies into a written work	Combine the top five idea organization methods, document design, and writing strategies into a written work	Advise and suggest learners	Combine the top five idea organization methods, document design, and writing strategies	Forum	Combination of idea organization methods, document design, and writing strategies
	4. Create a creative criticism written	To bring all the data to create a creative criticism written work	1. Create a creative criticism written 2 Evaluate organizing concept methods, document design, and writing strategies based on content in a written work	Advise and suggest learners	Create a written work to evaluate idea organization methods, document design, and writing strategies into a written work	Wiki	Written work

STAGE 5: Writing Assessment rubric's criteria and assess the ability of creative writing critiques of the test

Process	Method/Principle	Purpose	Learner's role	Lecturer's role	Activity	Social media	Result
Step 5 Evaluate (Week 4)	1. Present	To present a written work	1. Exchange web links with other groups for reading their own written works 2. Comment on other groups' written work	Join a discussion about the written works of the learners in each group	Present their opinion on the friends' written works	Wiki	Comment on the written works
	2. Decide	To consider, revise, edit a written work	Consider revise and edit their own written work to complete	Advise and suggest learners	Consider, revise, edit a written work	Wiki	Written works that have been fully considered, revised, and edited
	3. Assess	To assess the written works according to rubrics criteria	Listen to written work assessment results	1. Check written works of each group according to rubrics criteria 2. Announce written work scores	Acknowledge written work scores of one's own group and other groups	-	Written work scores of each group
	4. Test	To test the creative criticism writing ability	Write a short creative criticism written work individually	Check and announce a short creative criticism written work	Check and announce a short creative criticism written work of each learner	-	Short creative criticism written work score

best ideas into a single idea. Moreover, each group can provide a leader to choose the best idea but group members were respected and accepted that one.

The blended creative criticism writing instructional model is also the vital model that the higher education institutions should be applied to the activity plan that promotes qualifications frameworks based on learning outcomes in higher education in term of academic activity supporting desirable graduate's attributes in order to establish a standard for quality assurances of education in the 21st century.

ACKNOWLEDGEMENTS

I would like to express my gratitude to Professor Seifedine Kadry for his valuable opportunities. I also would love to thank Associate Professor Onjaree Natakuatoong, PhD and Theeravadee Thangkabutra, PhD at Department of Educational Technology and Communications, Faculty of Education, Chulalongkorn University for excellent academic and mental supports. Most importantly, this chapter would not be written successfully without great encouragements from my highly admirable family (Captain Terdsak Vibulsarin, Mrs. Nathasha Vibulsarin, and Ms. Kanokpan Wiboolyasarin) and splendid friend (Mr. Natthawut Jinowat). Finally, I am thankful for Suan Dusit Rajabhat University as well as Associate Professor Sirot Pholpuntin, PhD for all inspirations.

KEYWORDS

- **blended learning**
- **creative criticism writing**
- **instructional model**
- **participatory communication**
- **round table writing**
- **social media**

REFERENCES

1. Decaroli, J. What research say to the classroom teacher: Critical thinking. *Social Education.* 1973, 23, 67–69.
2. Divito, A. *Recognized and Assessing Creativity Developing Teacher Competencies*; Englewood Cliffs, New Jersey: Prentice-Hall, 1971.
3. Ennis, R. H. A logical basis for measuring critical thinking skill. *Journal of Educational Leadership.* 1985, 10, 45–48.
4. Furner, B. A creative writing through creative dramatics. *Journal of Elementary English.* 1973, 50.
5. Mayfield, A. What is Social Media? http://www.icrossing.co.uk/fileadmin/upload s/ eBooks/What_is_Social_Media_iCrossing_ebook.pdf (accessed Aug 17, 2014).
6. Osborn, A. F. *Applied Imagination*; New York: Charles Scribner's sons, 1963.
7. Quellmalz, E. S. Needed: Better methods for testing high-order thinking skills. *Educational Leadership.* 1985, 43, 29–34.
8. Sirichodok, T., Makaramani, R., Sangpum, W. A Guideline of Social Media Application for the Collaborative Learning in Higher Education, Proceedings of the International e-Learning Conference on Smart Innovations in Education and Life-long Learning, Nonthaburi, Thailand, June 14–15, 2012.
9. Torrance, E. P. *Scientific Views of Creativity and Factors Affecting Its Growth;* Kagan, J., Ed.; Creativity and Learning; Houghton Mifflin: Boston, 1989.
10. Wallach, M. A.; Kogan, N.; and Bern, D. J. Group influence on individual risk taking. *Journal of Abnormal and Social Psychology.* 1962, 65, 75–78.
11. Wiboolyasarin, W. Blended instructional model based on participatory communication with round table using social media to enhance creative criticism writing ability for undergraduate students: A synthesis and proposed model. *IJEEEE.* 2012, 2, 521–525.
12. Wiboolyasarin, W. Development of a Blended Instructional Model Based on Participatory Communication Approach Using Round Table Writing Technique on Social Media to Enhance Creative Criticism Writing Ability for Undergraduate Students. PhD Dissertation, Chulalongkorn University, Bangkok, 2013.

CHAPTER 2

ADAPTIVE E-LEARNING ACHIEVING PERSONALIZATION FOR STUDENTS

ROBERT COSTELLO

Graduate School, University of Hull, Cottingham Rd, Hull, Yorkshire HU6 7RX, United Kingdom

CONTENTS

2.1 INTRODUCTION

In this chapter, the main focus will be on personalization contributing to one of the major problems of e-learning, the fact that generally the learning content is not personalized to the learner's needs. The main drive of personalized e-learning is to adapt to the specific needs and influences of the learner through the implementation of supportive activities, content and group-learning-paradigms built upon a variety of different pedagogical approaches.

The challenges faced by academic institutions are who to adopt these complexities and tools to support not just the academic but also the individual when designing and improving the learning experience. Adaptive e-learning offers an important approach to the 'one size fits all' enabling individuals to be identified and categorized into different group-learning-paradigms to assist in the alignment of curriculum development, assessment, and instructional strategy, based upon learner-centricity.

There will be a variety of issues associated with this design of using group-learning-paradigms like:

- the homogeneous views, how to group individual views into clusters within an e-learning environment and what techniques/mechanism are available to achieve this; what problems are commonly associated with relationships and their groups.
- the concept drift refers to how individual learn differently depending on the extent of availability, reliability and the validity of psychometric measuring. Concept drift can occur either because the acquired learning style information needs to be adjusted or because the student simply changes his/her preferences.

One of the major problems of learning, is the fact that generally the learning content is not personalized to the learner's needs. Obviously, this is not a realistic idea in the physical form due to limitations of cost, time and the requirement for the teacher to gain experience of the learner's mechanisms for learning in different contexts.

The use of adaptive and personalized e-learning environments has demonstrated that tailored approaches to learning are achievable (Costello et al., 2013), for example; Intelligent Tutor System (aims at managing the teaching process; improving, evaluating and self-improving techniques for the students; providing guidance and support for the students). According to Treviranus et al. (2006), Gaeta et al. (2009) and Costello et al. (2013) a personalized learning environment should provide adaptable and accessible features to users' requirements. Costello and Nigel (2014) and Safran et al. (2006) agrees with Treviranus et al. (2006) that features of adaptive and personalized can be embedded into existing Virtual Learning Environments to act like a bridge between the learner, learning materials and equipment.

According to Safran et al. (2005) and McLoughlin and Lee (2010) the use of technologies can provide the necessary tools for enabling online learning to be learner centric, by giving the learner more control and autonomy. In the last several years there has been a significant growth in socially oriented web applications designed around embracing community annotation and recommendation, for example, social networking sites such as Facebook and social sharing sites such as YouTube (McLoughlin and Lee, 2010; Costello and Nigel, 2014).

According to Gutierrez and Rogoff (2003) by grouping individuals within a similar style approach, can introduce a new learning experience and explores different expectations of the individual. The author agrees with Gutierrez and Rogoff (2003) that group categorization can make a difference to the learning process, through the use of collective intelligence, when making a decision belonging to learning object.

This chapter provides a critical introduction to learning, e-learning 2.0, User Modeling and most relevant to this book the area of personalized learning. The chapter starts with an introduction to learning and how people learn. It follows this with a brief examination of e-learning, e-learning 2.0, User Modeling and the opportunities in e-learning for facilitating learners' needs through the concept drift.

2.2 LEARNING

Learning refers to the orientation of problem solving, decision making, and using embedded real-life tasks and activities to enable the learner to think, communicate, and build upon prior knowledge and experience (Warburton, 2003). Learning takes place with respect to content and context; you learn something somewhere (Edelson, 2001).

Teaching, according to Bereiter et al. (1989), and Grabinger et al. (1995), provides the learner with the opportunity to develop a firm conceptual base for the content of coherent knowledge structures. Building on this base the learner will develop effective ways of synthesizing, processing and transforming knowledge.

Learners have individual approaches to how they learn. There are many different types of learning theories that can be used to describe these approaches. The following sections analyze the following four learning theories thought to be of most direct relevance to this thesis:

- Social Learning Theory
- Experiential Learning
- Cognitive Behavioral Theory
- Learning Styles/Strategies

2.2.1 SOCIAL LEARNING THEORY

According to Bandura (1969) social learning theory reflects on how one person learns through the use of actions, feelings, and thoughts, after observing a learning experience. Ormrod (1999) suggests that social learning theory explains learning in terms of a continuous reciprocal interaction between cognitive, behavioral, and environmental influences. The social learning theory provides a model to describe how learners' often learn most effectively from observing other people.

Stahl et al. (2006) and Jones (2010) indicate that they have applied social learning theory in Computer-Supported Collaborative Learning (CSCL), to encourage students to learn together in small groups via the use of interactive software; allowing students to learn by expressing their questions; pursuing lines of inquiry together; teaching each other; and seeing how others are learning. Social learning theory in accordance with Dalgarno and Lee (2010), Saade & Kira (2009) and Jones (2010) could provide potential for providing learner personalization within electronic collaborative environments.

Literature has shown that by using social learning theories within e-learning can help to support: group activities; challenges and interest (Stahl et al., 2006); classification of group's dependent on behavior; and group problem solving (Ormrod, 1999). The research within this chapter encompasses elements of individuals learning through the learning experiences of others by enabling the capture of individual and group responses to learning content. However, this does not happen through observation, it happens through an individual reflective process, so individuals are essentially making decisions on learning content based on the experiences of others.

2.2.2 EXPERIENTIAL LEARNING

David Kolb in 1985 provided a cyclical model for experiential learning within the field of adult learning. There are four levels to the Kolb model, which characterize the learning process: concrete experience, reflective observation, abstract conceptualization, and active experimentation.

- **Concrete experience:** stresses that there needs to be an obvious relationship between the learner, knowledge gained and practical experience.
- **Reflective observation:** focuses on learners developing through watching others or developing observations about one's own experience that can be used to analyze the effect of what works and what does not, what was learned about the situation.
- **Abstract conceptualization:** uses theories to explain observations, concepts, principles, and/or generalized learning concepts. These concepts might include patterns, rules, methods, or the beliefs of the domain expert.
- **Active experimentation:** refers to taking on the general learning concept from the abstract conceptualization section, to demonstrate practically how that principle works within other areas.

Beard et al. (2007) suggest that the application of experiential learning theory within the context of computer-based learning can be supported through literature across multiple fields such as social and cognitive psychology and philosophy. Goodyear (2005) suggest that applying experiential learning theory to computer based learning focuses the domain expert on building exercises and tasks to suit the four different levels of the learning process.

Using the recommendations of Kolb (1985), Goodyear (2005) and Beard et al. (2007), the pedagogical model developed within chapter, will adhere to Kolb's cyclical model for identifying individual learner preferences to enable course-content to be matched to the individual, thus, hopefully improving individual performance and learning experiences while studying online.

McLoughlin and Luca (2002), indicates that encompassing the ideas of experiential learning in approaches to content production and delivery can provide the individual with a multitude of learner choices that can be tailored to their personalized learning classification. Dabbagh (2005) suggests that individual learning emphasizes on the systematic interaction between pedagogical theories and learning technologies.

2.2.3 COGNITIVE BEHAVIORIST THEORIES

There have been a variety of key researchers in the field of cognitive and behaviorist theory, for example: John B. Watson; Edwin R. Guthrie; and F. B. Skinner. Watson (1913) was the founder of behaviorism, in which his research led to techniques used in animal laboratory's to understand behavior being mapped to and applied to the analysis of the behavior of human beings. The goal of behaviorism within psychology—was to predict and control behavior, not to analyze consciousness into its elements or to study vague "functions" or processes like perception, imagery, and volition‖.

Clark (2005) suggests that Edwin R. Guthrie Jr. believed in the contiguity explanation of learning through the notion of the 'principle of association': if two events appear close together, in time or space, then they will become associated with each other. According to Skinner (1985), behaviorism theories are directly associated with the positivist and operationalist views belonging to methodology and philosophical sciences within the field of human behavior.—"*The Behaviorism theories were directly linked into: how a person remembers when tied into the learning experience (complex thinking and problem solving)*" (Skinner, 1985).

Skinner's (1985) theory focuses on attempts to provide behavioral explanations for a range of cognitive phenomena (learning is a function of change in behavior). There are a variety of other Cognitive and Behaviorist theories that can be applied to an educational setting, for example, Gestalt Cognitive learning theory. According to Soff (2013), the Gestalt Cognitive learning theory originated from three main researchers: Werthiemer, Kohler, and Koffa who did their early work in Germany. Gestalt Cognitive learning theory proposes that learning consists of the grasping of a structural whole and not just a mechanistic response to a stimulus. De Freitas and Martin (2006) indicates that cognitive behaviorist theories can be used to represent how informal and formal learning can support and reinforce one's learning.

2.2.4 LEARNING STYLES/STRATEGIES

Learning Styles (LS) are characterized as individual approaches to learning, for example, an individual may learn through seeing visual objects, hearing an oration, reflecting on past experiences and through practical problem solving. Felder et al. suggest that individuals have preferred approaches to learning. According to Felder and Silverman (1998), using a learning-style model can enable domain experts to classify individual students in relation to their learning approaches. Literature suggests that learning styles can be identified by the following; perceptual modality (how learners take in and perceive information), information processing, and personality patterns. Researchers like Kurt Lewin, Jean Piaget, David Kolb, Paul Sinclair, Benjamin Bloom, Phil Race, Peter Honey, Alan Mumford, and many more noted that identification of learning styles may lead to an influence on learner progression. Learning styles emphasize the fact that individuals perceive and process information in a variety of different ways this also implies that how much individuals learn can depend on whether the educational experience is geared towards their particular style of learning.

Learning styles can have beneficial influences within the educational system that can affect students' curriculum, assessments, and how particular modules are taught. It is these primary conceptual thoughts that could improve the students' learning experience if the domain expert considers the range of learning styles presented to them within a group of students.

According to Koper and Oliver (2004), Koper et al. (2004) and Marshall & Mitchell (2005) the results from the categorization process enable course content to be either more explanatory or more structured towards a majority of learner's needs. Karagiannidis and Sampson (2004), Canavan (2004) and Kanninen (2009) suggests that learning styles are used within computer based learning to enhance teaching by accommodating the students' learning preferences. Canavan (2004) indicates that the integration of learning styles, within computer-based learning can enable course content, exercises, discussions, and tasks to be developed to facilitate a variety of learner's needs and abilities. This chapter uses learning style categorization as a mechanism to enable personalization of learning content delivered to individual users and as a mechanism to group learners together.

In 1990, Eisenstadt et al. had suggested the notion of 'Neophytes' in learning, which translates to novice/beginner. Eisenstadt and Brayshaw (1990) suggested within their research that for some people, learning was difficult to grasp, and it was too hard to adjust to a learning experience even though attempts may be made to simplify and support them. According to Sadler-Smith (1996) the use of learning styles within education provides a vital tool to assist the individual and improve their learning experience by enabling the course content to be designed in accordance to how they learn. Learning styles can be used to allow the student to facilitate their acquisition of knowledge, skills or attitudes through study or experience in accordance to their preference learning style. Karagiannidis and Sampson (2004), Canavan (2004) and Kanninen (2009) agree with Sadler-Smith's views that using learning styles can have beneficial influences within the educational system. From the author's perspective; learning styles do provide a building block for which the domain expert has some knowledge on how that individual will function when processing new skills and concepts.

2.3   E-LEARNING

As identified in Section 2.2.3, traditional non-computer based approaches to the personalization of learning materials (based on learning styles) have involved domain experts in the production of multiple resources to enable a best fit to individual learners. In such a scenario either the learner, or the domain expert, needs to have knowledge of the learners approach to learning. The use of technology can assist in the profiling of the learner and the retrieval of resources which best fit their individual learning styles. Therefore, this section focuses on e-learning, moving through to the new wave of learning technologies based on the power of the social web, e-learning 2.0.

According to Alsultanny (2006) and Wang et al. (2012), e-learning is an efficient, effective way of providing a just-in-time learning approach by offering a dynamically changing technological environment that aims to replace old-fashioned time-place content learning.

2.3.1 E-LEARNING 2.0

The Web as platform' principle simply outlines the shift over the past of several years of previously desktop based application functionality to web and mobile based services. In addition, the growth of web and mobile based services which allow for the collection of shared resources, for example, Flickr, Facebook, Twitter, Dropbox and YouTube. This principle removes issues previously existing with the interoperability of applications across platform as now applications run via the web browser, which is cross platform compatible. According to Wang et al. (2012), Web 2.0 is more than just a web platform it is a service that offers more than just sharing content, tagging, wikis, blogs, and social networking but it offers features like enhancing student communications skills, engagement in real-time activities within collaborative development activities. Web 2.0 is an easy way for people to publish self-generated materials like music, videos and photos.

Also coined alongside Web 2.0 is the term e-learning 2.0, which according to Ullrich et al. (2008), Ghali and Cristea (2009), and Wang et al. (2012), refers to online learning environments that incorporate the idea of the Social Web making use of technologies such as collaborative authoring tools, rating tools, social identification (e.g., bookmarking) and annotation to improve motivation of the students online. According to Hamburg et al. (2008) e-learning 2.0 uses web-based tools to create new forms of learning materials (e.g., blogs, video sharing repositories, social networking spaces etc.) and to provide different ways of delivering learning materials. Hamburg et al. (2008) and Wang et al. (2012), suggest that incorporating social web concepts into online environments can assist with collaborative learning through the use of formal learning; the creation and construction of content; and the receiving and giving of feedback through discussion groups.

According to Safran et al. (2007), Ullrich et al. (2008) and Wang et al. (2012), e-learning 2.0 can be categorized or identified within three particular themes, these are: Technology, Social Networking and learner centricity. These link in to the Web 2.0 themes of '*web as platform*' and '*harnessing collective intelligence.*'

- **Technology:** (Ullrich et al., 2008; Wang et al., 2012; Costello et al., 2012) suggests that the use of technology within e-learning 2.0 can provide support for a variety of key educational features: Wiki-blogs, pod-casts, RSS (Rich Site Summary or Really Simple Syndication), e-portfolios and design platform technologies that can support 3D Virtual Environments, mobile technologies (tablets, and Smartphone's).
- **Social Networking:** According to Safran et al. (2007), Chatti et al. (2007) and Ghali and Cristea (2009) social interaction plays an important part within e-learning 2.0 because it allows students to interact, share ideas, communicate (E-mail, chat, video conference), and use forum's to discuss problems. Hamburg et al. (2008) and Brady et al. (2010) mentions that collaborative learning may provide a useful perspective on learning, knowledge creation and management from a social networking perspective.
- **Learner Centricity:** Pazos Arias et al. (2012) suggests that this is when the curriculum is built around the individual incorporating features of social networking and technology to assist in the educational experience.

Primarily on social networking within online learning by investigating practices of socialization; sharing and editing content within, wikis and social media. It is clear to see that this new generation of e-learning 2.0 is focused on applying some form of collaborative community learning through the use of collective intelligence and engaging in sophisticated dialogue models of education (Hamburg et al., 2008; Safran et al., 2007;

Ullrich et al., 2008; Ghali and Cristea, 2009; Alevizou et al., 2010 and Kramer and Bente, 2010).

Hamburg et al. (2008) does indicate that future trends must try to overcome other issues like:

- "Lack of immediate context of applying the learning for example by incorporating new learning in a personal knowledge schema or portfolio;
- Lack of time and lack of access to sufficient bandwidth to ensure high quality training, especially user-friendly tools and quality content;
- The attitude of managers – they often have not enough knowledge or are not convinced of the effectiveness of e-learning. Instead they put their trust in classroom-based training. Many of them prefer "learning from peers" Hamburg et al. (2008).

Expanding on the recommendations of Hamburg et al., further research, should aim at improving the motivation, group-based-paradigms and emotional responses of individuals/groups while studying online via rich media, personalization, and the use of Self-Regulated Learning (SRL) within Personal Learning (PL) (Costello and Mundy, 2009; Kramer and Bente, 2010; Costello and Shaw, 2013; Costello and Shaw, 2014).

Learning technologies offer new opportunities to meet the rapidly growing demand for new ways of learning (such as collaborative and community learning; group-learning-paradigms; matching and tailoring to the learners needs). These new ideas and concepts have the potential to act as a way of offering a personalized tailored approach to: exchange and reuse of learning objects; tailored learning activities; and matching content to individual preferences). Computer based research towards the personalization of learning experiences has been undertaken since the 1970s. The next few sections will introduce early forms of achieving this through User Modeling, and stereotyping, moving on in the final section of this chapter to outline research in personalized learning.

2.3.2 USER MODELING

When users interact with a computer, they provide a great deal of information about themselves. Even when they are not physically at a computer, users continuously radiate data, by walking, speaking, moving their eyes, and gesturing. User modeling enables architectures to be built to interpret this type of information and personalize learning experiences taking into account individual behaviors, habits, and knowledge. User modeling is an approach embedded in Human Computer Interaction (HCI) design to enable designers to understand how people use their soft/web-ware. A user model is a mechanism through which a user can be described and analyzed in relation to their use of a particular piece of soft/web ware. The approach enables designers to overcome problems linked to user perceptions reducing opportunities for error and improving the time taken by users to understand designed interfaces.

The design of modern systems is increasingly user-centered, with users now often involved from the planning stages of web and mobile development. Early user involvement can help prevent serious mistakes in systems designs. Benefits of a user-centered approach like user modeling are mainly related to time and cost savings during development, completeness of system functionality, repair effort saving, as well as user satisfaction.

User modeling is usually traced back to the late 1970s (Razmerita and Lytras, 2008; Kobsa, 2001), in which a lot of work was done in this area of research relating to how application systems were developed and how different types of information was collected from different users. According to Kobsa (2001), User Modeling System can be used to describe different scenarios; introduce rules and reasoning to help shape understanding and logic. Kobsa (2001) indicate that by applying user modeling to analysis human traits like (user interests; personality traits) can help interpret assumptions, beliefs, and introduce a variety of different strategies to support individuals.

Applications that integrate user-modeling capabilities are able to capture the assumptions of the users based on their use of the applications, features and user profiles (Gauch et al., 2007; Costello and Mundy 2009; Bettini et al., 2010). A number of systems have been constructed to use user modeling to personalize user experiences these are critiqued below.

Berendt in 2007 uses user modeling through the use of data mining to interpret and extract interests, behaviors and patterns, belonging to individual users to assist with the delivery of learning content. In addition, the system designed by Berendt used this information to group users (e.g., users with low IT literacy or users with limited language comprehension) and collect and share rating and tagging information from these user groups about particular learning resources. According to Berendt (2007) there is evidence to support this approach; however, further research is needed in the field of: semantics, data mining, pedagogy, system design, and finally privacy.

In 2011 Manouselis et al., introduces a Technology Enhanced Learning "TEL" environment that gathers data sets belonging to the learner's educational level, curriculum, and formal and informal training that they had. TEL works by supporting the learner to achieve a specific goal by providing context that is annotated and sequential the learning experience. Manouselis et al., suggests that lessons can be supported through additional goals and tasks to encourage the learner. Each task within the TEL, the learners can give each task or learning experience the student can rate the learning context to help monitor the educational materials and experience.

Shi et al. (2013) suggests a Social Personalized Adaptive E-Learning System called "Topolor" that provides and supports individuals through content adaptation, and learning path adaptation, which the capability of leaving individual remarks belonging to topic and exercises.

User modeling within online educational contexts will enable e-learning to provide more accurate information retrieval based on the profiling of behavioral, psychological and emotional states of individual users. According to Douce and Porch (2009) using User Modeling within VLE's can provide a way of incorporating pedagogical profiles to design more personalized approaches. Boticario and Santos (2006), Martins et al. (2008), Martins et al. (2008) and Costello and Mundy (2009) suggest educational systems that use user modeling as a way of identifying and extracting user traits have been successful in a variety of institutions. However, there is a need for a generic model that can offer the same success from trials belonging to small-scale institutions, which can be imported more readily into mainstream environments.

User modeling plays an important part within adaptive design and can provide a multitude of benefits ranging from: the personalization of learning materials; the use of pedagogical theories; support and guidance; and customization for learners. According to Akbulut and Cardak (2012) these new approaches to personalization will enable the future development of VLE, which support and provide provision for more accessible and personalized learning content and structures. The next section will expand on the notion of personalized learning, and how it directly relates to this chapter.

2.3.3 PERSONALIZED LEARNING ENVIRONMENTS (ADAPTIVE) – AN OVERVIEW

Personalized Learning Environments PLE's or Adaptive E-learning are developed to incorporate a variety of approaches that take into account different ways of learning. These different approaches comprise a variety of techniques: knowledge representation; cognitive learning styles; adaptation to the learner needs; search; and retrieval techniques (see Section 2.3.2).

Personalized learning encapsulates pedagogical approaches that adjust to individual learning preferences. Frameworks for personalized learning provide learners with the opportunity to support different goals and learning needs. Rosmalen et al. (2006) and Klašnja-Milićević et al. (2010) suggests that PLE's can provide extensive mechanisms to improve learner performance while studying online. There are a number of different approaches for achieving personalized learning, the most pertinent of which to this chapter are outlined below.

Hauren (2001) described a Personalized Continuing Medical Education Solution (PCMES), which focused on providing individually relevant learning materials based on expertise, interest and need. The algorithm at the heart of this solution needed to match personal attributes to the knowledge base to retrieve just in time materials. PCMES provided a learning environment that could handle real time response to a medical database, which would bring back learning materials that were appropriately associated with the academic level of the learner. The PCMES environment had several benefits, according to Hauren (2001), these were: delivery of low cost learning; up to date repositories and the provision of online support for students. Hauren (2001) indicated that the major limitation of the PCMES model was the complexity of the environment and the amount of staff training and staff time required in its usage.

Over time with the development of research and with the advancement of literature within the field of adaptive and personalized learning a multitude of different academic e-learning environments appeared that offered featured within Adaptive Hyper Architecture (AHA) or Adaptive Information Retrieval (AIR) techniques that would assist with online learning.

Once particular system was that heavily focused on AIR was the Adaptive Intelligent Personalized Learning (AIPL) (Costello and Mundy, 2009) environment that used a several learning styles to enable course context to match by identifying the learner traits and using intelligent algorithms to filter out learning materials that were not relevant to them with the aid of social rating and group-based-learning paradigms.

With the advancement of online technologies in 2010 Ghali et al., introduced an adaptive Web 2.0 e-learning tool called My Online Teacher (MOT), which was developed to support a variety of features and facilities like: collaborative authoring; group-learning-paradigm; social annotation (group rating, feedback, etc.); adaptive hypermedia recommendation facility (which provides learning context based on other people previous reading materials). Ghali and Cristea (2009) indicated that some of the features provided by MOT 2.0 such as grouping, subscriptions, communications, recommendations, accessing other people's material were useful and assisted with their learning experience.

One particular personalized e-learning environment that was introduced by Peter et al. (2010) was the iLearn project that used the VARK learning style to enhance platform's adaptability for the learner through using learning styles. This particular approach enabled rich media/digital assets to be developed in accordance to their individual needs.

In 2012 Costello et al., introduced several new features to the Graduate Virtual Research Environment (GVRE) that incorporated personalized learning at its core of the e-learning environment. The first concept was the introduction of a Dynamic Background Library (DBL), to enable EU/International students to incorporate categorized learning materials (Videos/ Websites) in their own languages and share them with English speaking students via a translation function to improve information retrieval filtering within the GVRE. The Second feature was the introduction of time-line, which has enabled the provision of research and transferable skills training to be linked directly to the Researcher Development Framework. This approach allows new researchers to conduct a guided search through a rich media repository, belonging to the individual year of study.

Klašnja-Milićević et al. (2010) has indicated that to improve the effectiveness of e-learning is to incorporate personalized learning. This can be achieved by using Adaptive e-learning system that incorporates a variety of different learning strategies and technologies to predict and recommend the preferred learning material. This can be achieved by recommending and adapting the appearance of hyperlinks or simply by recommending actions and resources.

According to Yarandi et al. (2013) they introduced an adaptive e-learning environment that focused on a various learner characteristics traits. The adaptive e-learning environment provides features similar to that of Ghali and Cristea (2009) by adding in social interaction, learner peer recommendation, and learner path adaptation. These particular types of adaptive e-learning environments (Adaptive Systems) enable learners to identify existing educational paths that pervious students have taken before, giving them guidance.

Adaptive Systems try to anticipate the needs and desires of the user. Any knowledge that the Adaptive Hypermedia Architecture (AHA) has belonging to learner is based on previous actions. The system may simply monitor what a user is doing or it may ask questions to enable the architecture to adapt to his or her needs.

Some of the technique associated with Personalized and Adaptive e-learning has been shown to improve the educational experience like having a social interaction layer (Ghali and Cristea, 2009), or identifying user traits and adapting them within the e-learning environment (Yarandi et al., 2013). As Manouselis et al., mentioned in 2011 by analyzing previous educational pathways could be one way of assisting the learner who is struggling to readjust there educational pathway either at college to university levels. Another way of improving and tacking issues of personalization and adaptive e-learning is using group-based-learning paradigms, which is discussed in the next section.

2.4   GROUP-BASED-LEARNING PARADIGMS

This section is written to introduce the challenges and complexities associated with grouping learners. The following layout will be used to represent a logical approach that is needed to deal with the complexities: homogenous views (Section 2.4.1); concept drift (Section 2.4.2); A Generic Solution to the Concept Drift (CD) (Section 2.4.3); Categorizing of Groups (CG) (Section 2.4.4).

2.4.1 HOMOGENOUS VIEWS

According to Spiro et al. (1996) grouping individuals can vary depending on subject matter and learning capabilities. The characteristics of the individuals within group settings can be varied depending on how they learn, for example: some learners might like to learn through orderly tasks; others might like complex challenges; and some learners might not like the pedagogical approaches adopted by the domain expert. It is these homogenous views of the individuals within a group setting that can make the group either succeed or fail. It is important that the homogenous views are compensated within any group setting by building on interests, traits and personal preferences of the learners. To overcome some of the issues associated with homogenous views, researchers like Severiens et al. (1994), Oxford (2003), Alexander et al. (2004), Costello and Mundy (2009), and Yarandi et al. (2013) are using learning traits to group students. Alexander et al. (2004) suggests that you can group learners through the use of cluster analysis. Cluster Analysis is where profiles can be grouped together on the basis of their expertise that is, participant's knowledge, interests and strategic processing.

Group categories can be categorized as the following:

- **Acclimatization cluster:** New students with low levels of domain knowledge and individual interest within specific areas.
- **Early competence cluster:** This particular cluster relates to individuals that seek knowledge in specific fields, interests and professions.
- **Mid-competence cluster:** Refers to knowledge and experiences that individuals already have which they can apply to deep-processing strategies or text-based strategies.
- **Proficiency cluster:** Students and staff members with a high level of academic and experience.

The research conducted by Alexander et al. (2004) through the use of cluster analysis, within grouping does work. However, grouping individuals into academic levels would not give a fair advantage to the students with very little knowledge, or mixed abilities.

Researchers like Oxford (2003) indicate that psychological and socio-cultural trends can be used as a way of grouping individuals; however, this is complex and requires a vast amount of knowledge and research into understanding the characteristics of the individuals, for example: anxiety, beliefs, support, assisting relationships and actual knowledge of the individual themselves. There are, however, more efficient ways of grouping individuals that have been demonstrated by Severiens et al. (1994), Gutierrez and Rogoff (2003), Costello and Mundy (2009), and Cristea and Ghali (2010).

Severiens et al. (1994), Gutierrez and Rogoff (2003), Costello and Mundy (2009) and Peter et al. (2010) indicate that by categorizing individuals through the use of learning styles/theories a balance can be created within education experiences between resources and approaches to learning. Eisenstadt and Brayshaw (1990), Leutner and Pass (1998) and Boyd and Murphrey (2004) would argue against the points of (Severiens et al., 1994; Gutierrez and Rogoff, 2003) that no matter what features applied by the domain expert to assist the individuals, the learner may still find it difficult, and even if the materials are designed accordingly this can still mislead them.

According to Oxford (2003) some of the major limitations associated with grouping are: how the learning activities are written in accordance to the behavior of the individual set within a group environment; matching the needs of the individual/group to the right learning materials; and finally, creating and exploring relationships between tasks to engage group learning. Whilst researchers like Severiens et al. (1994), Oxford (2003), Alexander et al. (2004), Wilson et al. (2007) and Cristea and Ghali (2010), have provided a great wealth of knowledge, it is still in the early stage of academic research about how to categorize groups of individuals in successful ways. Even though, there is not a correct procedure for grouping there is enough academic research and literature to support the idea that certain aspects of grouping can be applied to e-learning (Severiens et al., 1994; Oxford, 2003; Gutierrez and Rogoff, 2003; Costello and Mundy, 2009; Yarandi et al., 2013).

2.4.2   SOLUTIONS TO HOMOGENOUS VIEWS

To tackle the issues associated with homogenous views, the following research carried out by Oxford (2003), Severiens et al. (1994), Gutierrez and Rogoff (2003), and Zajac (2009) are used by using learning styles as a way of grouping individuals in accordance with how they learn most effectively. By grouping individuals in accordance with their own learning styles provides a way within the an online environment to overcome the following problems: unfair categorization of knowledge and expertise (i.e., all abilities are placed together); placing the learner with other learners that have similar learning traits as each other; placing the learner/s in an environment that was designed for them thus creating a more tailored learning experience.

This is further supported by Yalcinalp and Gulbahar (2010) that indicates by using homogenous views within personalized learning can aid in:

1.   encourage learners to learn by anticipating needs;
2.   make the interaction efficient and satisfying for both the organization and the participants; and
3.   build a relationship that stimulates learners to return for consistent and progressive learning" this overcoming some of the traditional issues of e-learning.

As Yalcinalp and Gulbahar (2010) mentions about homogenous views will enable personalized e-learning environment to adapt to the needs of the learners through either Adaptive Educational Hypermedia (AEH) Architectures/systems or even AIR techniques. However, it is important at this point to note that both AEH and AIR can be used to achieve personalized learning and provide one way of solving approaches to homogenous views. Within this chapter, we will be focusing on using learning styles.

To provide some insight into overcoming the issues associated with homogeneity within chapter a three level approach will be used:

- **Singular:** One student has only one category, i.e., Analytical;
- **Amalgamation:** One student can have many values, i.e., Analytical and Pragmatic or even Reflective with Analytical;
- **Concept Drift:** For more information see Section 3.2 for a definition and how it works within chapter.

Each one of these particular group-clustering methods (i.e., singular, amalgamation and concept drift) can create homogenous views within PAFS. The concept of homogenous views within this chapter can be represented through a three-layer triangle that enables students to change clusters over time, depending on their learning styles results. For more information, see Figure 2.1.

By grouping the individuals into clusters of learning styles, according to Severiens et al. (1994), Gutierrez and Rogoff (2003) and Zajac (2009) this will improve the learner's experience. To overcome the issues Oxford (2003) and Yalcinalp and Gulbahar (2010) had suggested modern personalized e-learning environment matching individuals with specific groups could be used to reduce some of the limitations associated with collaborative grouping, for example: inappropriate matching and creating relationships within a dynamic moving environment.

2.4.3 THE CONCEPT DRIFT (CD)

The concept drift is associated with the difference between the values on the Likert scale. The CD relies on the results from the psychometric measuring belonging to any particular Learning Process Questionnaire. The psychometric measuring systems used within this chapter have; however, divided the research community.

On the one hand, some researchers argue that learning styles do not effectively improve the learning experience (Eisenstadt and Brayshaw, 1990; Leutner and Pass, 1998; Boyd and Murphrey, 2004) and that psychometric measuring systems are ineffective. However, reading the Learning

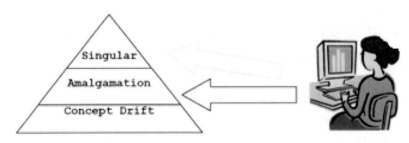

FIGURE 2.1 A three-layer approach.

Styles and Pedagogy in post-16 learning: A systematic and critical review report by Coffield et al. (2004) indicates that:

"The logic of lifelong learning suggests that students will become more motivated to learn by knowing more about their own strengths and weaknesses as learners. In turn, if teachers can respond to individuals" strengths and weaknesses, then retention and achievement rates in formal programmes are likely to rise and "learning to learn" skills may provide a foundation for lifelong learning" (Coffield et al., 2004).

The Coffield report does question the whole concept of using learning style and indicates that "whether a particular inventory has a sufficient theoretical basis to warrant either the research industry which has grown around it, or the pedagogical uses to which it is currently put" (Coffield et al., 2004).

Further reading into the Coffield report would indicate that their final assumption about learning styles would be that of "researchers and users alike will continue groping like the five blind men in the fable about the elephant, each with a part of the whole but none with full understanding" (Coffield et al., 2004).

From the authors perspective based on the Coffield et al. (2004) report it is clear that the impact of Learning Styles cannot either be proven or disproven due to the large amounts of literature supporting both claims. Due to the nature of learning styles being freely available from the Internet, it is possible for the domain expert to use them to find out quickly about ones learning type, so they can adapt coursework, and learning materials.

As stated above: "if teachers can respond to individuals" strengths and weaknesses, then retention and achievement rates in formal programmes are likely to rise and "learning to learn" skills may provide a foundation for lifelong learning" (Coffield et al., 2004).

Some researchers like Leutner and Pass (1998), Genovese (2004), Boyd and Murphrey (2004), Litzinger et al. (2005) have indicated that many models within pedagogical theories can improve instructional design. By using a variety of different models and pedagogical approaches such as learning strategies and learning styles can arguably help improve instructional design, through modifying teaching and student self knowledge awareness about how they learn best. However, their research suggested many of the investigations carried out on learning styles lack theoretical clarity and adequate measurement instruments.

Research conducted by Leutner and Pass (1998) suggested that individual learning differences depend on the extent of availability, reliability and the validity of psychometric measuring. Boyd and Murphrey (2004) furthers the debate initiated by Leutner and Pass (1998) by arguing that learning styles have several weaknesses in terms of the reliability, validity, and the identification of the different characteristics of learners needs. According to Leutner and Pass (1998), Genovese (2004) and Litzinger et al. (2005), psychometric measuring is a popular method for identification and analysis of the learner's needs, however, the scales used to capture individual needs sometimes lead to negative or low correlations between the needs of learners, and the actual outcome of results. Research conducted by Leutner and Pass (1998) has indicated that to sufficiently test the validity of psychometric measurements it would be necessary to implement the scale based upon behavioral observation, instead of using self-based learning process questionnaires to identify the student's needs.

Isaksen and Geuens (2007) suggested that psychometric measurements should be designed to assist the mediation between stimulus and response in relation to the scales that would best describe the characteristic ways in which individuals conceptually learn best within a learning environment. The research conducted by Leutner and Pass (1998), Genovese (2004), Boyd and Murphrey (2004) and Litzinger et al. (2005) has indicated that psychometric measurements used in identifying learning styles are ineffective, inefficient, and lack clarity in how they are applied; however, researchers like Duff and Duffy (2002), Zywno (2002), Carmona et al. (2007), and Pashler et al. (2008) believe that learning styles have been widely accepted within the academic world, even though limited evidence exists concerning the psychometric properties. Duff and Duffy (2002) have indicated those learning style questionnaires (LSQ) that use psychological factors serve as an indicator of how an individual interacts with and responds to the learning environment, and guarantees that some scales will be negatively correlated.

However, LSQ are designed to probe the relative strengths of individuals, therefore, it could be expected that students with a preference for particular learning activities would outperform those with preferences for other learning activities. Zywno (2002) agrees with Duff and Duffy (2002) by stating that learning styles are important to the individual learner, and that psychometric measurements have been rigorously tested over time. They have concluded that psychometric tools are statistically acceptable for characterizing individual learning preferences. Zywno (2002) suggested that instructors that have applied psychometric learning tools to the individual have shown/demonstrated a greater statistical significance between learning styles and performance based on results retrieved from the LPQ (Honey, 2001; Klasnja-Milicevic et al., 2011; Huang et al., 2012).

According to Kovar and Zekany (2001), Zywno (2002) and Huang et al. (2012) a considerable amount of research has been conducted in the area of learning styles and psychometric measurement tools, which has revealed that students learn better when using preferences with which they have success, and have the potential to be better learners. Kovar and Zekany (2001) suggests that it is important for the learner to be able to take steps to change their learning style to suit the situation by developing competence in a variety of learning styles' categories. Carmona et al. (2007) furthers the debate by arguing that psychometric measurement tools can be used to identify ways in which the individual will collect, process, and organize new knowledge/information. The research carried out by Carmona et al. (2007) suggests that the higher a psychometric value an individual obtains the closer the correlation to the learner needs.

For an example of student learning needs see Table 2.1.

According to the research of Carmona et al. (2007), the Table 2.1 would produce the following results:

Student A: would be classed as Theoretical
Student B: would be classed as Reflector

For a graphical representation of the table: See Graph 1.

However, Carmona et al. (2007) have indicated that a concept drift can occur with student **B**, where Analytical and Reflector are closely associated; in which case the student would try both possibilities to adjust accordingly to his/her needs. For more information see Table 2.2.

TABLE 2.1 Student Learning Needs

	Pragmatic	Analytical	Reflector	Theoretical
Student A:	10	8	9	12
Student B:	9	11	12	7

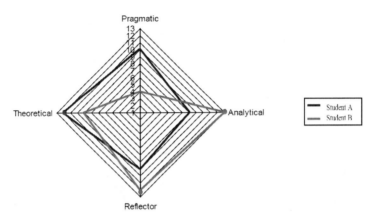

GRAPH 1.   Student learning needs.

TABLE 2.2   Concept Drift

	Pragmatic	Analytical	Reflector	Theoretical
Student A:	10	8	9	12
Student B:	9	11	12	7

The concept drift is concerned with the two top values within the scale. In the Table 2.2, concept drift is identified within the shaded sections.

In a concept drift, Genovese (2004) would suggest offering the student the chance to use both settings to see which one would effectively improve the learning experience. Genovese (2004) suggests that values that are low within the psychometric scales do not influence the results and the true result would be the highest number, depending on the scale used and the amount of research carried out to support the learning style in question.

Markham (2004) suggests that the psychometric scales would not have been accepted within the academic community without the consent of, or authorization from the American Psychological Association (APA), which is supported by local bodies such as the Australian Council for Educational Research (ACER).

According to Markham (2004), researchers that use learning styles to capture learners' behaviors must have a greater understanding, and should define how the scales within the psychometric testing have been conducted, by illustrating the consequences for, and benefits to the individual using them. Isaksen and Geuens (2007) suggests that learning styles seem to be more spontaneously applied without conscious deliberation, whereas strategies seem to be more a matter of choice and training.

2.4.4 A GENERIC SOLUTION TO THE CONCEPT DRIFT (CD)

The Concept Drift according to Carmona et al. (2007) is when the student has been identified as having two or more high numbers that are closely associated with each other.

One way of overcoming the issue of concept drift is to introduce a rule base that could be designed and implemented to calculate and categorize the psychometric values belonging to the learning process questionnaire. The algorithm could use a rule-base to retrieve the highest psychometric value before placing the learner into a community that best suits the learner's needs (Catherine et al., 2013); however, the rule-base must be capable of detecting a concept drift, which would flag/indicate that the learner has high similar values (Costello and Mundy 2009; Doctor and Iqbal, 2012).

To compensate for the concept drift within an algorithm design a 2-tier layer Learning Process Questionnaire could be used to enable the personalized learning environment to identify and average out the highest values within the psychometric values (see Figure 2.2).

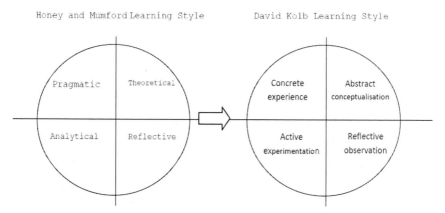

FIGURE 2.2 A 2-tier layer.

The Honey and Mumford learning style could be used as a primary learning style that would enable the algorithm to retrieve the psychometric measurements from the individual. The second learning style from David Kolb could be used as a secondary aid (or vice-versa) for allowing the algorithm to interrogate the concept drift by enabling the personalized learning environment to select the two highest similarities from both learning styles.

The second learning style is represented by the values of A, B & C, D to enable the matrix to calculate the difference. According to Kolb (1985) the learning style is divided into two sections, which are: Concrete Experience, and Abstract Conceptualization (A, B) and Active Experimentation, and Reflective Observation (C, D).

- The total of A's is computed as the Concrete Experience (A) score.
- Total of B's is computed as the Abstract Conceptualization (B) score.
- Total of C's is computed as the Active Experimentation (C) score.
- Total of D's is computed as the Reflective Observation (D) score.

For a tabular representation of the above details see Table 2.3.

Table 2.3 has indicated that Abstract conceptualization has the highest value within the psychometric scale. According to Kolb (1985) the learning style is capable of supporting four different concept drifts, which are:

- Abstract Conceptualization (AC) and Active Experimentation (AE).
- Concrete Experience (CE) and Reflective Observation (RO).
- Abstract Conceptualization (AC) and Reflective Observation (RO).
- Concrete Experience (CE) and Active Experimentation (AE).

By applying both learning styles together within a 2-tier system any e-learning environment could adjust content, context, curriculum structure to suit a multitude of learner needs. For a tabular example of the 2-tier layer, see Table 2.4:

TABLE 2.3   Learning Style

	Concrete experience	Abstract conceptualisation	Reflective observation	Active experimentation
David Kolb Learning Style	8	16	12	14

TABLE 2.4 Representation of Combining Learning Styles

		Total Value		
Pragmatic	11	Concrete experience	8	19
Theoretical	12	Abstract conceptualisation	16	28
Analytical	8	Active experimentation	14	22
Reflective	4	Reflective observation	12	16

Within the matrix, applying the two learning styles together can enable the two scales to be joined enabling a more specific psychometric value to be used. Several researchers (Brusilovsky and Vassileva, 2003; Dolog et al., 2004; Wang and Chen, 2008; Martins et al., 2008; Alves et al., 2008) have all used learning styles to adjust the learning content to suit the individual. As indicated by Alves et al. (2008) the use of learning styles within online learning can assist with the categorization of individuals into groups.

The concept drift is aimed at reducing mismatching between the individuals/groups to the learning materials by gathering knowledge about the individual into a knowledge base system that uses filtering techniques to reduce learning materials/learning activities that aren't suited to the learner's specification.

The benefits of blending these two learning styles together are:

- The Honey and Mumford learning style enables the domain expert to have an understanding of how the individual prefers to learn.
- The Kolb learning style enables the domain expert to have an understanding of the individual learning behavior.
- Using the blended learning styles will enable a precise group categorization to be performed within the online learning environment.
- The environment will be able to adjust to a variety of learning needs of the individual because of the blended learning styles together (how the individual prefers to learn; and what learning behaviors they have).

Each of the two learning styles was designed for a particular reason to study the behavior of the individual, and how that individual prefers to learn. By blending these two learning styles together the domain expert would be able to extract and create a more specific image on how that individual learns and what behavioral traits they have.

2.4.5 CATEGORIZING OF GROUPS (CG)

According to Tzouveli et al. (2005) and Subramaniam (2006), using groups of profiles has enabled environments to adapt to: similar groups' habits, interests, skills, projects, locations and personalized settings. The purpose of using CG within Chapter is to find any close correlations between the learning relationships of individual learners and fellow learners within a module. CG works by interrogating and comparing string parameters belonging to individual learning profiles and records them into a matrix. CG is derived from close interrogation of LPs belonging to all the students studying on a particular module, to enable comparisons to be made. Once similarities have been identified within the CG, the algorithm will group them into a matrix, for an example of the CG operational functionality, see Figure 2.3.

The CG reads in each parameter from left to right of the LP. Once all the values belonging to the LP have been read and recorded, the next step of the operation is to use a cycle that loops through a rule base until the highest value can be identified and it is this value, which the CG uses. The following four steps are required to enable the CG to work.

Step 1: Placing values into a rule base see Figure 2.4.
Step 2: The Rule Base will average out each conjoining value, see Figure 2.5.

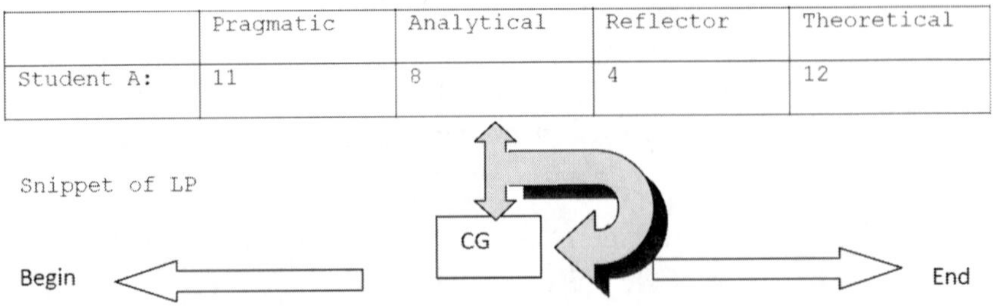

	Pragmatic	Analytical	Reflector	Theoretical
Student A:	11	8	4	12

FIGURE 2.3 (CG) procedure: (original).

Step 3: The highest value belonging to the matrix will be extracted in this case (14) with the category (Abstract Conceptualization and Theorist).
Step 4: A group will be formed or if a group already exists **Student A** will be placed into that particular cluster.

CG can be used to overcome some of the issues associated with grouping (like unfair matching depending on experience; abilities and capabilities). According to research conducted by Oxford (2003), Severiens et al. (1994), and Gutierrez and Rogoff (2003) matching groups based on learning preferences can improve the learning experience while studying online. Genovese (2004), Litzinger et al. (2005) and Mehigan and Pitt (2010) indicate that trying to match individuals into groups via learning preferences can assist with the development of students to gain knowledge

Learners Style Response **Student A**:

Active Experimentation	Reflective Observation	Abstract Conceptualization	Concrete Experience
14	12	16	8
Activist	Reflector	Theorist	Pragmatist
8	4	12	11

FIGURE 2.4   A Learners Matrix.

Active Experimentation	Reflective Observation	Abstract Conceptualization	Concrete Experience
14	12	16	8
Activist	Reflector	Theorist	Pragmatist
8	4	12	11
Average Calculation	Average Calculation	Average Calculation	Average Calculation
14 + 8 / 2 = 11	12 + 4 / 2 = 8	16 + 12 / 2 = 14	8 + 11 / 2 = 9.5

FIGURE 2.5   Averages belonging to User Profile.

of how other people learn. Spiro et al. (1996) does indicate that using learning styles within online learning environments can assist with matching learning materials to groups/individuals while studying online.

2.5 SUMMARY

This chapter through the analysis and design provides guidance theoretically to the field of adaptive e-learning by:

- Building upon research belonging to: uses a pedagogical and group-learning-paradigm approach through learner groups linked to learning styles.
- A pedagogical approach that was based upon the ideas and concepts of researchers (Riding and Sadler-Smith, 1997; Power et al., 2005; Cristea, 2005; Eze et al., 2007; and Melia and Pahl, 2009), which suggest that by applying pedagogical learning approaches to learning environments we can better support the learner.

Literature has shown that there are a variety of ways of personalizing online learning. Within this chapter the primary focus was on content retrieval. There are many different categories, which content retrieval can be placed into, for example: adaptive information retrieval, adaptive hyper retrieval, learning paths, intelligent tutoring systems, knowledge-base and finally clustering.

The chapter has exposed problems and issues with current models that are being used within universities, for example: not using pedagogical learning approaches to structure the materials; over use of technology to demonstrate course materials; mismatching of learning content to the users; information overload of learning resources; and not incorporating learner centricity as a center point for online learning.

KEYWORDS

- **Adaptive e-learning**
- **group-learning**

- **learner**
- **learning environments**
- **pedagogical learning approaches**
- **personalizing online learning**

REFERENCES

1. Agbonifo, O. Catherine, Adewale, O. Sunday, Alese, B. Kayode. (2013) "Design of a Neurofuzzy-based Model for Active and Collaborative Online Learning" International Journal of Education and Research, Vol. 1, No. 10, 2013.
2. Alevizou, P., Conole, G., Culver, J., Galley, R. (2010). Ritual performances and collective intelligence: theoretical frameworks for analyzing emerging activity patterns in Cloudworks, Proceedings of the 7th International Conference on Networked Learning 2010.
3. Alexander, P. A., Sperl, C. T., Buehl, M. M., Fives, H. (2004). Modeling Domain Learning: Profiles From the Field of Special Education. Journal of Educational Psychology 2004, Vol 96, No 3, 545–557.
4. Aleksandra, Klasnja-Milicevic, Boban, Vesin., Mirjana, Ivanovic., Zoran, Budimac. (2010). E-learning personalization based on hybrid recommendation strategy and learning style identification. Computers and Education 56.3 (2011), 885–899.
5. Alsultanny, Y. (2006). e-Learning System Overview based on Semantic Web, Electronic Journal of e-Learning Volume 4, Issue 2, 111–118.
6. Alves, P., Amaral L., Pires, J. (2008). Case-based reasoning approach to Adaptive Web-based Educational Systems The 8th IEEE International Conference on Advanced Learning Technologies, Santander, Spain, IEEE Computer Society.
7. Bandura, A. (1969). Social Learning theory of Identification Processes. From handbook of socialization theory and research, David A. Goslin, Ed Copyright 1969 by Rand McNally & Company Chapter 3.
8. Beard, C., Wilson, P. J., McCarter, R. (2007). Towards a Theory of e-Learning: Experiential e-Learning. Journal Of Hospitality, Leisure, Sport & Tourism Education. Vol. 6, No. 2. ISSN: 1473–8376.
9. Berendt, B. (2007). Data Mining for User Modeling Workshop (DM.UM'07) Corfu, June 2007 Context, (e)Learning, and Knowledge Discovery for Web User Modeling: Common Research Themes and Challenges.
10. Bereiter, C., Scardamalia, M. (1989). Intentional Learning As A Goal of Instruction Ontario Institute for Studies in Education issues 12 [Accessed, July 2014] [Online] Available: http://ikit.org/fulltext/1989intentional.pdf.
11. Bettini, C., Brdiczka, O., Henricksen, K., Indulska, J., Nicklas, D., Ranganathan, A., Riboni, D. (2010). A Survey of Context Modeling and Reasoning Techniques. Pervasive and Mobile Computing, Volume 6, Issue 2, April 2010, 161–180.

12. Boticario, J. G., Santos, O. C. (2006). Issues in developing adaptive learning management systems for higher education institutions: International Workshop on Adaptive Learning and Learning Design 2006.

13. Boyd, B. L., Murphrey, T. P. (2004). Evaluating the Scope of Learning Style Instruments Used in Studies Published in the Journal of Agricultural Education Journal of Southern Agricultural Education Research Volume 54, Number 1, 2004.

14. Brusilovsky, P., Vassileva, J. (2003). Course Sequencing techniques for large-scale web-based education, International Journal Engineering Education and Lifelong Learning Vol. 13 (1/2), pp. 75–93.

15. Canavan, J. (2004). Personalized E-learning Through Learning Style Aware Adaptive Systems [Accessed, July 2014] [Online] Available: http://www.scss.tcd.ie/publications/tech-reports/reports.05/TCD-CS-2005–08.pdf.

16. Carmona, C., Castillo, G., Millán, E. (2007). Discovering Student Preferences in E-learning Proceedings of the International Workshop on Applying Data Mining in e-learning 2007.

17. Chatti, M. A., Jarke, M., Frosch-Wilke, D. (2007). The future of e-learning: a shift to knowledge networking and social software, Int. J. Knowledge and Learning, Vol. 3, Nos. 4/5, 2007.

18. Cristea, A. (2005). Authoring of Adaptive Hypermedia. Journal of Educational Technology & Society, 8 (3), 6–8.

19. Clark, O. D. (2005). From Philosopher to Psychologist: The Early Career of Edwin Ray Guthrie, Jr. History of Psychology, Volume 8, Issue 3, August 2005, 235–254.

20. Coffield, F., Moseley, D., Hall, E., Ecclestone, K. (2004). Learning Styles and Pedagogy in post-16 learning: A systematic and critical review. Learning and Skills research center. [Accessed, July 2014] [Online] Available: http://sxills.nl/lerenlerennu/bronnen/Learning%20styles%20by%20Coffield%20e.a.pdf.

21. Costello, R., Mundy, D.P. (2009). The Adaptive Intelligent Personalized Learning Environment, ICALT, 606–610, 2009 Ninth IEEE International Conference on Advanced Learning Technologies, 2009.

22. Costello, R., Mundy, D. P. (2009). A Move towards personalized learning e-Teaching and Learning Workshop 2009, The eCenter, University of Greenwich and Higher Education Academy ICS.

23. Costello, R., Shaw, N. (2014). Personalized Learning Environments HEA STEM (Computing), Learning Technologies Workshop, University of Hull (2014).

24. Costello, R., Shaw, N., Mundy D., and Allan, B. (2012). Tailoring e-learning 2.0 to facilitate EU/International Postgraduate student needs within the (GVRE) Graduate Virtual Research Environment. International Journal for e-Learning Security (IJeLS), Volume 2, Issues 3/4, September/December 2012, 173–180.

25. Costello, R., Shaw, N. (2013). The Development of a Blended Learning Experience to Enable a Personalized Learning Approach to Researcher Development for Research Students The International Journal of Learning in Higher Education, Volume 19, Issue 3, 75–82.

26. Cristea, I A., Ghali, F. (2010). Towards Adaptation in E-Learning 2.0 [Accessed, July 2014] [Online] Available: http://www.dcs.warwick.ac.uk/~acristea/Journals/NRHM10/NRHM-aic-2010.pdf.

27. Dabbagh, N. (2005). Pedagogical models for E-learning: A theory-based design framework. International Journal of Technology in Teaching and Learning, 1(1), 25–44.
28. Dalgarno, B., Lee J. W. M. (2010). What are the learning affordances of 3-D virtual environments? British Journal of Educational Technology Vol 41, No 1, 10–32.
29. De Freitas C., Martin O. (2006). How can exploratory learning with games and simulations within the curriculum be most effectively evaluated? Computers and Education 46 (3) 249–265.
30. Doctor, F., Iqbal, R. (2012). An intelligent framework for monitoring student performance using fuzzy rule-based Linguistic Summarization International Conference on Fuzzy Systems (FUZZ/IEEE), 10–15 June 2012, 1–8.
31. Dolog, P., Henze, N., Nejdl, W., Sintek, M. (2004). Personalization in Distributed e-learning environment, proceedings of WWW 2004, May 17–22, 2004, New York.
32. Dolog, P., Henze, N., Nejdl, W., Sintek, M. (2005). Personalization in Distributed eLearning Environments [Online] Available: WWW 2004, May 17–22, 2004, New York, New York, USA. ACM 1–58113–912–8/04/0005 Proceedings of the 13th international World Wide Web 2004.
33. Douce, C., Porch, W. (2009). Evaluating accessible adaptable e-learning, Towards User Modeling and Adaptive Systems for All (TUMAS-A) workshop, AIED'09, Brighton.
34. Duff, A., Duffy, T. (2002). Psychometric properties of Honey & Mumford's Learning Styles Questionnaire (LSQ) Personality and Individual Differences Volume 33, Issue 1, 5 July 2002, 147–163.
35. Edelson, D. C. (2001). Learning-for-use: A Framework for the Design of Technology-Supported Inquiry Activities. Journal of Research in Science Teaching Vol. 38, No. 3, 355–385.
36. Eisenstadt, M., Brayshaw, M. (1990). A fine-grained account of Prolog execution for teaching and debugging Instructional Science 19, 407–436 (1990).
37. Eugenia, Y, H., Sheng, W, L., Travis, K., H. (2012). What type of learning style leads to online participation in the mixed-mode e-learning environment? A study of software usage instruction. Computers & Education 58 (2012) PP. 338–349.
38. Eze, E., Ishaya, T., Wood, D. (2007). Contextualizing multimedia semantics towards personalized eLearning. Journal of Digital Information Management, April, 2007 Source Volume: 5 Source Issue: 2.
39. Felder, R., Silverman, L. (1998). Learning and Teaching Styles In Engineering Education 78(7), 674–681 (1988).
40. Gaeta, M., Orciuoli, F., Ritrovato. P. (2009). Advanced ontology management system for personalized e-Learning. Knowledge-Based Systems 22 (2009) 292–301, Journal Elsevier.
41. Gauch, S., Speretta, M., Chandramouli, A., Micarelli, A. (2007). User Profiles for Personalized Information Access, The Adaptive Web Lecture Notes in Computer Science Volume 4321, 2007, 54–89.
42. Genovese, J. E. C. (2004). The Index of Learning Styles: An Investigation of its Reliability and Concurrent Validity with the Preference Test - Individual Differences Research; Dec 2004, Vol. 2 Issue 3, 169–174, 6p.

43. Ghali, F., Cristea, I. A. (2009). MOT 2.0: A Case Study on the Usefulness of Social Modeling for Personalized E-Learning Systems. In the 14th Int. Conference of Artificial Intelligence in Education (AIED'09) (2009), IOS Press.

44. Grabinger, R. S., Dunlap, J. C. (1995). Definition of Real Rich Environments for Active Learning: A Definition [Accessed, July 2014] [Online] Available: www. researchinlearningtechnology.net%2Findex.php%2Frlt%2Farticle%2Fdownload%2 F9606%2F11214&ei=IX21U-uuEuqJ7Aax-4CoCA&usg=AFQjCNE3kVVOv2OW zWVl4NNQQ9dSSe7hUQ&bvm=bv.70138588,d.ZGU.

45. Gutierrez, K. D., Rogoff, B. (2003). Cultural Ways of Learning: Individual Traits or Repertoires of Practice – Educational Researcher 2003.

46. Goodyear, P. (2005). Educational design and networked learning: Patterns, pattern languages and design practice. Australasian Journal of Educational Technology 2005, 21(1), 82–101.

47. Hamburg, I, Engert, S, Anke, P., Marin, M. (2008). Improving E-Learning 2.0-Based Training Strategies of SMEs Through Communities of Practice, Proceedings of the Seventh Lasted International Conference, Web-based Education, March 17–19, 2008 Innsbruck, Austria.

48. Harun, M. H. (2001). Integrating e-learning into the workplace‖, The Internet and Higher Education, Volume 4, November 3, 2001, pp. 301–310(10).

49. Honey, P. (2001) E-Learning: A Performance appraisal and some suggestions for improvement. The Learning Organization, Volume 8, Number 5, 2001, 200–202.

50. Isaksen, S. G., Geuens, D. (2007). Exploratory Study of the Relationships between an Assessment of Problem Solving Style and Creative Problem Solving [Accessed, July 2014] [Online] www.viewassessment.com/media/761/VIEW-CPS-KJTPS.pdf.

51. Jones, E. (2010). A CSCL Approach to Blended Learning in the Integration of Technology in Teaching, Interdisciplinary Journal of E-Learning and Learning Objects Volume 6, 2010.

52. Kanninen, E. (2009). Learning Styles and E-Learning [Accessed, July 2014] [Online] Available: http://hlab.ee.tut.fi/video/bme/evicab/astore/delivera/wp4style.pdf.

53. Karagiannidis, C., Sampson, D. (2004). Adaptation Rultes Relating Learning Styles Research and Learning Objects Meta-Data. Workshop on Individual Differences in Adaptive Hypermedia, 3rd International Conference on Adaptive Hypermedia and Adaptive Web-based Systems (AH2004), Eidhoven, Netherlands.

54. Klašnja-Milićević, A., Vesin, B., Ivanović, M., Budimac, Z. (2010). Integration of Recommendations and Adaptive Hypermedia into Java Tutoring System [Accessed July 2014] [Online] Available:http://www.doiserbia.nb.rs/img/doi/1820–0214/2010%20OnLine-First/1820–02141000021K.pdf.

55. Kobsa, A.,(2001). Generic User Modeling Systems, User Modeling and User-Adapted Interaction 11: 49–63, 2001. 49 2001 Kluwer Academic Publishers.

56. Kolb, D. A. (1985). Learning style inventory, version 3.1, Boston, MA: McBer and Company [Accessed, July 2014] [Online] Available: http://www.haygroup.com/tl/ Questionnaires_Workbooks/Kolb_Learning_Style_Inventory.aspx.

57. Koper, R., Pannekeet, K., Hendrikes, M., Hummel, H. (2004). Building communities for the exchange of learning objects: theoretical foundations and requirements.

Educational Technology Expertise Center (OTEC), ALT-J: Research in Learning Technology, v. 12, n. 1, 21–35, Mar 2004.

58. Koper, R., Olivier, B. (2004). Representing the Learning Design of Units of Learning. Educational Technology & Society, 7 (3), 97–111.

59. Kovar, S., Zekany, K. (2001). Teaching in a Multimedia Cost Accounting Classroom: Do Learning Styles Matter? Accounting Educators' Journal Volume XIII, 2001.

60. Kramer, C. N., Gary, B. (2010). Personalizing e-Learning. The Social Effects of Pedagogical Agents. Springer Science and Business Media, LLC 2010 10.1007/s10648-010-9123-x.

61. Litzinger, T. A., Lee, S. H., Wise, J. C., Felder, R. M. (2005). A Study of the Reliability and Validity of the Felder-Soloman Index of Learning Styles Proceedings of the 2005 American Society for Engineering Education Annual Conference and Exposition, American Society for Engineering.

62. Manouselis, N., Drachsler, H., Vuorikari, R., Hummelv, H., Koper R. (2011). Recommender Systems in Technology Enhanced Learning Recommender Systems Handbook 2011, 387–415.

63. Markham, S. (2004). Learning Styles measurement: a cause for concern, Technical Report Computing Educational Research Group [Accessed, July 2014] [Online] Available: http://cerg.csse.monash.edu.au/techreps/learning_styles_review.pdf.

64. Marshall, S., Mitchell, G. (2005). Applying SPICE to e-learning: an e-learning maturity model? Proceedings of the sixth conference on EDUCAUSE in Australasia, 2005.

65. Martins, A. C., Faria, L., Vaz de Carvalho, C., Carrapatoso, E. (2008). User Modeling in Adaptive Hypermedia Educational Systems. Educational Technology and Society, 11 (1), 194–207.

66. Martins, C., Faria, L., Carrapatoso, E. (2008). Constructivist Approach for an Educational Adaptive Hypermedia Tool, The 8th IEEE International Conference on Advanced Learning Technologies, Santander, Spain, IEEE Computer Society.

67. McLoughlin, C., Luca, J. (2002). A learner-centered approach to developing team skills through web-based learning and assessment British Journal of Educational Technology, 2002 Vol. 33 No 5 2002 Pages 571–582.

68. McLoughlin, C., Lee, W. J. M. (2010). Personalized and self-regulated learning in the web 2.0 era: International exemplars of innovative pedagogy using social software. Australasian Journal of Educational Technology, 2010, 26(1), 28–43.

69. Mehigan, J. T., Pitt, I. (2010). Chapter 11 Individual Learner Styles Inference for Development of Adaptive Mobile Learning Systems - Guy R. (Ed) Mobile Learning: Pilot Projects and Initiatives. Santa Rosa, California: Informing Science Press. (167–183).

70. Melia, M, Pahl, C. (2009). Validation of Adaptive e-Learning Courseware, IEEE Transactions on Learning Technologies, vol. 2, no. 1, January–March 2009.

71. Ormrod, J. E. (1999). Human Learning (3rd ed.) Upper Saddle River, NJ: Merrill/Prentice Hall.

72. Oxford R. L. (2003). Language learning styles and strategies: Concepts and relationships International Review of Applied Linguistics in Language Teaching, 2003.

73. Pashler, H., McDaniel, M., Rohrer D., and Bjork, R. (2008). Learning Styles: Concepts and Evidence . Psychological Science in the Public Internet. Volume 9 Number 3.

74. Pazos Arias, J. J. et al. (2012). Recommender Systems for the Social Web ISRL 32, 195–207 Springer-Verlag Berlin Heidelberg, 2012; Rebeca P. Diaz Redondo, Ana Fernandez Vilas, Jose J. Pazos Arias, Alberto Gil Solla, Manuel Romos Cabrer, and Jorge Garcia Duque. Chapter 10 SCORM and Social Recommendation: A Web 2.0 Approach to E-learning.

75. Peter, E. S., Bacon, E., Dastbaz, M. (2010). Adaptable, personalized e-learning incorporating learning styles," Campus-Wide Information Systems, Vol. 27 Iss: 2, 91–100.

76. Power, G., Davis, H. C., Cristea, A. I., Stewart, C., Ashman, H. (2005). Goal Oriented Personalization with SCORM [Accessed, July 2014] [Online] Available: http://eprints. ecs.soton.ac.uk/10735/01/icalt05_gui.pdf#search='Goal%20Oriented%20Personalization%20with%20SCORM'.

77. Razmerita, L., Lytras, D. M. (2008). Ontology-Based User Modeling Personalization: Analyzing the requirements of a semantic learning portal emerging technologies and information systems for the knowledge society, lecturer notes in Computer Science, 2008, Vol 5288/2008, 354–363.

78. Riding, R. J., Sadler-Smith E. (1997). Cognitive style and learning strategies: some implications for training design International Journal of Training and Development 1997 1:3 ISSN 1360–3736.

79. Rosmalen, P. Van., Vogten, H., Es R Van., Passier, H., Poelmans P., Koper R. (2006). Authoring a full life cycle model in standards-based, adaptive e-learning, Educational Technology and Society, 9(1), 72–83.

80. Saadé, R., Kira, D. (2009). Computer Anxiety in E-Learning: The Effect of Computer Self-Efficacy. Journal of Information Technology Education: Research. 8 (1), pp. 177– 191.

81. Sadler-Smith, E. (1996). Learning styles: a holistic approach. Journal of European Industrial Training 20/7 [1996] 29–36.

82. Safran, C., García-Barrios, V. M., Gütl, C. (2006). Concept-based Context-Modeling System for the Support of Teaching and Learning Activities Proceedings of Society for Information Technology and Teacher Education International Conference 2006, 2395– 2402.

83. Safran, C., Helic, Gütl, C. (2007). E-learning practices and Web 2.0 Conference ICL 2007 September 26–28, 2007 Villach, Austria 1(8).

84. Scott, W., Oleg, L. L., Mark, J., Beauvoir, P., Sharples, P. (2007). Personal Learning Environments: Challenging the dominant design of educational systems. Journal of e-Learning and Knowledge Society — Vol. 3, n. 2, June 2007 (pp. 27–38).

85. Severiens, S. E., Dam, G., Ten, T. M. (1994). Gender differences in learning styles: a narrative review and quantitative meta-analysis: Higher Education 27: 487–501, 1994.

86. Shi, L., Al Qudah, D., Qaffas, A., Cristea, I, A. (2013). Topolor: A Social Personalized Adaptive E-Learning System. User Modeling, Adaptation, and Personalization. Lecture Notes in Computer Science Volume 7899, 2013, 338–340.

87. Skinner, F. B. (1985). Cognitive Science and Behaviorism, British Journal of Psychology (1985). 76, 291–301 (c) 1985 The British Psychological Society.

88. Soff, M. (2013). Gestalt Theory in the Field of Educational Psychology: An Example Gestalt Theory – 2013 (ISSN 0170-057 X) Vol. 35, No.1, 47–58.

89. Spiro, R. J., Feltovich P. J., Coulson R. L. (1996). Two Epistemic World-Views: Prefigurative Schemas And Learning In Complex Domains: Applied Cognitive Psychology, Vol. 10, S51-S61 (1996).

90. Stahl, G., Koschmann, T., Suthers, D. (2006). Computer-supported collaborative learning: An historical perspective. [Accessed, July 2014] [Online] Available:http://citeseerx.ist.psu.edu/viewdoc/download?doi=10.1.1.73.6085&rep=rep1&type=pdf.

91. Subramaniam, G. (2006). "Stickability" in Online Autonomous Literature Learning Programmes: Strategies for Sustaining Learner Interest and Motivation Malaysian Journal of ELT Research ISSN: 1511–8002.

92. Treviranus, J., Nevile, L., Heath, A. (2006). Individualized Adaptability and Accessibility in E-learning, Education and Training Part 1: Framework ISO/IEC JTC1 SC36 Information Technology for Learning, Education, and Training.

93. Tzouveli, P., Mylonas, P., Kollias, S. (2005). Spero a Personalized Integrated E-Learning System. [Accessed, July 2014] [Online] Available: http://manolito.image.ece.ntua.gr/papers/305.pdf.

94. Ullrich, C., Tan, X., Luo, H., Borau, K., Shen, L., Shen, R. (2008). Why Web 2.0 is Good for Learning and for Research: Principles and Prototypes World Wide Web Conference Committee (IW3C2).

95. Wang, F., Chen, D. (2008). A Knowledge Integration Framework for Adaptive Learning System Based on Semantic Web Languages, The 8th IEEE International Conference on Advanced Learning Technologies, Santander, Spain, IEEE Computer Society.

96. Wang, X., E. D. Peter, E. D. L., Curtin, J., Klinc, Kim J. Mi., Davis, R, P. (2012). ITCon – Journal of Information Technology in Construction – ISSN 1874–4753.

97. Kevin Warburton, K., (2003). Deep learning and education for sustainability, International Journal of Sustainability in Higher Education, Vol. 4 Iss: 1, pp. 44–56.

98. Watson, J. B. "Psychology as the behaviorist views it." (1913). Psychological Review, Vol 20(2), Mar 1913, 158–177. doi: 10.1037/h0074428.

99. Yavuz, A., Cigdem, S. C. (2012). Adaptive educational hypermedia accommodating learning styles: A content analysis of publications from 2000 to 2011, Computers & Education, Volume 58, Issue 2, February 2012, Pages 835–842.

100. Yarandi, M., Jahankhani, H., Addel-Rahman H. T. (2013). A Personalized Adaptive E-learning approach based on semantic web technology, Volume 10, Number 2, December 2013.

101. Yalcinalp, S., Gulbahar, Y. (2010). Ontology and taxonomy design and development for personalized web-based learning systems" British Journal of Educational Technology. Volume 41, Issue 6, 883–896, November 2010.

102. Zajac, M. (2009). "Using learning styles to personalize online learning." Campus-wide information systems 26(3), 256–265.

103. Zywno, M. (2002). Instructional Technology, Learning Styles and Academic Achievement Proceedings of the 2002 American Society for Engineering Education Annual Conference and Exposition, American Society for Engineering Education.

CHAPTER 3

THE TREND AND DELIVERY ARCHITECTURE OF E-LEARNING SYSTEM

CHI MAN MUI

Head, Department of Information Technology, Chinese YMCA College, Hong Kong SAR, China, Tel.: (852) 26419588, E-mail: cmmui@cmmui.net

CONTENTS

The rapid advancement of technology has not only improved the living standard, but also the education quality and self-learning opportunities. Nowadays, teaching and learning are not limited to teacher-centered lessons, blackboard and chalk, pens and workbooks, which are boring in students' eyes. Instead, technology enables students to learn everywhere with shrinking computer devices, rising Internet access and programs.

By definition, "e-learning" is a way to use of new technologies and learn through the Internet or other electronic media. E-learning was originated in 1980s. At that time, some courseware published on a CD while some of them published on web. The use of technology was the "technology trigger" stage for e-learning.

The term "e-learning" was introduced in 1999 during a CBT Systems seminar in Los Angeles. This was known as the "Peak of Inflated Expectations" of e-learning. At that time, people think e-learning is everything and the education sector promotes e-learning system like open source student information system (SIS), education tablets, mobile-learning system, etc.

Not surprisingly, following the typical "hype" development, there was a big depression on e-learning in 2000s. People stayed away for a while, and it took some time before the lasting results became visible.

In the past few years, the education sector gradually witnessed a shift back to e-learning again. We may nearly be in the final stage of the hyper cycle now. We use E-Textbooks, Learning Management System (LMS) and Massive Open Online Courses (MOOCs) in many countries. E-learning provides a low costs and faster delivery learning method for students.

These e-learning systems help students learn and explore new knowledge by themselves efficiently. Nowadays, schools and parents emphasize more interaction between teachers and students, self-learning opportunities and e-learning. This chapter also talks about how traditional education can be incorporated into computer and web-based education by e-learning system.

3.1   INTRODUCTION

The term e-learning means 'Electronic Learning' where 'E' stands for Electronic. Thus, e-learning refers to the use of electronic media and information and communication technology (ICT) in education. E-learning means use of computer to impart knowledge or courses in or outside the classroom. E-learning can be used in schools, colleges, business or corporate trainings and distance learning. It has been widely used from K-12 education to corporate trainings.

If one can stop time and economically bring together all the learners as well as the Trainers or Experts at one place, along with all the required resources, then there won't be a need for e-learning. But in today's fast paced world, it is almost impossible to give enough time and spend an adequate amount of money, as budgets are limited. E-learning helps eliminate the constraints of time, resource and distance.

E-learning comprise of all the educational activities that can be performed by individuals or groups, both online or offline, either synchronously or asynchronously. E-learning is also referred as virtual education, online learning, computer-based training, web-based training and Internet-based training. But, e-learning is somewhat different from the above educational methods.

E-learning is available in various types including text, audio, images, animation, and video. It is available in various media like audio or video tapes, CD's, satellite TV, local intranet or extranet, computer and web based learning.

3.1.1 ADVANTAGES OF E-LEARNING

• *Flexible and Self-paced Method of Education.*

E-learning is a flexible and self-paced method of learning and education. Many of the students find it difficult to give time from their daily routine and jobs to attend college for higher education. E-learning helps such students pursue their choice of higher education without compromising much on their daily or household activities, jobs, etc. There is no biding of place, time or speed of learning.

• *Cost Effective and Available 24/7.*

E-learning is saves the costs incurred on printed material and travel cost. E-learning can be used anytime as it is available 24 hours a day and all 7 days a week throughout the year.

• *Open Access to Various Programs.*

E-learning have a given an opportunity to the individual or a group to opt for choice of program from the list of various courses available. With the help of e-learning, one can pursue courses available in different countries sitting at one place.

• *Ease in Reach for Expert Knowledge.*

It is not feasible to bring all the experts in a particular subject at one place. E-learning makes it easy to reach out to experts in their fields located at various different locations.

• *Access to Various Tools and Learning Methods.*

E-learning gives an opportunity to access various tools and learning styles or methods depending upon individual's choice. In traditional classrooms, it is not possible to work on various tools related to the same topic which is a plus point in e-learning where various tools can be used for same topic as per individual's choice and understanding.

• *No Age Constraint and Is Discreet.*

Many a time, student finds it difficult to learn in large group, especially when he finds it difficult to understand particular topic whereas others finds it easy. E-learning helps students learn at their own pace and has no age constraint.

3.1.2   DISADVANTAGES OF E-LEARNING

• *Lack of Social Interaction & Communication.*

E-learning results in lack of communication and social interaction among the students. It also has a disadvantage of lack of social interaction among the student and the teacher which generally does not happen in classrooms.

• *Lack or Delay in Feedback.*

When a student use an option of e-learning, there a might be a case where some doubt arise or the student find it difficult to understand certain topic. Since it is not classroom training, doubts cannot be cleared instantly. There might be a delay in explanation or feedback from the teacher or the trainer.

• *Problem in Understanding Language.*

Some student (who is not a native English speaker) may find it difficult to understand or cope with the language of teaching. This may create hindrance to understand the concepts and to communicate with the teacher.

3.2   E-LEARNING: CATALYST FOR CHANGE IN PEDAGOGY

E-learning with its' growing popularity in several directions plays a crucial role for learners, educators as well as trainers. With the increase in demand for distance education, e-learning has gained a lot of importance to the people. As more and more people are using e-learning, it is essential to understand its importance and role in the paradigm shift in pedagogy (Figure 3.1).

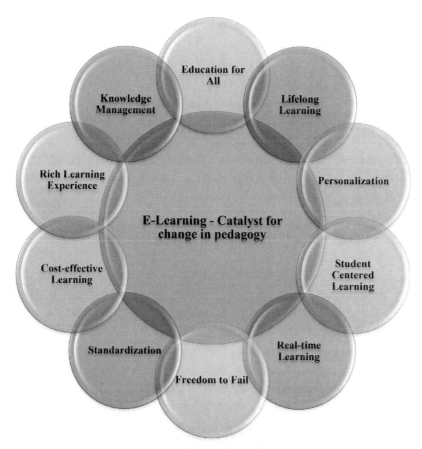

FIGURE 3.1 E-learning: catalyst for change in pedagogy.

3.2.1 EDUCATION FOR ALL

E-learning has a power to reach individuals or groups residing anywhere in the world through computer and Internet. With the help of E-learning, the knowledge is no more limited to the people who belong to that particular area. E-learning provides equal education opportunity to all and provides learners with knowledge and skills.

3.2.2 LIFELONG LEARNING

Earlier people used to learn at particular age only. With the introduction of E-learning, people learn and teach at any age. There is no age bar. E-learning gives an opportunity to improve skills, knowledge and competencies lifelong.

3.2.3   PERSONALIZATION

E-learning gives an opportunity to the individuals or groups to learn through computers or mobile as per their requirements or choice. The Learner can learn at its own pace and make use of the tools or learning styles that suits him/her the best. E-learning has made it easy and comfortable for the learner, which in turn results in effective learning.

3.2.4   STUDENT-CENTERED LEARNING

Teacher-centered and whole class instruction method is no longer dominant. With the help of E-learning, the teacher has become a facilitator, coach or a mentor to the learner, where teacher guide the learners on how to understand the question, find information and analyze the information. E-learning gives confidence to the learners and make them self-dependent.

3.2.5   REAL-TIME LEARNING

E-learning gives an opportunity to the learner to learn anytime and anywhere through the use of computer. Unlike traditional classrooms, learner can learn anytime when it is suitable for them. Also, for live or classroom learning it is important that everyone should be online or available at the same time. But with the help of E-learning, one does not have any time binding. Learner can also learn offline and upload the assignments or results when online.

3.2.6   FREEDOM TO FAIL

Learners many a times find it humiliating to fail in the classroom learning. The leaner may hesitate to ask the question related to the topic he finds it difficult to understand when all others don't find it difficult to understand the topic. E-learning improves learner's confidence and gives opportunity to ask questions and failure to understand the topic.

3.2.7 STANDARDIZATION

E-learning courses are same across the sessions, which may not be possible when the same topic is been taught in different countries by different teachers or tutors. There is consistency in teaching method and the course content.

3.2.8 COST-EFFECTIVE LEARNING

Creation and design of new training content every time is not feasible and it incurs lot of money and time. With the help of E-learning, the course content readily available for anyone and anytime. One does not need to invest time and money to develop the same course content. Also, with the use of E-learning, costs incurred on travel and study or course material is reduced.

3.2.9 RICH LEARNING EXPERIENCE

E-learning combines text, audio, video, animation, images, etc. This combination of multimedia and instructional design gives rich learning experience to the learners and creates an impact on the minds of the learners. The learners find it interesting and easy to understand.

3.2.10 KNOWLEDGE MANAGEMENT

E-learning can be readily used for knowledge management purposes for the individual, groups or the organizations. One can combine and collaborate all the content for the use of future learners. Forums, online community groups, wiki spaces, etc. are effective E-learning solutions.

3.3   THE TREND AND THE HYPE CYCLE FOR E-LEARNING

Every innovation starts in a particular way and performs or goes through some specific process. Gartner's Hype Cycle of E-learning (Figure 3.2) helps us understand the maturity and adoption of E-learning and its applications and how they help benefit the learners.

Five phases of E-learning's life cycle are as follows.

3.3.1   TECHNOLOGY TRIGGER

During the first phase of E-learning life cycle – Podcasting and Mobile E-learning triggered interest in the E-learning technology and its concepts. Every new technology attracts interest of the masses and the media. Similarly these products triggered interest and also proved to be commercially viable.

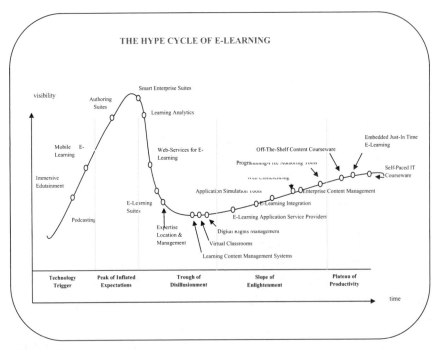

FIGURE 3.2   Hype cycle of e-learning.

3.3.2   PEAK OF INFLATED EXPECTATIONS

Authoring suites, Small Enterprise Suites and Learning Analytics had various success stories especially Small Enterprise Suites and Learning Analytics and proved to be popular among the learners. New ideas on improvement of E-learning got implemented.

3.3.3   TROUGH OF DISILLUSIONMENT

Web services for E-learning, E-learning Suites, Expertise Location and Management, Virtual Classrooms and Digital Rights Management played a crucial role in this stage. Even though Virtual Classrooms proved to be popular and useful, few of these products failed to deliver as per the expectations. Due to the popularity and high demand investments in virtual classrooms still continue.

3.3.4   SLOPE OF ENLIGHTENMENT

People realized the importance and benefits of E-learning with the help of E-learning Application Service Providers, E-learning Integration, Application Simulation Tools, Web Conferencing, Enterprise Content Management, etc. and used widely all over the world. More number of E-learning tools are created to give better learning experience.

3.3.5   PLATEAU OF PRODUCTIVITY

Mainstream adoption of E-learning with the help of Embedded Just-in time E-learning, Off-The-Shelf Content Courseware and Self-Paced IT Courseware started to grow and reached new heights. People became more aware of the importance of e-learning and started analyzing using these applications widely. E-learning's market applicability and relevance are clearly paying off and people are moving towards E-learning.

3.4   DELIVERY ARCHITECTURE

E-learning proves to be an excellent way to achieve high quality results in short time duration. With the introduction of new technologies and concepts, it becomes essential for everyone to update their skills and knowledge from time to time so as to remain in the competition and not get replaced by the competitors.

With the growing industry trends and introduction of new technologies, e-learning market is growing globally. Latest trends of e-learning for the year 2014 are inspiring and very effective.

3.4.1 CLOUD-BASED LEARNING MANAGEMENT SYSTEMS

Cloud-based Learning Management Systems is a most preferred e-learning tool by large enterprises. It is a powerful tool that can improve the e-learning experience. Cloud-based Learning Management Systems are cost-effective and offer superior quality data security, improved accessibility, more storage space, faster deployment and easy maintenance. This tool is fully customizable. Cloud-based Learning Management Systems are hosted on Internet and can be accessed by logging into service provider website. It does not require any course design and management software to be installed on the computers. One can directly use their Internet browsers to use these Cloud-based Learning Management Systems.

3.4.2 MOBILE LEARNING

With increasing use of mobile phones and tablets, people prefer to use their mobile devices or tablets for e-learning. The flexibility to use preferred device for e-learning make people happy and comfortable. Everyone wish to learn on their own terms. Thus, more and more websites are been designed and developed such that they are compatible to any device, be it mobile phones or tablets. With the help of Mobile Learning, e-learning has become more popular and comfortable to the users. With the growth in Mobile Leaning demand, more mobile applications and tools are been designed for e-learning purpose.

3.4.3 GAME-BASED LEARNING

Game based learning a new and interesting way of e-learning where game thinking and mechanics are used to teach or explain various concepts and technologies. People spend hours together to play games. Similar principles are applied to e-learning of certain concepts or technologies such that it maintains interest of the learner in the topic. Game based learning reduces stress in learning and keeps the learner engaged. The learners find Game based learning challenging and interesting.

3.4.4 SOCIAL LEARNING

Social Learning, an effective e-learning tool is growing in popularity. Social Learning is more effective and the results are more productive. People find Social learning encouraging and motivating. This way of e-learning, facilitate communication, enable sharing and helps grow network. When compared to formal learning, social learning, where learning takes place through collaboration, proves to be better. Social learning helps connect talent pools and expertise and find just-in time solutions for the issues.

3.4.5 SCENARIO-BASED LEARNING

Scenario-based Learning is very effective tool for e-learning where learners are put through real life or scenario-based situations. Scenario-based Learning helps learners gain skills and knowledge before the actual situation arises. This will prepare the learners better similar scenario in future. The learners get more involved and interactive in such type of learning. Scenario-based Learning has realism and motivates the learner to face and tackle the situation effectively in future.

3.4.6 JUST-IN-TIME LEARNING

Just-In-Time Learning is a concept where e-learning is been provided through webinars, web-based tutorials, games and quizzes, when needed. Whenever anyone faces a problem or issue in the project, he can contact the expert for help to solve the issue. Just-In-Time Learning eliminates the need to refresh the training. This saves time of training and helps find solutions to the problems when they arise.

3.5   TECHNOLOGY DEVELOPMENT

With growing technology and online trends, e-learning is not just limited to Internet. E-learning can happen on various technologies and not just a computer. With the help of new technology innovations, it becomes necessary that e-learning supports those technologies, so as to make the learner comfortable.

3.5.1   CD OR DVD

CD or DVD in e-learning is used as a storage medium, where all the information on the CD or DVD can be shared with others. CD and DVD are static and its content cannot be replaced or updated easily as compared to online courses. Usually the e-learning course content remains same over the years. Also, all countries do not have good Internet connection all the time and hence they face problem to access the data available on Internet. In such situation, CD and DVD are beneficial.

CD or DVD can also be preserved for years. Even though it has a drawback that e-learning cannot become interactive with the use of CD or DVD as compared to Internet-based course, but still a very popular device used for e-learning.

3.5.2   PERSONAL COMPUTER

Personal Computer is widely used for Internet browsing and e-learning purpose. Though e-learning through personal computer is an old technology, but still it is most popular. People can make use personal computer for e-learning in office as well as at home. Even though it limits the mobility of the user, it is still the most used technology for e-learning. E-learning through personal computers has various advantages over the traditional classroom learning. The learner can take courses at his own convenience and is a cost-effective solution. With the use personal computer, one can connect with anyone residing in other countries.

3.5.3   TABLET

Mobile Learning technique includes learning through devices like mobile phone, notebook and tablet. With growing demand of tablets, people prefer to make use of device they are most comfortable with, for the purpose of learning. Tablet learning can take place anywhere and anytime. Tablet learning allows mobility of the learner, as they are easy to carry. With the more number of Tablet users, more e-learning courses are designed such that they are compatible to devices like mobile phones, tablets and notebooks.

3.5.4   CLOUD

Cloud is the most popular and preferred technology in today's time. It is not feasible for people located in different countries, to remain online at the same time for the purpose of e-learning. Cloud-based e-learning allows the individuals or groups to access information anytime and anywhere using Internet. There is no location or time limitation for Cloud-based e-learning. Cloud-based e-learning is popular among the corporate. With the help of cloud, one can interact with its team members located at different locations, create assignments and review the progress of the project and the team. Cloud-based e-learning provides data security and is a cost-effective solution, mainly for corporate.

3.5.5   MOBILE

Mobile Learning, popularly known as 'M-Learning,' has proved to be beneficial for many users. Generally, people look for comfort more than the knowledge. They are happier when learning takes place with the use of their preferred technology. With the growing number of mobile users over the world, it becomes essential that technology should be compatible with the mobile devices. Thus developers have made the applications compatible with mobile devices with the use of HTML5. Now, one can access or collect information available on Internet using mobile phones. The use of mobile for the purpose of e-learning has saved lots of time for the users.

3.6   TYPES OF DELIVERY METHOD

E-learning can take place in various forms and one can make a choice from the available forms of e-learning.

3.6.1   ONLINE LEARNING

Online Learning does not involve face-to-face meetings between the tutor and the learner. Online courses are available on the Internet. One who opts for online learning can access these courses through web browsers. This learning will be purely online.

3.6.2   BLENDED OR FACE-TO-FACE LEARNING

Blended or face-to-face learning is a combination of classroom learning and online learning. It implies that some part course learning takes place in classroom and other part of course learning takes place online.

3.6.3   SYNCHRONOUS E-LEARNING

Synchronous e-learning refers to the learning where tutors or teachers conduct classes over the Internet. Synchronous e-learning happens in real-time where all the participants interact during the same period of time. Synchronous e-learning includes:

3.6.3.1   Virtual Classrooms

Virtual classroom is just like real-world classroom, but online. Virtual classroom allows participants can interact with each other and with the tutor, view presentations or videos and share documents amongst the participants.

3.6.3.2   Audio or Video Conference

Synchronous e-learning can take place with the help of audio or video conferencing. Audio or Video conferencing can be conducted using Skype where all the participants are online at the same time.

3.6.3.3   Chat

Chat messengers or even Skype can be used for group chat where all the participants are online and learning takes place through text chat.

3.6.3.4   Shared Whiteboard

A Shared Whiteboard allows the participants to share and communicate through comments, drawing, pointing and highlighting. With the help of Shared Whiteboard, all the study material can be shared with all the participants.

3.6.3.5   Application Sharing

With the help of Application Sharing, tutor can demonstrate how to use particular software or give an opportunity to the participant to have hands-on experience on the software.

3.6.4   ASYNCHRONOUS E-LEARNING

Asynchronous e-learning refers to a learning where all the participants do not need to be online at the same time. It is self-paced learning and can take place anytime. There is no dependency on any other participant in asynchronous e-learning.

3.6.4.1   Self-Paced E-Learning

Self-paced e-learning refers to the learning that can take place anytime depending on the convenience of the learner. Self-paced e-learning can happen through CD or DVD, Internet and Intranet or Local Area Network. One can also communicate with the experts, ask questions, track the performance, etc. using self-paced e-learning.

3.6.4.2   Discussion Groups

Various online discussion groups like forums, message boards and bulletin boards allow asynchronous e-learning to the participants, where the participant can discuss and communicate with various experts and take their opinions on the same. It's not necessary that all participants remain online at the same time. Discussion group is in fact a collection or conversation that takes place over the time.

3.6.5 WEB-BASED LEARNING

Web-based learning refers to the learning that can take place anywhere and anytime. The instruction of study material is available on Internet or corporate intranet. Web-based learning can take place either synchronously (instructor-led or virtual classrooms) or asynchronously (self-paced e-learning). Web-based learning can happen through tutorials, portals, audio or video, etc.

3.6.6 COMPUTER-BASED LEARNING

Computer-based learning can refer to the learning that can take place either online or offline through computers. Computer-based learning can also happen with the use of CD or DVD to understand the particular topic.

3.6.7 VIDEO OR AUDIO TAPE LEARNING

E-learning also includes use of audio or video tapes on a particular topic. One can either listen to the audio or have a look at the videos available to understand the topic better.

3.6.8 LEARNING MANAGEMENT SYSTEM (LMS)

Learning Management Systems (LMS) are widely used to manage the process of e-learning. LMS manages the learning, provide reports and manage access to self-paced or instructor-led courses. LMS can be used for the purpose of administration, training and employee management.

3.6.9 LEARNING CONTENT MANAGEMENT SYSTEMS (LCMS)

Learning Content Management Systems (LCMS) supports and manage team-based development of self-paced courses. It includes development tools, project management and quality assurance tools.

3.6.10 KNOWLEDGE MANAGEMENT

Knowledge Management provides support to employees in their jobs. Knowledge management includes document management, knowledge capture and information portals.

3.7 THE INTEGRATION OF E-LEARNING AND TRADITIONAL EDUCATION

E-learning plays a very important role in current scenario, both for teachers as well as students. Just relying on traditional classroom learning is not enough. One needs to update their knowledge and skills from time-to-time. Also, the time duration for classroom training is limited. It becomes difficult for teachers to cover all the topics in the class. Hence there comes an importance of e-learning. E-learning can be integrated with traditional teaching and learning techniques (Figure 3.3). Teachers/students integrate e-learning in various ways.

3.7.1 BLENDED LEARNING

Blended learning is one of the most effective ways of learning. Blended learning is a combination of online and face-to-face or classroom learning. A teacher can conduct few topics in the traditional classroom and other topics can be taught online through online courses, chat, audio or video conferencing, etc.

3.7.2 AUDIO/VIDEO LEARNING

Traditional classroom learning can be made interesting by including various different methods of learning. Teachers can make use of audios or

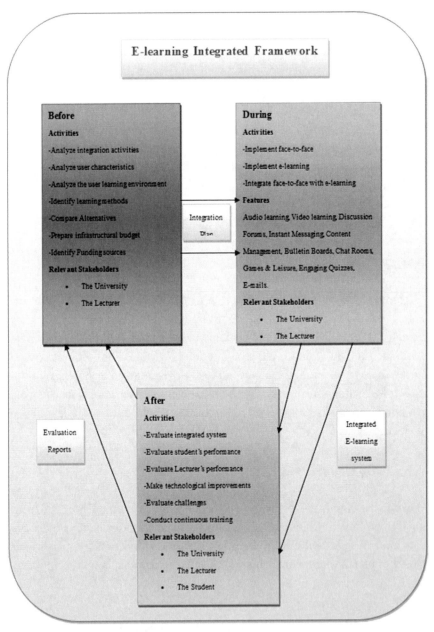

FIGURE 3.3   E-learning integrated framework.

videos with the traditional way of teaching, just to develop interest of the students in the particular topic. Audio or Video learning can also be used to support and understand the topics covered in the classrooms.

3.7.3 COMPUTER-BASED LEARNING OR WEB-BASED LEARNING

Teachers can integrate computer-based or web-based learning with traditional classroom training in order to save time and to explain the concept with practical examples. Teacher can also make use of presentations for the purpose of teaching and learning.

3.7.4 CD/DVD

Teachers or students can also make use of CD/DVD along with traditional classroom training. E-learning courses can be stored in CD/DVD. It can also be used to support the concept and technology. Many a times, Internet connection is limited. In such cases, it is easy to make use of offline CD/DVD for the purpose of learning.

3.7.5 SCENARIO-BASED LEARNING

Teachers can include various scenario-based activities to prepare the student for future. The scenario-based learning can also give an opportunity to make mistakes before the actual situation arise.

3.8 CASES IN HONG KONG SCHOOL

In Hong Kong, most teachers use traditional teaching method. For the science education in most schools, teachers will show the theories to students and proving the correctness of the theories by formulae. However, this method cannot motivate students to explore and learn.

In science education, some experiments are dangerous, for example, the dilution of concentrated acid and alkaline, and cannot be done in school laboratory. Teachers usually demonstrate the experiment in school. However, students cannot visualize the process. Therefore, some e-learning mobile application can be applied; the application simulated the dangerous experiment, and the students could experience the simulated explosion for a wrong procedure (Figures 3.4 and 3.5).

E-learning is popular among the young generation as it is interesting and is a cost-effective solution. Below listed some e-learning methods that are commonly used in Hong Kong:

3.8.1 PRINT

E-learning can take place through the use of print media including – eBooks, e-zine articles, Blog, etc. Some language teachers are blogger in

FIGURE 3.4   Virtual Experiment: Dilution of concentrated acid and alkaline.

FIGURE 3.5   Virtual Experiment: Simulated explosion for a wrong procedure.

Hong Kong and they teach language by uploading articles every day and encourage students to improve their reading skills.

3.8.2   AUDIO AND VIDEO

Audio and Video plays a very important role in e-learning in Hong Kong. Students prefer audios and videos to understand the topic instead of text. They create long lasting impression on the minds and easily understandable.

The concept of flipped classroom and micro-teaching are commonly use in Hong Kong. Teachers always upload videos to the Internet and students learn the topic before and after the lesson.

3.8.3   REVIEW AND EXAMS

The performance of the participant can be checked by conducting exams or quizzes related to the topic. Review and exams can be performed either electronically or by interaction.

3.9   CONCLUSION

The use of e-learning method can accelerate student access to information to improve the availability or reality of learning materials. At the same time, students can be motivated by the e-learning materials and starts learning by themselves.

KEYWORDS

- **delivery architecture**
- **e-learning**
- **hype cycle**
- **Technology Development in Education**
- **traditional education**

REFERENCES

1. Bergin, D. (1999). Influences on classroom interest. Educational Psychologist, 34(2), 87–98.
2. Hung, K. M., Mui, C. M. "An Action Research in Hong Kong: How Does e-Learning Act as a Catalyst for Change in Pedagogy?" International Journal of e-Education, e-Business, e-Management and e-Learning, 2014, vo. 4, no. 1, pp. 67–71.
3. Leidner, D. E., Jarvenpaa, S. L. (1995). The use of Information Technology to Enhance Management School Education: A Theoretical View. MIS Quarterly, 19(3), 265–292.
4. Pachler, N. (1999). 'Theories of Learning and ICT,' In: M. Leask, N. Pachler (Eds.) Learning to Teach Using ICT in the Secondary School, London, Routledge.
5. Prensky, M. (2001). Digital Natives, Digital Immigrants, On the Horizon, Vol. 9 No. 5. NCB University Press.
6. Renzulli, J. (1977). The enrichment triad model: a guide for developing defensible programs for the gifted. Mansfield Center, CT: Creative Learning Press, Inc.
7. Rosas, R., Nussbaum, M., Cumsille, P., Marianov, V., Correa, M., Flores, P., et al. (2003). Beyond Nintendo: Design and Assessment of Educational Video Games for First and Second Grade Students. Computers and Education, 40(1), 71–94.
8. Wade, S. E. (2001). Research on importance and interest: Implications for curriculum development and future research. Educational Psychology Review, 13(3), 243–261.

CHAPTER 4

E-LEARNING 3.0: TECHNOLOGIES, APPLICATIONS, BENEFITS, TRENDS AND CHALLENGE

TE FU CHEN

Assistant Professor, Department of Business Administration,
Lunghwa University of Science and Technology, Taoyuan Taiwan,
E-mail: phd2003@gmail.com

CONTENTS

4.1   INTRODUCTION

"Universities won't survive. The future is outside the traditional campus, outside the traditional classroom. Distance learning is coming on fast."

—Peter Drucker, 1997

The perspicacious Peter Drucker, the man who invented management, said that 16 years ago, and he couldn't have been more right.

Traditionally, the fields of education and training move slowly. The adoption of technology has been relatively quick – taking place in two decades rather than the usual five decades. Christopher (2013) indicated E-learning has gone through rapid and unprecedented changes, following the profound advances in technology and, according to Education Sector Factbook 2012, E-learning is expected to grow at an average of 23% in the years 2013–2017. Distance has long stopped being a barrier, due to the plethora of new devices and APIs, as well as to the super fast Internet speeds available around the world, and the considerable improvement of mobile networks. According to Comscore, in 2014 "the number of mobile users will exceed the number of desktop users for the first time." This means that the default content access points will be mobile phones, laptops, and tablets, which are bound to become more powerful and smarter than ever. Our vocabulary has been enriched with words, phrases and acronyms, such as mLearning, Tin Can, xAPI, HTML5, Gamification, and MOOC, and every day there is a new E-learning trend and a new E-learning technology designed around it (Christopher, 2013).

Nowadays, more than 50% of companies use E-learning and most universities and school systems have adopted technology for their distance learning programs. Also, the adoption of specific new applications takes place over several years rather than in a single year (McIntosh, 2014).

E-learning is an inclusive term that describes educational technology that electronically or technologically supports learning and teaching. Tavangarian et al. (2004) indicated E-learning is the use of electronic media, educational technology and information and communication technologies (ICT) in education. Information and communication systems, whether free-standing or based on either local networks or the Internet in networked learning, undergo many e-learning processes.

E-learning includes numerous types of media that deliver text, audio, images, animation, and streaming video, and includes technology applications and processes such as audio or video tape, satellite TV, CD-ROM, and computer-based learning, as well as local intranet/extranet and web-based learning (Tavangarian et al., 2004). E-learning is broadly synonymous with multimedia learning, technology-enhanced learning (TEL), computer-based instruction (CBI), computer managed instruction, computer-based training (CBT), computer-assisted instruction or computer-aided instruction (CAI), Internet-based training (IBT), flexible learning, web-based training (WBT), online education, virtual education, virtual learning environments (VLE), m-learning, and digital educational collaboration, distributed learning, computer-mediated communication, cyber-learning, and multi-modal instruction. (Day and Payn, 1987). These alternative names individually emphasize a particular digitization approach, component or delivery method, but conflate to the broad domain of e-learning. For example, m-learning emphasizes mobility, but is otherwise indistinguishable in principle from e-learning. Every one of these numerous terms has had its advocates, who point up particular potential distinctions. In practice, as technology has advanced, the particular "narrowly defined" aspect that was initially emphasized has blended into "e-learning" (Udaya and Vamsi, 2014).

Luskin (2010) advocates that the "e" should be interpreted to mean "exciting, energetic, enthusiastic, emotional, extended, excellent, and educational" in addition to "electronic." This broad interpretation focuses on new applications and developments, and also brings learning and media psychology into consideration. Parks suggested that the "e" should refer to "everything, everyone, engaging, easy"(Parks, 2013).

Government Technology (2013) and Army.mil (2013) indicated virtual learning in a narrowly defined semantic sense implies entering the environmental simulation within a virtual world (YouTube, 2013), for example in treating posttraumatic stress disorder (PTSD). In practice, a "virtual education course" refers to any instructional course in which all, or at least a significant portion, is delivered by the Internet. "Virtual" is used in that broader way to describe a course that not taught in a classroom face-to-face but through a substitute mode that can conceptually be associated "virtually" with classroom teaching, which means that people do not have to go to the physical classroom to learn (Karl, 2001). Accordingly, virtual education refers to a form of distance learning in which course content is delivered by various methods such as course management applications, multimedia resources, and video conferencing. Students and instructors communicate via these technologies (Karl, 2001).

The worldwide e-learning industry is economically significant, and was over $48 billion in 2000 according to conservative estimates (EC, 2000).

Developments in Internet and multimedia technologies are the basic enabler of e-learning, with consulting, content, technologies, services and support being identified as the five key sectors of the e-learning industry (Nagy, 2005). Information and communication technologies (ICT) are used extensively by young people (John and Catherine Foundation, 2005). E-learning expenditures differ within and between countries. Finland, Norway, Belgium and Korea appear to have comparatively effective programs (Aristovnik, 2012).

Downes (2005) described the next stage in the development of e-Learning, namely e-Learning 2.0. He noted that "e-learning as we know it has been around for ten years or so. During that time, it has emerged from being a radical idea – the effectiveness of which was yet to be proven – to something that is widely regarded as mainstream." The shift to e-Learning 2.0 was due to the emergence of Web 2.0 and the resulting emphasis on social learning and use of social software such as blogs, wikis, podcasts and virtual worlds (Ebner, 2007; Redecker et al., 2009; Wiki, 2010). The Web 2.0 has enabled large-scale user-generated content production. However, large parts of this data are simply stored away and are rarely, if ever, utilized by others (e.g., 97% of users never look beyond the top three search results (Research, 2006), so millions of carefully crafted documents are never even looked at). Clearly, this situation needs to be remedied, we believe that one of the main objectives of the Web 3.0 would be to enable utilization of this data, and AI will be a perfect tool for accomplishing this (Rubens et al., 2011).

Just like the original concept of e-Learning, e-Learning 2.0 has become a mature and widely accepted paradigm, and like its predecessor before it, is now due for change. If the past is any indicator of the future, then the emergence of e-Learning 3.0 will also be strongly influenced by the technologies that will bring forth the Web 3.0. The concept of Web 3.0 is still in its infancy, but we are starting to see a number of early indicators that Artificial Intelligence (AI) will become an integral part of the Web 3.0 (Rubens et al., 2011).

The chapter is organized as follows. First of all, the study briefly outline how the web technologies influence on the evolution of e-Learning, and discusses the literatures from E-learning 1.0 to 3.0 via the perspectives of technology and applications, then survey the views of other researchers about what e-learning 1.0, 2.0 and 3.0 might be like. Secondly, the chapter summarizes content in e-learning, advantages and disadvantages with regards to motivation in e-learning, explores potential AI utilization by learning methods, outline new developments in AI and their potential influence on the evolution of e-Learning 3.0, the principles of student-centered learning. Thirdly, the study discusses e-learning technology and learning management system, future e-learning trends and technologies in the global and analyzes the applications for E-learning which includes: Preschool, K–12, Online schools, Higher education, MOOCs, DOCC, Coursera, Corporate and professional, Public health, ADHD, Classroom 2.0, Virtual classroom, Virtual learning environment (VLE), Augmented reality (AR). Fourthly, the study summarizes benefits of E-learning. Finally, the study analyzes global trends and technologies in the E-learning Industry.

4.2   FROM E-LEARNING 1.0 TO 3.0

4.2.1   THE EVOLUTION OF E-LEARNING

Rupesh (2009) indicated "E-learning is a technological infrastructure with applications and software that manage courses and users." The software that facilitates e-learning may be called a Learning Management System (LMS) and supports course creation, content delivery, user registration, monitoring, and certification.

Gonella and Panto (2008) have traced the following four stages in the evolution of e-learning:

1.   Web-based Training
2.   E-learning 1.0
3.   Online Education
4.   E-learning 2.0

Gonella and Panto's discussion of those four stages is outlined as Table 4.1.

E-learning has been evolving alongside the World Wide Web. As new web technologies become available, they find their way into the education domain, which by applying these new technologies makes it possible to both utilize new learning methods, and to enhance the use of the existing ones. Simply put, new web technologies enable the application of

TABLE 4.1   Four Stages in the Evolution of E-Learning

Stages	Contents
Web-based Training	Web-based training emerged in the 1990s in business. It was based on the "online distribution of autonomously used learning materials." The emphasis was on "training" rather than on education or learning. The contents were mainly multimedia pages, which users would consult for information. With these web-based training systems, it was not possible to track the use of learning materials.
E-learning 1.0	The web-based training model evolved into the e-learning architecture, which can be referred to as "E-learning 1.0." E-learning uses and LMS to create, design, and manage courses, as well as supporting content delivery, user registration, monitoring, and certification. The focus of the system is on content and learning objects, with less consideration for the learning process. There is not much scope for communication and collaboration. Even though tools for collaboration are available, their application in learning is negligible.
Online Education	Earlier learning infrastructures had little or no provision for interaction. In the late 1990s, educators began emphasizing the active role of students in the learning process. Collaboration and communication tools assumed greater importance and teachers and students began using simple technologies such as mailing lists and newsgroups for interaction. Later, more sophisticated tools like conferencing systems were introduced. Learning consisted not only of materials delivered by the teacher but also of interactions and discussions among students, making learning a social process.
E-learning 2.0	The learning process is transformed when courses are interactive. User (student) contribution is not limited to newsgroups and mailing lists. "Social software" has revolutionized online learning. Web 2.0 has given birth to e-learning 2.0. The influence of new practices on the Web has resulted in a new array of services, which can be collectively termed "E-learning 2.0."

Source: Gonella and Panto (2008).

learning theories to E-learning. If we can understand this trend, then we can better understand the future of e-Learning. The best way to understand this trend, is to first look at it historically, by taking a quick look at where it has been. Then we can better assess existing predictions about where it is heading, and make our own predictions as well (Rubens et al., 2011).

4.2.2   E-LEARNING 1.0

Web 1.0 made content available online. This was a very significant development since it allowed (at least in principle) for easy access to view (or read) information. However, this "access" is often seen as the staple functionality available with the Web 1.0, which is why it is often referred to as the "read-only Web" (Richardson, 2005). E-Learning 1.0 quickly adapted this new technology. The motto "anytime, anywhere and anybody" emphasized characteristics of e-Learning 1.0 – providing easy and convenient access to educational contents (Ebner, 2007). Therefore, e-Learning 1.0 mostly focused on creating and administering content for viewing online. To ensure high quality and usefulness of created "read-only" materials, the concept of a "learning object" was developed.

Learning objects could be thought of as Lego blocks that allow for sequencing and organizing bits of content into courses, and to package them for delivery as though they were books or training manuals (Downes, 2005). In turn, to support the utilization of learning objects the concept of a learning management system (LMS) was introduced. The learning management system takes learning content and organizes it in a standard way, such as a course divided into modules and lessons, supported with quizzes, tests and discussions, and integrated into university's information system (Downes, 2005). These frameworks allowed not only to provide access to educational materials but also to log and analyze their usage. This in turn, allowed for an application of a number of learning theories and methods, including instructivism, behaviorism, and cognitivism, each briefly summarized below.

Instructivism focuses on transferring content from teacher to learner (Redecker et al., 2009). Utilizing the web for content distribution provides an alternative channel to lectures and textbooks. This theory requires the student to passively accept information and knowledge as presented by the instructor, which made it a particularly good fit for rather passive LMS of the Web (and e-Learning) 1.0. Behaviorism treats learning as a "black box" process, that is, only the inputs and corresponding outputs are observable quantitatively, and the in networking of the learning process are assumed to be unknown. It is based on the principle of stimulus-response, and learning is viewed as an acquisition of new behavior through either classical conditioning, where the behavior becomes a reflex response to stimulus as in the case of Pavlov's Dogs, or operant conditioning, where there is reinforcement of the behavior by a reward or punishment (Meetoo-Appavoo, 2011).

Behaviorism emphasizes performance rather than the reasons that the learner performs a certain way (Chen, 2003). The interactive capabilities of LMS 1.0 systems were primarily based on the behaviorism. LMS logs were used to observe: "input" through the access logs of learning materials, "output" through the advancement and performance measures, as well as system's attempts at conditioning. Once this data is analyzed the necessary adjustments could be made as to condition the learning process more efficiently. Cognitivists were not satisfied with treating learning as a black-box process, and by analyzing learning logs tried to gain better understanding of the inner workings of the mind during the learning process (Meetoo-Appavoo, 2011) . The obtained knowledge was then incorporated into LMS as to take into consideration what has became known about the processes of learning, such as thinking, memory, knowing, and problem-solving.

4.2.3   E-LEARNING 1.0 VS. E-LEARNING 2.0

One may say that e-learning 1.0 is based on Web 1.0, and e-learning 2.0 is based on Web 2.0. E-learning is found in the following forms (Hart, 2008):Online courses with a blend of online teaching and face-to-face interaction (to a limited extent). Live learning systems that support the delivery of scheduled online sessions. The factor most lacking in e-learning 1.0 is the social and collaborative approach to learning. With the fusion of e-learning 1.0 and Web 2.0, a new set of services has emerged, which makes the learning process more creative and learning experience more enduring. Hart (2008) says that, E-learning 1.0 was all about delivering content, primarily in the form of online courses and produced by experts, that is, teachers or subject matter experts. E-learning 2.0 is about creating and sharing information and knowledge with others using social media tools like blogs, wikis, social bookmarking and social networks within an educational or training context to support collaborative approach to learning. Tracing the changes in the world of e-learning, Downes (2005) remarks that the model of e-learning as being a type of content, produced by publishers, organized and structured into courses, and consumed by students, is turned on its head. Insofar as there is content, it is used rather than read— and is, in any case, more likely to be produced by students than courseware authors. And insofar as there is structure, it is more likely to resemble a language or a conversation rather than a book or a manual.

4.2.4   E-LEARNING 2.0

The Web has undergone a metamorphosis, bringing radical changes in the information industry. "Collaboration" is the hallmark of Web 2.0. Here are some notable changes that have occurred on the Web over the past few years, as observed by Downes (2005). These may also be perceived as the characteristics of Web 2.0 shown as Table 4.2.

4.2.4.1   Definition of E-Learning 2.0

Gonella and Panto (2008) present the following definition: e-learning 2.0 refers to a second phase of e-learning based on Web 2.0 and emerging trends in e-learning. The term suggests that the traditional model of e-learning as a type of content, produced by publishers, organized and structured into courses, and consumed by students, is reversed; so that content is used rather than read and is more likely to be produced by students than courseware authors.

4.2.4.2   Significant Features of E-Learning 2.0

Table 4.3 shows a few significant features of e-learning 2.0.

In addition to the ability of reading contents online (provided by Web 1.0), Web 2.0 introduced a capability to write (or save content) and is therefore commonly referred to as the "read-write web" (Downes, 2005; Rubens et al., 2011). The capability to write content allowed for the transformation of the web from a passive provider of information to a social platform that allowed people to interact and collaborate with each other by expressing their thoughts and opinions online. Utilizing these new

TABLE 4.2   The Characteristics of Web 2.0

Characteristics	Contents
"Read-Only Web" to "Read-Write Web"	A few years ago, the practice was that "hundreds would publish and millions would read." The Web was more "readable" than "writable." There was no or little scope for "ordinary" user to write on the web. Web 2.0 has made it a "Read-Write Web," with as many writers as readers. Web 2.0 has provided a democratic publishing space.
Web of Documents to Web of Data	The Web was previously a repository of static documents. Now, it is more dynamic with new data being added constantly. With the new set of tools, people syndicate and remix existing content in new and useful ways. It is not the documents but the data, which rules Web 2.0. "Web as a Medium" to "Web as a Platform." The Web was used as a medium by publishers to transmit information and by users to consume it. Now, it is a platform to create, share, organize, and distribute content, and anyone can use this platform.
Reactive User to Proactive User	Web users had a very limited role in the Web 1.0 environment. They were expected to consume readymade sources of information and react. In the Web 2.0 environment, the availability tools for communication and collaboration has made the user proactive. These tools empower users to initiate communication and collaborate among peers to share their views and reviews. Users stay connected to each other and inform each other of happenings on the Web. Thus, the user is no longer a spectator, but a dynamic actor on the Web 2.0 platform.
Technological Revolution to Social Movement	Web 2.0 enables users to connect, communicate and collaborate with each other, forming online communities and socializing. Web 2.0 encourages participation through open applications and services. Everyone has the right to create content, use it, and reuse it. Web 2.0 is not a technological revolution, but a social one.

Source: Downes (2005).

capabilities allowed e-Learning 2.0 to incorporate the social aspects of learning theories (Mondahl et al., 2009; Anderson, 2007).

Constructivism views learning as a process of knowledge construction rather than absorption (Glasersfeld, 1996; Chen, 2003). It is often associated with pedagogic approaches that promote active learning or learning by doing (Meetoo-Appavoo, 2011). The use of Web 2.0 technologies allowed incorporation into LMS 2.0 the capability to allow students not only to passively read educational materials, but also to express opinions and to socialize (Meetoo-Appavoo, 2011).

TABLE 4.3   Significant Features of e-Learning 2.0

Features	Contents
People-Centered Learning	E-learning 1.0 paid attention to the delivery of content. Delivering content authored by subject matter experts will not stimulate learning. E-learning 2.0 brings people together in the learning process. It focuses on "people-centered learning" rather than on "content-centered learning."
Bottom Up Learning Approach	In a conventional e-learning system, an instructor prepares material according to a particular curriculum and uploads it to the LMS where it is consumed by learners. There is a top–down approach in such a system. The autonomy of creating and publishing contents lies with the tutor. E-learning 2.0 adopts a bottom–up approach. It gives learners the opportunity to participate in courseware creation.
Content Creation	The traditional e-learning system limits the role of student to that of content consumer, with no role in content creation. A successful learning process is one in which there is scope for interaction between teachers and learners at every stage. E-learning 2.0 provides tools for collaboration among students. Students discuss the course and its contents, interact with teachers and help them design learning resources. Learners have their say in deciding and getting what is most useful to them.
Dynamic Content Publishing	Content publishing is instant and dynamic in an e-learning 2.0 environment. With tools like blogging, one can publish content on the fly. Students can read each other's blog posts, comment and interact, thus forming a social network among them. Further, with tools like wikis, students may create and edit content collectively, which promotes collective authorship and harnesses collective intelligence.
Folksonomy	The LMS organizes learning objects in a standard way, in terms of modules and lessons, tests and discussions. This system of organizing content is rigid. Students have no liberty to organize contents in their own way. With the use of "tagging" tools, e-learning 2.0 facilitates organization of information in a personalized way. This enables quick access to the learning resources. More emphasis is given to folksonomy than taxonomy.

Source: Gonella and Panto (2008).

This functionality made it possible to include ambiguous situations and open-ended questions, which further promoted extensive dialogues among students (Meetoo-Appavoo, 2011; Wikipedia, 2014). Overall, this allowed for the embedding into learning environments the corner stones of constructivist learning: context, construction, and collaboration (Jonassen and Land, 2000; Wikipedia, 2014), unlike social constructivism which focuses on an individual's learning that takes place because of their interactions in a group (Zhong et al., 2000). This motivated the incorporation of wikis and social networking services (SNS) into LMS 2.0, which enabled both the collaborative construction of artifacts and the corresponding examination and analysis of created artifacts (Wikipedia, 2014).

4.2.4.3   The Challenges in E-learning 2.0

The success of an e-learning 2.0 platform depends on many factors. One important factor is the preparedness of students to use the platform for their benefit. A highly ambitious e-learning 2.0 project may not deliver the expected results and may end up being a failure if its chief beneficiaries, i.e., the students, lack familiarity with Web 2.0 and its tools. In other words, an e-learning 2.0 environment expects a student to be a "student 2.0" and demands more responsibility and accountability. Therefore, it cannot be expected to deliver success at all levels of education. The level of expertise among peers is a matter of concern in an e-learning 2.0 environment. As Calvani, et al. (2008) have observed, "E-learning 2.0 is mostly based on peer-to-peer learning; as any other environment-based on collaboration, the principle remains valid as regards people that interact: the more expert they are, the better the chances that interactions are mutually profitable." Another challenge is resources and learning cultures. In organizations where resources are limited and learning cultures vary, the implementation of e-learning 2.0 is a hard task.

4.2.5   E-LEARNING 3.0

Rubens et al. (2011) indicated "The concept of e-Learning 2.0 has become well established and widely accepted. Just like how e-Learning 2.0 replaced its predecessor, we are again on the verge of a transformation. Both previous generations of e-Learning (1.0 and 2.0) closely parody the prevalent technologies available in their kin Web versions (1.0 and 2.0, respectively).

In order to acquire a better perspective to assess what technologies will be available in the Web 3.0 and therefore e-Learning 3.0, Rubens et al. (2011) take a historical glance at the previous generations of e-Learning and the Web. They then survey some existing predictions for e-Learning 3.0 and finally provide their own. Previous surveys tend to identify educational needs for e-Learning, and then discuss what technologies are required to satisfy these needs. Educational needs are an important factor, but the required technologies may not reach fruition. Gauging past trends Rubens et al. (2011) take the reverse approach by first identifying technologies that are likely to be brought forth by the Web 3.0, and only then looking at how these technologies could be utilized in the learning domain. In particular, Rubens et al. (2011) pinpoint Artificial Intelligence (more specifically Machine Learning and Data Mining) as a major driving force behind the Web 3.0. They therefore examine the influence that AI might exert on the development of e-Learning 3.0."

Rubens et al. (2011) indicated the predictions of the future of e-Learning vary due to the differences in opinions of what the Web 3.0 will be like, and which technologies will best suit the needs in the learning domain. Wheeler (2011) considers that the Web 3.0 will be the "Read/Write/ Collaborate" web. E-Learning 3.0 will have at least four key drivers: distributed computing, extended smart mobile technology, collaborative intelligent filtering, 3D visualization and interaction. Distributed computing in combination with smart mobile technology will enable learners to come closer to "anytime anyplace" learning and will provide intelligent solutions to web searching, document management and organization of content. It will also lead to an increase in self-organized learning, driven by easier access to the tools and services that enable us to recursively personalize our learning. Collaborative intelligent filtering performed by intelligent agents will enable users to work smarter and more collaboratively. 3D visualization and interaction will promote rich learning, by making a whole range of tasks easier including fine motor-skill interaction, exploration of virtual spaces and manipulation of virtual objects. Halpin (2007) considers that e-Learning 3.0 will be both "collaborative" and "intelligent."

Rubens et al. (2011) indicated intelligent agents will "facilitate the human thinking greatly." Collaboration will be further improved by tools like Twitter due to a number of its "communicative conceptual characteristics"; a place to share and consume information, a new real-time search engine, a service for Web users, a platform of debate, a tool for listening and analyzing, a perfect traffic generator, an excellent means to meet new people and create new connections, and talk about what you are doing right now. Moravec (2008) suggests that in e-Learning 3.0, meaning will be socially constructed and contextually reinvented, and teaching will be done in a co-constructivist manner. The focus of learning will shift from "what to learn" to "how to learn." The technology will play a central role, however it will do so in the background and become invisible. Technology will connect knowledge, support knowledge brokering, and enable translation of knowledge to beneficial applications.

Rego et al. (2010) considers that e-Learning concept of "anytime, anywhere and anybody" will be complemented by "anyhow," that is, it should be accessible on all types of devices. Virtual 3D worlds such as Second Life are expected to become a common feature of the 3D web, facilitated by the availability of 3D visualization devices. As a result, e-Learning 3.0 will be able to reach a wider range and variety of persons being available on different kinds of platforms/systems, through different tools, where users will have the possibility to personalize their learning and have an easier access to comprehensive information. This situation may turn e-Learning into a cross-social learning methodology since it will be possible to be applied in all contexts, making collaboration easier. To enable this view of e-Learning 3.0, Rego et al. (2010)considers that LMS systems need to be capable of representing information through metadata, granting semantics to all contents in it, giving them meaning.

Chen (2003) considers that machine-understandable educational material will be the basis for machines that automatically use and interpret information for the benefit of authors and educators, making e-Learning 3.0 platforms more adaptable and responsive to each individual learner. A number of researchers express concerns about the issues that will arise with the advent of e-Learning 3.0. Silva et al. (2011) warns that the evolution of e-Learning management systems significantly enhances ethical dilemmas, and advocate for the adoption of an extension of the

"Three Ps" model of pedagogy to become the "P3E" model: personalization, participation, productivity, lecturer's ethics, learner's ethics, and organizational ethics. (Lukasiewicz, and Straccia, 2008; Glasersfeld, 1996) are concerned that e-Learning will be impacted by some of the challenges of the Semantic Web including vastness, vagueness, uncertainty, inconsistency, and deceit. Alkhateeb et al. (2010) expresses concerns about the privacy and loss of control, as university integrates into its infrastructure services that are located in a variety of countries, with different privacy laws and principles.

The existing forecasts about where e-Learning 3.0 is heading tend to focus on educational aspects, and only briefly mention technical aspects. The study takes a different approach; since from past transformations (e-Learning 1.0 and e-Learning 2.0), its close relationship with the available technologies of the current generation of the World Wide Web. The challenge then is to correctly identify which technologies will be brought forth by the Web 3.0. We then hypothesize how these technologies may be utilized in e-Learning 3.0 (Rubens et al., 2011).

4.2.6 CONTENT

While there are a number of means of achieving a rich and interactive E-learning platform, one option is using a design architecture composed of the "Five Types of Content in E-Learning." Content normally comes in one of five forms (Clark, Mayer, 2007):

1. Fact – unique data (e.g., symbols for Excel formula, or the parts that make up a learning objective);
2. Concept – a category that includes multiple examples (e.g., Excel formulas, or the various types/theories of Instructional Design);
3. Process – a flow of events or activities (e.g., how a spreadsheet works, or the five phases in ADDIE);
4. Procedure – step-by-step task (e.g., entering a formula into a spreadsheet, or the steps that should be followed within a phase in ADDIE);
5. Strategic Principle – task performed by adapting guidelines (e.g., doing a financial projection in a spreadsheet, or using a framework for designing learning environments).

Content is a core component of e-learning and includes issues such as pedagogy and learning object re-use as Table 4.4.

4.2.7   ADVANTAGES AND DISADVANTAGES WITH REGARDS TO MOTIVATION IN E-LEARNING

There are several advantages and disadvantages with regards to motivation in e-learning. For many students, e-learning is the most convenient way to pursue a degree in higher education. A lot of these students are attracted to a flexible, self-paced method of education to attain their degree. It is important to note that many of these students could be working their way through college, supporting themselves or battling with serious illness (Webster, 2013). To these students, it would be extremely difficult to find

TABLE 4.4   Content Issues

Issues	Description
Pedagogical elements	Pedagogical elements are defined as structures or units of educational material. They are the educational content that is to be delivered. These units are independent of format, meaning that although the unit may be delivered in various ways, the pedagogical structures themselves are not the textbook, web page, video conference, Podcast, lesson, assignment, multiple choice question, quiz, discussion group or a case study, all of which are possible methods of delivery.
Pedagogical approaches	Various pedagogical perspectives or learning theories may be considered in designing and interacting with e-learning programs. E-learning theory examines these approaches, including social-constructivist, one application of which was One Laptop Per Child, (Wiki.laptop.org, 2013) Laurillard's conversational model (Macs.hw.ac.uk, 2013) including Gilly Salmon's five-stage model (Salmon, Kogan, 2000), and cognitive (Bloom and Krathwohl, 1956), emotional (Bååth, 1982), behavioral (Areskog, 1995) and contextual perspectives (Black and McClintock, 1995). In 'mode neutral' learning online and classroom learners can coexist within one learning environment, encouraging interconnectivity (Smith et al., 2008). Self-regulated learning refers to several concepts that play major roles in e-learning. Learning courses should provide opportunities to practice these strategies and skills. Self-regulation and structured supervision both enhance e-learning (Williams and Hllman, 2004).

TABLE 4.4   Continued

Issues	Description
Learning object standards	Much effort has been put into the technical reuse of electronically based teaching materials and in particular creating or re-using learning objects. These are self-contained units that are properly tagged with keywords, or other metadata, and often stored in an XML file format. Creating a course requires putting together a sequence of learning objects. There are both proprietary and open, non-commercial and commercial, peer-reviewed repositories of learning objects such as the Merlot repository.
	Sharable Content Object Reference Model (SCORM) is a collection of standards and specifications that applies to certain web-based e-learning. Other specifications such as Schools Framework [dead link] allow for the transporting of learning objects, or for categorizing metadata (LOM). These standards themselves are early in the maturity process with the oldest being 8 years old. They are also relatively vertical specific: SIF is primarily pK-12, LOM is primarily Corp, Military and Higher Ed, and SCORM is primarily Military and Corp with some Higher Ed.
	PESC – the Post-Secondary Education Standards Council- is also making headway in developing standards and learning objects for the Higher Ed space, while SIF is beginning to seriously turn towards Instructional and Curriculum learning objects. In the US pK12 space there are a host of content standards that are critical as well- the NCES data standards are a prime example. Each state government's content standards and achievement benchmarks are critical metadata for linking e-learning objects in that space.
	An excellent example of e-learning that relates to knowledge management and reusability is Navy E-Learning, which is available to Active Duty, Retired, or Disable Military members. This online tool provides certificate courses to enrich the user in various subjects related to military training and civilian skill sets. The e-learning system not only provides learning objectives, but also evaluates the progress of the student and credit can be earned toward higher learning institutions. The Internet allows for learning to be directed at one's current objectives (Armstrong, 2012). This reuse is an excellent example of knowledge retention and the cyclical process of knowledge transfer and use of data and records.

time to fit college in their schedule. Thus, these students are more likely and more motivated to enroll in an e-learning class. Moreover, in asynchronous e-learning classes, students are free to log on and complete work any time they wish. They can work on and complete their assignments at the times when they think most cogently, whether it be early in the morning or late at night (Cull et al., 2013). However, many teachers have a harder time keeping their students engaged in an e-learning class. A disengaged student is usually an unmotivated student, and an engaged student is a motivated student (Dennen and Bonk, 2013). One reason why students are more likely to be disengaged is that the lack of face-to-face contact makes it difficult for teachers to read their students' nonverbal cues (Cull et al., 2013), including confusion, boredom or frustration. These cues are helpful to a teacher in deciding whether to speed up, introduce new material, slow down or explain a concept in a different way. If a student is confused, bored or frustrated, he or she is unlikely to be motivated to succeed in that class.

Table 5.5 summarizes the key advantages and disadvantages of e-learning.

4.3 ARTIFICIAL INTELLIGENCE

In more open-ended domains such as education, the success of AI has been limited. This limitation primarily comes from the fact that open-ended domains are inherently more complex and therefore an AI system needs to contain a lot of parameters, which in turn require a lot of data for estimation and as a result require significant amounts of computational power (Rubens et al., 2011).

4.3.1 THE COMPARISONS OF BIG DATA, LINKED DATA, CLOUD COMPUTING AND DATA-DRIVEN SCIENCE

Rubens et al. (2011) indicated crucial components needed for the AI to succeed in more general open-ended domains starting to fall in place. There is a vast amount of data of available, importantly a lot of this data is "open" to a wide audience, that is, not hidden behind the corporate or institutional walls (Big Data). No matter how vast the dataset is it tends to provide a limited view on the problem. New technologies are allowing establishing links between these datasets as to obtain a more complete picture (Linked Data). The significant infrastructure needed to store and intelligently process this data is now becoming easily accessible and affordable

TABLE 4.5   Key Advantages and Disadvantages of E-Learning

Advantages and disadvantages	Contents
Key advantages of e-learning	1. Improved open access to education, including access to full degree programs (Zameer, 2013).
	2. Better integration for non-full-time students, particularly in continuing education (Zameer, 2013).
	3. Improved interactions between students and instructors (Christian, 2013).
	4. Provision of tools to enable students to independently solve problems (Christian, 2013).
	5. Acquisition of technological skills through practice with tools and computers.
	6. No age-based restrictions on difficulty level, that is, students can go at their own pace.
Key disadvantages of e-learning	1. Ease of cheating.
	2. Bias towards tech-savvy students over non-technical students,
	3. Teachers' lack of knowledge and experience to manage virtual teacher-student interaction (Illinois University, 2008).
	4. Lack of social interaction between teacher and students, (E-learning-companion.com, 2013).
	5. Lack of direct and immediate feedback from teachers, (E-learning-companion.com, 2013).
	6. A synchronic communication hinders fast exchange of question, (E-learning-companion.com, 2013).
	7. Danger of procrastination. (E-learning-companion.com, 2013).

(Cloud Computing). The new scientific framework is becoming available for supporting AI in the process of scientific discovery (Data-Driven Science). The chapter reviews the comparisons of Big Data, Linked Data, Cloud Computing and Data-Driven Science in Table 4.6.

4.3.2   POTENTIAL AI UTILIZATION BY LEARNING METHODS

Rubens et al. (2011) indicated the traditional approach to constructing e-Learning systems has been a time consuming process requiring an explicit implementation of all the assumptions and rules. Adapting AI in e-Learning

TABLE 4.6   The Comparisons of Big Data, Linked Data, Cloud Computing and Data-Driven Science

Data	Contents
Big Data	Recently, vast datasets are becoming openly available due to increase in user-generated data brought forth technologies provided by Web 2.0. The type of data that is being generated also differs, larger portions of it are user generated such as blogs, tweets, and wikis. To emphasize the importance of the role that data is expected to play, Tim Berners-Lee has suggested that the next generation of Web should be referred to as "Data Web" (King, 2011). Currently web data is severely underutilized, for example, 97% of users never look beyond the top three search results (Research, 2006) so other millions of carefully crafted documents are never even looked at. Web data contains a precious resource –intelligence and is therefore often referred to as "Web Intelligence" (Zhong et al., 2000). This intelligence needs to be extracted and utilized, and AI is a perfect tool for accomplishing this objective. We consider that the role of Web 2.0 was to enable data production, and the role of Web 3.0 will be to enable utilization of this data.
Linked Data	Web 2.0 Data exhibits different characteristics, it is no longer stored in a central well-structured databases, but is in a free-form, fragmented and is spread across the Internet. One of the objectives of the next generation web was defined as to create "a web of data that can be processed directly and indirectly by machines" (King, 2011). Semantic Web is often considered a popular choice for accomplishing this task (King, 2011). However, we along with many others, for example, Lukasiewicz and Straccia (2008) believe that semantic linking is overly ambitious and is yet hard to achieve on the wide and general scale due to inherent ambiguity of natural language. However, this does not mean that the data could not be linked and utilized. In order to widen the linking objectives the concept of "Linked Data" has been recently developed (Fischetti, 2010). There has been a number of success of using AI to produce the needed links that even captures some of the semantics, for example, folksonomy (Goroshko and Samoilenko, 2011).
Cloud Computing	Processing and analyzing large quantities of data requires significant computational resources as well as frameworks to make these resources easily accessible. A variety of competitively priced cloud computing services are becoming available, for example, Amazon's AWS, Google's App Engine, Microsoft's Azure to name a few. In addition a number of supporting frameworks has been developed that made the power of computational clouds easily accessible, for example, a widely adopted Hadoop/Map Reduce, and more specialized ones such as Mahout, Hive, Pig, Oozie, and Rhipe (Rubens et al., 2011).

TABLE 4.6 Continued

Data	Contents
Data-driven Science	Large data sets can potentially provide a much deeper understanding of both nature and society (Challenges and opportunities, 2011). As a result, social scientists are getting to the point in many areas at which enough information exists to understand and address major previously intractable problems (Jonassen and Land, 2000). As a result, science is becoming data-driven at a scale previously unimagined and is fundamentally transforming the scientific process and is driving new innovations in science (Moravec, 2008). The traditional scientific process has followed a top–down approach of starting with a hypothesis, collecting the needed data, and finally evaluating the hypothesis. Data collection has traditionally been a very expensive and time-consuming process. Therefore, starting collecting data without having a hypothesis in mind was a very risky endeavor. Recently large number of datasets has become available for little or no cost, in addition collecting your own data has become very inexpensive (e.g., Rubens et al. 2011 have assembled a dataset of 20 million tweets on various aspects of learning for under $10). Having vast amount of data available, allowed to make scientific process a bottom–up approach, that is, by first gathering data, and then performing analysis as to discover hypotheses hidden within the data. Data-driven science (Nelson, 2009) is starting to gain a foothold in the education, as indicated by the rapid development and increasing applications in the new areas of educational data mining (EDM) (Baker and Yacef, 2009) and learning analytics (Siemens, 2010).

may allow us to concentrate more on modeling rather than a tedious implementation of all the rules. Maintenance and adaptation cost will also be reduced, since by using the data-driven approach, the model is able to adapt to the new users and contents, as well as potential changes. However, Marshall and Shipman (2003) assumed due to the open-ended nature of learning, applying AI is not always straightforward. Out of all learning theories, application of AI to behaviorism is one of the most straightforward ones. This is because behaviorism treats learning as a "black box" process, that is, only the inputs and corresponding outputs are observable quantitatively, and the inner workings of the learning process are assumed to be unknown. Chen (2003) proposed behaviorism emphasizes performance rather than the reasons that the learner performs a certain way. Rubens et al. (2011) summarized this paradigm fits perfectly with AI, as a main goal of AI is to model the dependency between inputs and outputs. Once the model is obtained the inputs needed for the desired response could be obtained by applying methods based on stochastic optimization or control theory.

According to Wiki (2010), cognitivisim in a way similar to behaviorism takes an objective approach by assuming that knowledge and learning tasks can be identified and performance can be measured, and the objective of education is to analyze and influence thought processes. Constructivism considers that learning and teaching is an open-ended process, unlike objective methods such as behaviorism and cognitivism. Rubens et al. (2011) indicated open-ended and vague nature of constructivism makes it difficult to analyze. Since there are no clear objectives, traditional supervised learning methods are not applicable. However, unsupervised learning methods do not require an explicit objective and therefore are applicable in these settings. By using unsupervised methods it is possible to find and analyze patterns that may allow to further enriching constructivist view of learning.

Rubens et al. (2011) summarized:

1. Given these similarities, AI methods could be applied to cognitivism in manner that is similar to behaviorism, but in addition, the explicit assumptions of cognitivism on how we store and manipulate information's need to be incorporated which could be achieved by utilizing existing cognitive models. Unlike behaviorism, many of the learning theories aim at understanding the inner workings of the learning process; therefore applying black-box models is not suitable.

2. There are a number of white-box machine learning methods whose inner workings could be easily examined, analyzed, verified and extended, for example, graphical models such as decision trees and Bayesian networks, or rule based methods such as inductive logic or association rule learning. Social aspects of learning could be thoroughly studied thanks to the data available from e-Learning and Web 2.0 tools.

3. Since collaborative tools are often used in an informal, self-driven manner, it allows to better understand collaborative behaviors outside of confines of classrooms and formal educational institutions; which previously were difficult if not impossible to do.

4. There is a multitude of methods, which could be applied to better understand collaborative behaviors. Network analysis could be used for examining structural dynamics of collaboration; natural language processing along with network flow models could be used to better understand knowledge diffusion.

5. Having e-Learning systems that are AI and data-driven, by no means precludes from utilizing and including existing pedagogical knowledge. Quite the opposite, existing knowledge can and should be incorporated into e-Learning systems. However, its influence on systems decision and its interdependency with others parts of the model should be estimated and verified based the data, rather than try setting these parameters via "educated" guessing.

6. A possible way of incorporating existing knowledge is by providing data on features that are considered important in the learning process (e.g., difficulty level of learning materials, student's learning type, etc.). If the data on the feature is already available it could be simply fed into the AI system; if not, feature mapping too can be learned from the data.

4.4 E-LEARNING TECHNOLOGY

4.4.1 EDUCATIONAL APPROACH

Bates and Poole (2003) and OECD (2005) indicated the extent to which e-learning assists or replaces other learning and teaching approaches is variable, ranging on a continuum from none to fully online distance learning. A variety of descriptive terms have been employed (somewhat inconsistently) to categorize the extent to which technology is used. For example, 'hybrid learning' or 'blended learning' may refer to classroom aids and laptops, or may refer to approaches in which traditional classroom time is reduced but not eliminated, and is replaced with some online learning (Celia, 2014; Career FAQs State, 2013; Valerie, 2013). 'Distributed learning' may describe either the e-learning component of a hybrid approach, or fully online distance learning environments (Bates and Poole, 2003). Another scheme described the level of technological support as 'web enhanced,' 'web supplemented,' and 'web dependent' (Sloan Commission).

4.4.2   SYNCHRONOUS AND ASYNCHRONOUS LEARNING

E-learning may either be synchronous or asynchronous, the chapter reviews the details in Table 4.7.

TABLE 4.7   Synchronous and Asynchronous of E-learning

Approaches	Contents
Synchronous learning	Synchronous learning occurs in real-time, with all participants interacting at the same time, while asynchronous learning is self-paced and allows participants to engage in the exchange of ideas or information without the dependency of other participants involvement at the same time. Synchronous learning refers to the exchange of ideas and information with one or more participants during the same period of time. Examples are face-to-face discussion, online real-time live teacher instruction and feedback, Skype conversations, and chat rooms or virtual classrooms where everyone is online and working collaboratively at the same time.
Asynchronous learning	Asynchronous learning may use technologies such as email, blogs, wikis, and discussion boards, as well as web-supported textbooks (Loutchko et al., 2002) hypertext documents, audio video courses, and social networking using web 2.0. At the professional educational level, training may include virtual operating rooms (Bernard Luskin, 2010). Asynchronous learning is particularly beneficial for students who have health problems or have childcare responsibilities and regularly leaving the home to attend lectures is difficult.

They have the opportunity to complete their work in a low stress environment and within a more flexible timeframe (Johnson and Henry, 2013). In asynchronous online courses, students proceed at their own pace. If they need to listen to a lecture a second time, or think about a question for a while, they may do so without fearing that they will hold back the rest of the class. Through online courses, students can earn their diplomas more quickly, or repeat failed courses without the embarrassment of being in a class with younger students. Students also have access to an incredible variety of enrichment courses in online learning, and can participate in college courses, internships, sports, or work and still graduate with their class. Both the asynchronous and synchronous methods rely heavily on self-motivation, self-discipline, and the ability to communicate in writing effectively (Gc.maricopa.edu, 2013).

4.4.3 LINEAR LEARNING

Computer-based learning or training (CBT) refers to self-paced learning activities delivered on a computer or handheld device such as a tablet or smartphone. CBT often delivers content via CD-ROM, and typically presents content in a linear fashion, much like reading an online book or manual. For this reason, CBT is often used to teach static processes, such as using software or completing mathematical equations. Computer-based training is conceptually similar to web-based training (WBT), the primary difference being that WBTs are delivered via Internet using a web browser (Vangie, 2014).

Assessing learning in a CBT is often by assessments that can be easily scored by a computer such as multiple-choice questions, drag-and-drop, radio button, simulation or other interactive means. Assessments are easily scored and recorded via online software, providing immediate end-user feedback and completion status. Users are often able to print completion records in the form of certificates. CBTs provide learning stimulus beyond traditional learning methodology from textbook, manual, or classroom-based instruction. For example, CBTs offer user-friendly solutions for satisfying continuing education requirements. Instead of limiting students to attending courses or reading printed manuals, students are able to acquire knowledge and skills through methods that are much more conducive to individual learning preferences. For example, CBTs offer visual learning benefits through animation or video, not typically offered by any other means (Ryan, 2012).

CBTs can be a good alternative to printed learning materials since rich media, including videos or animations, can easily be embedded to enhance the learning.

However, CBTs pose some learning challenges. Typically the creation of effective CBTs requires enormous resources. The software for developing CBTs (such as Flash or Adobe Director) is often more complex than a subject matter expert or teacher is able to use. In addition, the lack of human interaction can limit both the type of content that can be presented as well as the type of assessment that can be performed. Many learning organizations are beginning to use smaller CBT/WBT activities as part of a broader online learning program, which may include online discussion or other interactive elements (Muljadi, 2014).

4.4.4 COLLABORATIVE LEARNING

Computer-supported collaborative learning (CSCL) uses instructional methods designed to encourage or require students to work together on learning tasks. CSCL is similar in concept to the terminology, "e-learning 2.0" and "networked collaborative learning" (NCL) (Trentin, 2010). Collaborative learning is distinguishable from the traditional approach to instruction in which the instructor is the principal source of knowledge and skills. For example, the neologism "e-learning 1.0" refers to the direct transfer method in computer-based learning and training systems (CBL). In contrast to the linear delivery of content, often directly from the instructor's material, CSCL uses blogs, wikis, and cloud-based document portals (such as Google Docs and Dropbox). With technological Web 2.0 advances, sharing information between multiple people in a network has become much easier and use has increased (Beverly, 2009). One of the main reasons for its usage states that it is "a breeding ground for creative and engaging educational endeavors" (Beverly, 2009).

Using Web 2.0 social tools in the classroom allows for students and teachers to work collaboratively, discuss ideas, and promote information. According to Sendall et al. (2008), blogs, wikis, and social networking skills are found to be significantly useful in the classroom. After initial instruction on using the tools, students also reported an increase in knowledge and comfort level for using Web 2.0 tools. The collaborative tools also prepare students with technology skills necessary in today's workforce. Locus of control remains an important consideration in successful engagement of e-learners. According to the work of Cassandra B. Whyte, the continuing attention to aspects of motivation and success in regard to e-learning should be kept in context and concert with other educational efforts. Information about motivational tendencies can help educators, psychologists, and technologists develop insights to help students perform better academically (Whyte and Kurt, 1980).

4.4.5 *THE PRINCIPLES OF STUDENT-CENTERED LEARNING*

The traditional approach to e-learning is too often driven by the needs of the institution rather than the individual (O'Hear, 2006). E-learning platforms should focus on building student-centered learning environments. Lea et al. (2003) summarize the principles of student-centered learning:

1. the reliance on active rather than passive learning,
2. an emphasis on deep learning and understanding,
3. increased responsibility and accountability on the part of the student,
4. an increased sense of autonomy to the learner,
5. an interdependence between teacher and learner,
6. mutual respect within the learner teacher relationship,
7. a reflexive approach to the teaching and learning process on the part of both teacher and learner.

4.4.6 E-LEARNING TECHNOLOGY TOOLS AND MEDIA

Various types of E-learning technology tools and media are used to facilitate e-learning as Table 4.8.

TABLE 4.8 E-learning Technology Tools and Media

Technology tools and media	Contents
Media	To understand educational technology one must also understand theories in human behavior as behavior is affected by technology. Media and the family is another emerging area affected by rapidly changing educational technology (Luskin, 1996). Educational media and tools can be used for: task structuring support: help with how to do a task (procedures and processes), access to knowledge bases (help user find information needed) alternate forms of knowledge representation (multiple representations of knowledge, e.g., video, audio, text, image, data). Numerous types of physical technology are currently used: (Forehand, 2012) digital cameras, video cameras, interactive whiteboard tools, document cameras, and LCD projectors. Combinations of these techniques include blogs, collaborative software, ePortfolios, and virtual classrooms.
Audio and video	Radio offers a synchronous educational vehicle, while streaming audio over the Internet with webcasts and podcasts can be asynchronous. Classroom microphones, often wireless, can enable learners and educators to interact more clearly. Video technology (Lisa et al., 2011) has included VHS tapes and DVDs, as well as on-demand and synchronous methods with digital video via server or web-based options such as streamed video from YouTube, Teacher Tube, Skype, Adobe Connect, and webcams. Telecommuting can connect with speakers and other experts. Interactive digital video games are being used at K-12 and higher education institutions (Michael, 2011). Podcasting allows anybody to publish files to the Internet where individuals can subscribe and receive new files from people by a subscription (Reeves, 2013).
Computers, tablets and mobile devices	Computers and tablets enable learners and educators to access websites as well as programs such as Microsoft Word, PowerPoint, PDF files, and images. Many mobile devices support m-learning. Mobile devices such as clickers and smartphones can be used for interactive feedback (Eric, 2010). Mobile learning can also provide performance support for checking the time, setting reminders, retrieving worksheets, and instruction manuals (Melody and Ramsay, 2012; Liesbeth and Kirschner, 2014).
Open Course Ware (OCW)	OCW gives free public access to information used in undergraduate and graduate programs at institutions of higher education. Participating institutions are MIT (The Magazine for Database Professionals, 2006; Iiyoshi and Kumar, 2008) and Harvard, Princeton, Stanford, University of Pennsylvania, and University of Michigan (Lewin, 2012).

TABLE 4.8 Continued

Technology tools and media	Contents
Social networks	Group webpages, blogs, and wikis allow learners and educators to post thoughts, ideas, and comments on a website in an interactive learning environment (Courts and Tucker, 2012). Social networking sites are virtual communities for people interested in a particular subject or just to "hang out" together. Members communicate by voice, chat, instant message, video-conference, and blogs, and the service typically provides a way for members to contact friends of other members (Murray& Rhonda, 2007). The National School Boards Association found that 96% of students with online access have used social networking technologies, and more than 50% talk online specifically about schoolwork. These statistics support the likelihood of being able to bring these technologies into our classrooms and find successful teaching methods to employ their use in an educational setting. Social networking inherently encourages collaboration and engagement (Martha and Hudges, 2014). Social networking can also be used as a motivational tool to promote self-efficacy amongst students. In a study by Bowers-Campbell (2008). Facebook was used as an academic motivation tool for students in a developmental reading course (McCarroll and Kevin, 2013). Group members may respond and interact with other members (Pilgrim and Christie, 2011). Student interaction is at the core of constructivist learning environments and Social Networking Sites provide a platform for building collaborative learning communities. By their very nature they are relationship-centered and promote shared experiences. With the emphasis on user-generated-content, some experts are concerned about the traditional roles of scholarly expertise and the reliability of digital content. Students still have to be educated and assessed within a framework that adheres to guidelines for quality. Every student has his or her own learning requirements, and a Web 2.0 educational framework provides enough resources, learning styles, communication tools and flexibility to accommodate this diversity (McCarroll and Kevin, 2013).
Webcams	Webcams and webcasting have enabled creation of virtual classrooms and virtual learning environment (Dennis, 2013)

TABLE 4.8 Continued

Technology tools and media	Contents
Whiteboards	Interactive whiteboards and smart boards allow learners and instructors to write on the touch screen. The screen markup can be on either a blank whiteboard or any computer screen content. Depending on permission settings, this visual learning can be interactive and participatory, including writing and manipulating images on the interactive whiteboard (Wiki, 2014).
Screen casting	Screen casting allows users to share their screens directly from their browser and make the video available online so that other viewers can stream the video directly (Ipark.hud.ac.uk, 2013). The presenter thus has the ability to show their ideas and flow of thoughts rather than simply explain them as simple text content. In combination with audio and video, the educator can mimic the one-on-one experience of the classroom and deliver clear, complete instructions. Learners also have an ability to pause and rewind, to review at their own pace, something a classroom cannot always offer.

4.4.7 LEARNING MANAGEMENT SYSTEM

A learning management system (LMS) is software used for delivering, tracking and managing training and education. For example, an LMS tracks attendance, time on task, and student progress. Educators can post announcements, grade assignments, check on course activity, and participate in class discussions. Students can submit their work, read and respond to discussion questions, and take quizzes (Courts and Tucker, 2012). An LMS may allow teachers, administrators, students, and permitted additional parties (such as parents if appropriate) to track various metrics. LMSs range from systems for managing training/educational records to software for distributing courses over the Internet and offering features for online collaboration. The creation and maintenance of comprehensive learning content requires substantial initial and ongoing investments of human labor. Effective translation into other languages and cultural contexts requires even more investment by knowledgeable personnel (Sarasota et al., 2013).

Internet-based learning management systems include Canvas, Blackboard Inc. and Moodle. These types of LMS allow educators to run a learning system partially or fully online, asynchronously or synchronously. Blackboard can be used for K-12 education, Higher Education, Business, and Government collaboration (Blackboard.com, 2012). Moodle is a free-to-download Open Source Course Management System that provides blended learning opportunities as well as platforms for distance learning courses (Moodle.org, 2012). Eliademy is a free cloud based Course Management System that provides blended learning opportunities as well as platforms for distance learning courses (eliademy.com, 2014).

4.4.8 LEARNING CONTENT MANAGEMENT SYSTEM

A learning content management system (LCMS) is software for author content (courses, reusable content objects). An LCMS may be solely dedicated to producing and publishing content that is hosted on an LMS, or it can host the content itself. The Aviation Industry Computer-Based Training Committee (AICC) specification provides support for content that is hosted separately from the LMS. A recent trend in LCMSs is to address this issue through crowd sourcing (Sören, 2013).

4.4.9   COMPUTER-AIDED ASSESSMENT

Computer-aided assessment, also but less commonly referred to as e-assessment, ranges from automated multiple-choice tests to more sophisticated systems. With some systems, feedback can be geared towards a student's specific mistakes or the computer can navigate the student through a series of questions adapting to what the student appears to have learned or not learned (Technologyenhancedlearning.net, 2014). The best examples follow a formative assessment structure and are called "Online Formative Assessment." This involves making an initial formative assessment by sifting out the incorrect answers. The author of the assessment/teacher will then explain what the pupil should have done with each question. It will then give the pupil at least one practice at each slight variation of sifted out questions. This is the formative learning stage. The next stage is to make a summative assessment by a new set of questions only covering the topics previously taught (Ryan, 2012).

Learning design is the type of activity enabled by software that supports sequences of activities that can be both adaptive and collaborative. The IMS Learning Design specification is intended as a standard format for learning designs, and IMS LD Level A is supported in LAMS V2. E-learning and has been replacing the traditional settings due to its cost effectiveness (Ryan, 2012).

4.4.10   ELECTRONIC PERFORMANCE SUPPORT SYSTEMS (EPSS)

An Electronic Performance Support System is, according to Barry Raybould, "a computer-based system that improves worker productivity by providing on-the-job access to integrated information, advice, and learning experiences" (Raybould, 1992). Gloria Gery defines it as "an integrated electronic environment that is available to and easily accessible by each employee and is structured to provide immediate, individualized online access to the full range of information, software, guidance, advice and assistance, data, images, tools, and assessment and monitoring systems to permit job performance with minimal support and intervention by others" (Gery, 1989). Data system Student data systems have a significant impact on education and students (Cho and Wayman, 2009). Over-the-counter data (OTCD) refers to a design approach which involves embedding labels, supplemental documentation, and a help system and making key package/display and content decisions (Rankin, 2013).

4.5 APPLICATIONS

The study summarizes the applications for E-learning as Table 4.9.

4.6 *BENEFITS OF E-LEARNING*

The study shows benefits of incorporating technology into the classroom in Table 4.10.

4.7 *GLOBAL TRENDS AND TECHNOLOGIES IN THE E-LEARNING INDUSTRY*

Based on the new E-learning technologies that keep evolving, Christopher (2013) indicated the following E-learning trends will be established or further developed in the near future and proposed Top 8 Future E-learning trends of the Global E-learning Industry as Table 4.11.

4.7.1 Top 7 Future E-learning technologies of the E-learning Industry

Christopher (2013) indicated the most influential E-learning technologies of tomorrow as Table 4.12:

4.7.2 Top 9 E-Learning Predictions for 2014

Ravi (2014) indicated 2013 was an exciting year for e-Learning professionals with many discussions around learning paradigms and technologies.

TABLE 4.9 The Applications for E-learning

Applications	Contents
Preschool	Various forms of electronic media are a feature of preschool life (Rideout et al., 2003). Although parents report a positive experience, the impact of such use has not been systematically assessed (Rideout et al., 2003). The age when a given child might start using a particular technology such as a cellphone or computer might depend on matching a technological resource to the recipient's developmental capabilities, such as the age-anticipated stages labeled by Swiss psychologist, Jean Piaget (Buckleitner, 2013) Parameters, such as age-appropriateness, coherence with sought-after values, and concurrent entertainment and educational aspects, have been suggested for choosing media (Meidlinger, 2011).
K–12	E-learning is utilized by public K–12 schools in the United States as well as private schools. Some e-learning environments take place in a traditional classroom, others allow students to attend classes from home or other locations. There are several states that are utilizing virtual school platforms for e-learning across the country that continue to increase. Virtual school enables students to log into synchronous learning or asynchronous learning courses anywhere there are an Internet connection. Technology kits are usually provided that includes computers, printers, and reimbursement for home Internet use. Students are to use technology for school use only and must meet weekly work submission requirements. Teachers employed by K-12 online public schools must be certified teachers in the state they are teaching in. Online schools allow for students to maintain their own pacing and progress, course selection, and provide the flexibility for students to create their own schedule. Virtual education in K-12 schooling often refers to virtual schools, and in higher education to virtual universities. Virtual schools are "cyber charter schools" with innovative administrative models and course delivery technology (Allen and Seaman, 2008).

TABLE 4.9 Continued

Applications	Contents
Online schools	E-learning is increasingly being utilized by students who may not want to go to traditional brick and mortar schools due to severe allergies or other medical issues, fear of school violence and school bullying and students whose parents would like to homeschool but do not feel qualified (Edweek.org, 2012). Online schools create a safe haven for students to receive a quality education while almost completely avoiding these common problems. Online charter schools also often are not limited by location, income level or class size in the way brick and mortar charter schools are (KQED, 2014). E-learning also has been rising as a supplement to the traditional classroom. Students with special talents or interests outside of the available curricula use e-learning to advance their skills or exceed grade restrictions (Cavanaugh, 2009). Some online institutions connects students with instructors via web conference technology to form a digital classroom. These institutions borrow many of the technologies that have popularized online courses at the university level. National private schools are also available online. These provide the benefits of e-learning to students in states where charter online schools are not available. They also may allow students greater flexibility and exemption from state testing (Wiki, 2014).
Higher education	Enrollments for fully online learning increased by an average of 12–14 percent annually between 2004 and 2009, compared with an average of approximately 2 percent increase per year in enrollments overall (Allen and Seaman, 2003; Ambient Insight Research, 2009). Almost a quarter of all students in post-secondary education were taking fully online courses in 2008 (Allen and Seaman, 2003). In 2009, 44 percent of post-secondary students in the USA were taking some or all of their courses online, this figure is projected to rise to 81 percent by 2014 (Ambient Insight Research, 2009). During the fall 2011 term, 6.7 million students enrolled in at least one online course (Babson Research Study, 2013). Over two-thirds of chief academic officers believe that online learning is critical for their institution (The Learning Curve, 2013). The Sloan report, based on a poll of academic leaders, indicated that students are as satisfied with online classes as with traditional ones.

Although a large proportion of for-profit higher education institutions now offer online classes, only about half of private, non-profit schools do so. Private institutions may become more involved with online presentations as the costs decrease. Properly trained staff must also be hired to work with students online (Repetto and Trentin, 2011) These staff members need to understand the content area, and also be highly trained in the use of the computer and Internet. Online education is rapidly increasing, and online doctoral programs have even developed at leading research universities (Hebert, 2007). |

TABLE 4.9 Continued

Applications	Contents
MOOCs	Although massive open online courses (MOOCs) may have limitations that preclude them from fully replacing college education (Youngberg, 2012), such programs have significantly expanded. MIT, Stanford and Princeton University offer classes to a global audience, but not for college credit (Pappano, 2013). University-level programs, like edX founded by Massachusetts Institute of Technology and Harvard University, offer wide range of disciplines at no charge. MOOCs have not had a significant impact on higher education and declined after the initial expansion, but are expected to remain in some form (Kolowich, 2014). Private organizations also offer classes, such as Udacity, with free computer science classes, and Khan Academy, with over 3,900 free micro-lectures available via YouTube. There already is at least one counter stream to MOOC.
DOCC	Distributed open collaborative course or DOCC challenges the role of the Instructor, the hierarchy, the role of money and role of massiveness. DOCC recognizes that the pursuit of knowledge may be achieved better by not using a centralized singular syllabus that expertise is distributed throughout all the participants in a learning activity, and does not just reside with one or two individuals (Jaschik, 2013).
Coursera	Coursera, an online-enrollment platform, is now offering education for millions of people around the world. A certification is consigned by Coursera for students who are able to complete an adequate performance in the course. Free online courses are administered by the website-fields like computer science, medicine, networks and social sciences are accessibly offered to pursuing students. The lectures are recorded into series of short videos discussing different topics and assignments in a weekly basis. This virtual curriculum complement the curriculum taught in the traditional education setting by providing equality for all students, despite disability, and geographical location and socioeconomic status. According to Fortune magazine, over a million people worldwide have enrolled in free online courses (Mansour, 2013).
Corporate and professional	E-learning has now been adopted and used by various companies to inform and educate both their employees and customers. Companies with large and spread out distribution chains use it to educate their sales staff about the latest product developments without the need of organizing physical onsite courses. Compliance has also been a big field of growth with banks using it to keep their staff's CPD levels up. Other areas of growth include staff development, where employees can learn valuable workplace skills (Interconf, 2014).

TABLE 4.9 Continued

Applications	Contents
Public health	There is an important need for recent, reliable, and high-quality health information to be made available to the public as well as in summarized form for public health providers (Warner, Procaccino, 2004). Providers have indicated the need for automatic notification of the latest research, a single searchable portal of information, and access to Grey literature (Simpson, Prusak, 1995). The Maternal and Child Health (MCH) Library is funded by the U.S. Maternal and Child Health Bureau to screen the latest research and develop automatic notifications to providers through the MCH Alert. Another application in public health is the development of MHealth (use of mobile telecommunication and multimedia into global public health). mHealth has been used to promote prenatal and newborn services, with positive outcomes. In addition, "Health systems have implemented mHealth programs to facilitate emergency medical responses, point-of-care support, health promotion and data collection." (Tamrat, Kachnowski, 2012). In low and middle-income countries, MHealth is most frequently used as one-way text messages or phone reminders to promote treatment adherence and gather data (Källander et al., 2013).
ADHD	There has also been a growing interest in e-learning as a beneficial educational method for students with Attention Deficit Hyperactivity Disorder (ADHD). With the growing popularity in e-learning among K-12 and higher education, the opportunity to take online classes is becoming increasingly important for students of all ages (Grabinger et al., 2008). However, students with ADHD and special needs face different learning demands compared to the typical developing learner. This is especially significant considering the dramatic rise in ADHD diagnoses in the last decade among both children and adults (Center For Disease Control and Prevention. n.d., 2014). Compared to the traditional face-to-face classroom, e-learning and virtual classrooms require a higher level of executive functions, which is the primary deficit associated with ADHD (Madaus et al., 2012). Wolf (2001) lists 12 executive function skills necessary for students to succeed in postsecondary education: plan, set goals, organize, initiate, sustain attention/effort, flexibility, monitor, use feedback, structure, manage time, manage materials, and follow through. These skills, along with strong independent and self-regulated learning, are especially pronounced in the online environment and as many ADHD students suffer from a deficit in one or more of these executive functions, this presents a significant challenge and accessibility barrier to the current e-learning approach (Cull et al., 2013; Parker and Banerjee, 2007). Some have noted that current e-learning models are moving towards applying a constructivism learning theory (Sajadi and Khan, 2011) that emphasizes a learner-centered environment

TABLE 4.9   Continued

Applications	Contents
	(Keengwe et al., 2009) and postulates that everyone has the ability to construct their own knowledge and meaning through a process of problem solving and discovery (William, 2010). However, some of the principles of constructivism as required for e-learning may not be appropriate for ADHD learners; these principles include active learning, self-monitoring, motivation, and strong focus (Sajadi and Khan, 2011).
	Despite the limitations, students with special needs, including ADHD, have expressed an overall enthusiasm for e-learning and have identified a number e-learning benefits, including: availability of online course notes, materials and additional resources; the ability to work at an independent pace and spend extra time spent formulating thoughtful responses in class discussions; help in understanding course lecture/content; ability to review lectures multiple times; and enhanced access to and communication with the course instructor (Fichten et al., 2009; Cull et al., 2013).
Classroom 2.0	Classroom 2.0 refers to online multi-user virtual environments (MUVEs) that connect schools across geographical frontiers. Also known as "eTwinning," computer-supported collaborative learning (CSCL) allows learners in one school to communicate with learners in another that they would not get to know otherwise, (Sero.co.uk, 2013; Ite. educacion.es, 2013; Scuola-digitale.it, 2013) enhancing educational outcomes and cultural integration. Examples of classroom 2.0 applications are Blogger and Skype (Paolo, 2012).
Virtual classroom	A virtual classroom also provides the opportunity for students to receive direct instruction from a qualified teacher in an interactive environment. Learners can have direct and immediate access to their instructor for instant feedback and direction. The virtual classroom also provides a structured schedule of classes, which can be helpful for students who may find the freedom of asynchronous learning to be overwhelming. In addition, the virtual classroom provides a social learning environment that replicates the traditional "brick and mortar" classroom. Most virtual classroom applications provide a recording feature. Each class is recorded and stored on a server, which allows for instant playback of any class over the course of the school year. This can be extremely useful for students to review material and concepts for an upcoming exam. This also provides students with the opportunity to watch any class that they may have missed, so that they do not fall behind. It also gives parents and auditors the conceptual ability to monitor any classroom to ensure that they are satisfied with the education the learner is receiving (Muljadi, 2014).

TABLE 4.9 Continued

Applications	Contents
Virtual learning environment (VLE)	A Virtual Learning Environment (VLE), also known as a learning platform, simulates a virtual classroom or meetings by simultaneously mixing several communication technologies. For example, web conferencing software such as GoToTraining, WebEx Training or Adobe Connect enables students and instructors to communicate with each other via webcam, microphone, and real-time chatting in a group setting. Participants can raise hands, answer polls or take tests. Students are able to whiteboard and screen cast when given rights by the instructor, who sets permission levels for text notes, microphone rights and mouse control (Wiki, 2014).

In higher education especially, the increasing tendency is to create a virtual learning environment (VLE) (which is sometimes combined with a Management Information System (MIS) to create a Managed Learning Environment) in which all aspects of a course are handled through a consistent user interface throughout the institution. A growing number of physical universities, as well as newer online-only colleges, have begun to offer a select set of academic degree and certificate programs via the Internet at a wide range of levels and in a wide range of disciplines. While some programs require students to attend some campus classes or orientations, many are delivered completely online. In addition, several universities offer online student support services, such as online advising and registration, e-counseling, online textbook purchases, student governments and student newspapers (Muljadi, 2014). |
| **Augmented reality (AR)** | Augmented reality (AR) provides students and teachers the opportunity to create layers of digital information that includes both virtual world and real world elements, to interact with in real time. There are already a variety of apps, which offer a lot of variations and possibilities (Wiki, 2014). |

TABLE 4.10   Benefits of E-Learning

Benefits	Contents
Defray travel costs	Easy-to-access course materials. Course material on a website allows learners to study at a time and location they prefer and to obtain the study material very quickly (Nsba.org, 2014). Student motivation. According to James Kulik, who studies the effectiveness of computers used for instruction, students usually learn more in less time when receiving computer-based instruction and they like classes more and develop more positive attitudes toward computers in computer-based classes (Electronic-school.com, 2014). Teachers must be aware of their students' motivators in order to successfully implement technology into the classroom (Guo et al., 2012). Students are more motivated to learn when they are interested in the subject matter, which can be enhanced by using technologies in the classroom and targeting the need for screens and digital material (Gu et al., 2013) that they have been stimulated by outside of the classroom.
More opportunities for extended learning	According to study completed in 2010, 70.3% of American family households have access to the Internet (Warschauer and Matuchniak, 2010). According to Canadian Radio Television and Telecommunications Commission Canada, 79% of homes have access to the Internet (CRTC, 2014). This allows students to access course material at home and engage with the numerous online resources available to them. Students can use their home computers and Internet to conduct research, participate in social media, email, play educational games and stream videos. Using online resources such as Khan Academy or TED Talks can help students spend more time on specific aspects of what they may be learning in school, but at home (Awesome Inc., 2014). These online resources have added the opportunity to take learning outside of the classroom and into any atmosphere that has an Internet connection. These online lessons allow for students who might need extra help to understand materials outside of the classroom. These tutorials can focus on small concepts of large ideas taught in class, or the other way around (Wiki, 2014). Schools like MIT have even made their course materials free online so that anybody can access them. Although there are still some aspects of a classroom setting that are missed by using these resources, they are still helpful tools to add additional support to the already existing educational system (Wiki, 2014).
Wide participation	Learning material can be used for long distance learning and are accessible to a wider audience (Nsba.org, 2014).
Improved student writing	It is convenient for students to edit their written work on word processors, which can, in turn, improve the quality of their writing. According to some studies, the students are better at critiquing and editing written work that is exchanged over a computer network with students they know (Nsba.org, 2014).

TABLE 4.10 Continued

Benefits	Contents
Differentiated Instruction	Educational technology provides the means to focus on active student participation and to present differentiated questioning strategies. It broadens individualized instruction and promotes the development of personalized learning plans in some computer programs available to teachers. Students are encouraged to use multimedia components and to incorporate the knowledge they gained in creative ways (Smith and Stephanie, 2004). This allows some students to individually progress from using low ordered skills gained from drill and practice activities, to higher level thinking through applying concepts creatively and creating simulations (Wenglinsky, 2006). In some cases, the ability to make educational technology individualized may aid in targeting and accommodating different learning styles and levels. Overall, the use of Internet in education has had a positive impact on students, educators, as well as the educational system as a whole. Effective technologies use many evidence-based strategies (e.g., adaptive content, frequent testing, immediate feedback, etc.), as do effective teachers (Ross, 2010). It is important for teachers to embrace technology in order to gain these benefits so they can address the needs of their digital natives (Hicks, 2011).
Additional Benefits	The Internet itself has unlocked a world of opportunity for students. Information and ideas that were previously out of reach are a click away. Students of all ages can connect, share, and learn on a global scale. Using computers or other forms of technology can give students practice on core content and skills while the teacher can work with others, conduct assessments, or perform other tasks (Ross, 2010). Using technology in the classroom can allow teachers' to effectively organize and present lessons.
	Multimedia presentations can make the material more meaningful and engaging. "Technology's impact in schools has been significant, advancing how students learn, how teachers teach and how efficiently and effectively educational services can be delivered," said Carolyn April, director, industry analysis, CompTIA." With emerging technologies such as tablets and netbooks, interactive whiteboards and wireless solutions gaining ground in the classroom, the reliance on IT by the education market will only grow in the years ahead" (Comptia.org, 2014). Studies completed in "computer intensive" settings found increases in student-centric, cooperative and higher order learning, students writing skills, problem solving, and using technology (An and Reigeluth, 2011). In addition, positive attitudes toward technology as a learning tool by parents, students and teachers are also improved.

TABLE 4.11 Top 8 Future E-learning Trends of the Global E-Learning Industry

Trends and Technologies	Contents
1. Massive Open Online Courses are the hottest trend right now in E-learning.	This flexible and diverse concept sounds simple-online videos of real-life lectures-, but not when it involves the astronomic number of 36,000 students, which is how many people enrolled in one of Harvard's first massive online courses. And it's not just Harvard that jumped on the MOOC bandwagon. Other world famous Ivy League universities, such as MIT, Caltech, Berkeley and Princeton, have similarly climbed aboard. I believe that the changes and disruptions to traditional form of university teaching will be wide and profound, since institutions that mainly rely on disseminating information in traditional classrooms will have fewer resources to such an evolution and therefore will be pressured to cut costs. So the only way to deal with that will be by investing in blended learning, using the flipped classroom model, which is getting wildly popular all over the world, by the way. Here are some effective blended design approaches for your consideration – Effective Blended Learning Design Approaches.
2. Credits and fees for MOOCs	It's only natural that MOOCs will eventually stop being a free service, since they do require the presence of an instructor, the use of technology, and quite soon content providers will have to enrich them with webinars, discussions, wikis, etc. All this will also lead to the necessity of official grades, as well as credits that can also be transferred from one school to another. How would you like a Master's degree in a growing field, from a well-regarded university, all for less than $7,000? On first glance, that doesn't sound too bad at all (Georgia Tech unveils first all-MOOC computer science degree). Also, check out the following informative articles to learn more about the Business opportunities around MOOC. Last but not least, if you are wondering about corporate MOOCs you may want to read the MOOCs and the e-Learning Industry article.
3. Micro-Learning, or in other words mini bytes of learning content.	This is yet another trend gaining growing popularity in the sector and will most probably be a big hit in the future, especially in the corporate world. Five minute videos, one page documents, focused lessons, small chunks of information, and other flexible activities that will be easily incorporated in a busy person's daily life, since the cognitive load is considerably lighter. And let's not forget that micro-learning is perfectly suited for mobiles, something that partially explains its popularity.

TABLE 4.11   Continued

Trends and Technologies	Contents
4. The importance and greater recognition of informal learning	Accessibility and availability of social media tools enable and encourage people to gain the information they need. This is bound to continue and evolve in the future, thanks to the plethora of free learning resources, such as podcasts, videos, blogs, webinars, etc. You may want to read the Social Learning Best Practices for the Workplace and the 5 Sexy Steps to Informal Learning, Content Curation and Knowledge Hustling (Robinson and Ritzko, 2009).
5. The role of the instructor will change Open Educational Resources (OER)	OER are freely accessible documents and media, quite often written by the world's best authorities on any subject and sector. This can only mean one thing. The role of the traditional educator will be transformed. And all this knowledge available should be used wisely, creatively and effectively to support learning in or outside the traditional classrooms.
6. The concept of research will be upgraded.	This is linked to the above-mentioned role of the instructor/educator. Tedious publications, worn lectures, and absence of updated material will soon come to their very end, if they have not already come to an end! Students' participation in knowledge building is the new comparative advantage for those who want to stand out in the education field. Knowledge is easily accessible and we need to let our students/learners create knowledge autonomously. To truly involve and engage them in active learning. To encourage and believe in student generated content. I highly encourage you read the Progressive Education: The Rising Power Of Student Voice and the 3 Reasons To Encourage Student-Generated Content.
7. The majority of students will be overseas.	There is no longer need to study in another country, when you may as well receive the same – or even better – level of education through distance learning, even from an Ivy League university. Sure, the experience won't be the same, but the cost savings will be substantial.
8. Growing influence of learning communities.	The term is multifaceted, implying extending classroom practice, curriculum enhancement, student tasks, engagement of students, teachers and administrators, etc. They support learning, promote collective creativity and shared leadership, and unite learning groups with shared values, vision and practices in a global perspective.

Source: Christopher (2013).

TABLE 4.12 Top 7 Future E-learning Technologies of the E-learning Industry

Technologies	Contents
1. mLearning with a native app	Just to be clear, mobile learning is not E-learning in a mobile device, since the proper content conversion requires skillful instructional design and development skills. The differences between mobile web apps and native apps is that the latter requires development for multiple platforms, that is, specific operation systems and machine firmware, the application is stored locally on the device, and user data can be stored on the device, in the cloud, or in both. According to Deloitte, "the cost of developing an app for 2 OSs is 160% higher than for 1 OS." The concept of Mobile – combined with the BYOD trend – changes the way the next generation of learning experiences will be designed, since learning will no longer target people who are chained to their desk in front of a PC. The continuous development of mobile devices, which are equipped with digital compasses, dual cameras, incredible audio, etc., coupled with their obvious advantage of mobility will lead E-learning to a whole new level.
2. Cloud-based learning	Cloud-based learning has a dual effect; on a school level and on a corporate level. One-time downloads and installs of course materials will no longer be the default methods of obtaining a course's content. Providers will be able to offer cloud-stored individual E-learning modules, or even full E-learning courses as packets that can be purchased and downloaded on demand. On the other hand, the increasing demand for affordable, global training will be addressed by cloud-based technology, which will streamline corporate training processes and create tailor-made solutions for smaller businesses. You may want to check Sugata Mitra's TED Talk on How Can We Build A School In The Cloud.
3. Use of game-play mechanics for non-game applications, aka Gamification	Gamification is not a new trend, but rather one that will certainly evolve. It's a powerful tool that enables technological innovation, develops student/learner skills, crafts behaviors and enhances problem solving. Gamification has proven to be an invaluable instrument to improve employee performance, upgrade education, customer engagement, as well as personal development. And I believe its possibilities and applications are endless.
4. SaaS authoring tools	Yet another hot trend on the rise. SaaS is basically enterprise software hosted in a cloud, which translates to easily downloadable software, virtual updates, massive savings in costs and time and so much more. According to a recent survey by Mint Jutras, more than 45% of all software will be SaaS by the year 2023. I highly encourage you to check the following list of 54 Cloud Based E-learning Authoring Tools.

TABLE 4.12   Continued

Technologies	Contents
5. Notification systems in LMS	These systems begin to become the core of distributed mobile and omnipresent learning support. They are used to draw attention to important events, give instructions and information, raise awareness regarding various activities, or to directly and instantly provide information related to the user's training/learning material. You may want to check the LMSs Comparison Checklist of Features.
6. HTML5	According to Gartner, within 2014 "improved JavaScript performance will begin to push HTML5 and the browser as a mainstream enterprise application development environment." The benefits include, but are not limited to, better performance, multimedia and connectivity. Built With .com's trends data indicate that "in 2013 of the top 1 million websites worldwide, there was a 100% increase in the use of HTML5 compared to 2012." HTML5 seems to be the favorite platform of mobile developers, because it eradicates the need for multiple apps. The very existence of one single HTML5 app guarantees, inter alia, better maintainability and quicker updates.
7. Tin Can API, aka xAPI	One of the most exhaustively analyzed topics amongst learning professionals today. It enables the collection of data about a wide range of learning experiences a person goes through. It relies on a Learning Record Store, and it overcomes the majority of limitations of SCORM, which was the previous standard. For more information about what you can do with the Tin Can API see Discover Simple Communication with Tin Can API. I believe it has innumerable practical implementation aspects and will evolve even further, revolutionizing the way we learn, creating more personal and richer learning environments. Maybe is the technological solution for the Individualized Learning Plans in E-learning. Last but not least, I highly encourage you to read the How Tin Can is Making Tools Better, Together article.

Source: Christopher (2013).

The year saw many interesting developments in areas such as Learning Standards/Specifications (xAPI), Programming Languages (HTML 5), Gamification, Game-based Learning, etc. They believe that 2014 will see more powerful changes, owing to the maturing and convergence of some of these technologies. If 2013 was a year of 'contemplations,' 2014 will be a year of 'decisions'! Table 4.13 shows the Top 9 e-Learning Predictions for 2014.

According to Ravi (2014), the study summarizes the key points as the follows:

1. More tools adopting Tin Can API
2. Tools continuing to invest in HTML5 and mobile (Responsive) learning
3. Tools moving toward sophisticated graphics and rich media
4. Tools continuing to evolve to meet the market's needs (such as Gamification, GBL, Collaborative Development Support, and Video-based Course Development Support)
5. Tools evolving further to offer instructional models suited for multi-screen learning
6. More tools offering cloud-based authoring options
7. Tools moving towards support for developing just-in-time learning and Performance Support systems
8. Tools supporting Learning Analytics for video-based learning content Bottom Line.

4.7.3   Top 10 E-learning Industry Trends for 2013

Thanekar (2013) indicated as per the market predictions by GSV advisors, the global E-learning market is estimated to grow at a Compound Annual Growth Rate (CAGR) of 23% over 2012–2017. In dollars, this translates into $90 bn to $166.5 bn in 2012 and $255 bn in 2017. That's a very healthy growth rate. As the E-learning market continues to grow from strength to strength, it's only natural that E-learning evolves too. Here are our predictions about the upcoming trends in this industry. While there is possibly some overlap between a few of them, the growing importance of each calls for a separate acknowledgement.

Table 4.14 shows the top 10 E-learning industry trends for 2013:

TABLE 4.13   Top 9 e-Learning Predictions for 2014

Predictions	Contents
1. Courseware Authoring Tools	1. LMSs, LCMSs, and Learning Management Platforms
	2. Learning Standards/Specifications (Experience API)
	3. Learning Analytics
	4. Learning Styles (Personalized and Adaptive)
	5. Web Design Techniques (Responsive/Adaptive E-learning Design)
	6. Mobile Learning
	7. Learning Modes (Informal Learning/Social Learning)
	8. Programming Languages (HTML 5)
	9. Gamification and Game-based Learning
	10. Video-based Learning
	11. Cloud Computing
2. e-Learning Forecasts for 2014	1. Wider adoption of mobile learning in workplaces to improve productivity
	2. Richer and more dynamic learning experiences due to the evolution of mobile phones and tablets
	3. Integration of mobile learning with e-learning, and with other emerging technologies such as Learning Analytics
	4. Capabilities such as Mobile Augmented Reality, Near Field Communication (NFC), and QR Codes powering up 'contextual mobile learning'
	5. Adoption of Business App Stores within organizations to help employees work smarter
3. Social Learning/ Informal Learning	1. Major disruption in the thinking of organizational learning
	2. Incorporation of social media into core training strategy
	3. Wider adoption of the 70:20:10 framework for organizational development
	4. Opening up of learning content that is presently locked up behind firewalls or LMSs to enable access to learners for just-in-time learning or performance support
	5. More LMSs and learning platforms enabling options for managing and tracking informal learning
	6. More authoring tools, LMSs, and learning platforms embracing collaboration workflows to encourage social learning around formal learning courses

TABLE 4.13   Continued

Predictions	Contents
4. Gamification and Game-based Learning	1. An increased use of gamification for learning in a variety of subject areas and for all types of training
	2. More application software and Learning Management Systems integrating gamification elements as part of their applications (user level operations)
	3. Social networking systems integrating more gamification elements as part of their functions
	4. Wider adoption of gamification for encouraging employee/client/ partner/customer participation and involvement in organizational areas other than learning
	5. The emergence of more tool and consulting service providers to help clients integrate gamification into their business activities
	6. The emergence of more accrediting agencies around gamification and badging
	7. The evolution of new strategies for gamification and GBL across screens and devices
	8. Evolution of GBL into more sophisticated and serious games for learning complex subjects and topics
	9. More focused initiatives (such as Qualcomm's Enterprise Mobile Store) to power up gamification and GBL in organizations worldwide
	10. Gamification amalgamating further with the power of other emerging technologies such as Big Data Analytics and Internet of Things (IoT), providing us with more powerful gamified learning experiences
5. HTML5	1. Wider acceptance as the standard Enterprise
	2. Development Platform (combined with CSS and JavaScript) for building web-based applications
	3. More content authoring tools adopting HTML5 (including support for powerful interactivities and media)
	4. Evolution of the language towards standardization (to address the challenges regarding rendering issues with different browsers and devices)
6. Responsive e-Learning Design	1. More authoring tools adopting 'Responsive output'
	2. Tools evolving further allowing us to create 'Adaptive' content for just-in-time learning and performance support

TABLE 4.13   Continued

Predictions	Contents
7. Tin Can API	1. Wider acceptance of this learning standard as the key resource to harness the power of Big Data for e-Learning
	2. More tools, LMSs, and delivery platforms supporting the standard to track informal learning experiences
	3. More vendors, developers, and organizations embracing Tin Can in their learning designs, and extending its power to Talent Analytics and other broader parts of the enterprise
	4. More mobile learning apps utilizing the power of Tin Can API to deliver personalized learning experiences
	5. Evolution of best practices for statement structures to ensure interoperability among various systems
8. Learning Analytics	1. More tools powered up by the possibilities of Learning Analytics to provide personalized and adaptive learning solutions
	2. More LMS vendors and learning platforms integrating Learning Analytics features to their products to help training administrators track learner behaviors
	3. Wider adoption of Learning Analytics in organizational areas other than learning (such as Human Resources, Talent Management, Sales, Marketing, Customer Support, and Program Management)
	4. Further evolution of Learning Analytics by tapping deeper into Big Data to provide better insights
	5. Integration of Learning Analytics with other systems (such as Human Capital Management System) to power up organizational performance
9. Video-based Learning	1. Increased acceptance of byte-sized video-based learning content
	2. More rapid development tools that enable development of interactive video-based learning content
	3. Tools evolving further to enable creation of highly effective video snippets for performance support and just-in-time learning

Source: Ravi (2014).

TABLE 4.14 Top 10 E-learning Industry Trends for 2013

Trends	Contents
E-learning On Tablets and mEnablement	From 'just another type of PC' to 'an interesting media consumption device,' tablets have come a long way in a short time. Overall, the tablet market is showing exponential growth with tablet sales touted to overtake that of notebook PCs with an estimated volume of 240 million units sold worldwide by end of 2013 (by Tech-Thoughts). In the enterprises too, the usage of tablets for business related activities and enterprise mobility is on the rise, making E-learning on tablets almost a necessity. As we mentioned in The Question Of Why (Not) E-learning On iPads Or Tablets?, while E-learning on tablets is not 'real' mLearning, it serves as a bridge to Mobile Learning. This has given birth to another trend most commonly found in organizations these days – mEnablement. mEnablement is the conversion of existing (legacy) E-learning courseware into a tablet compatible format. Know How To mEnable Your E-learning.
Pervasive Learning and Embedded Ubiquitous Learning	'Pervasive Learning,' as described by Dan Pontefract, is learning at the speed of need through formal, informal and social learning modalities. The idea of pervasive learning makes perfect sense as more and more of us become concept workers, and work and learning merge. This is further driven by the emerging technology, which is helping pervasive learning to happen more effectively and will continue to impact positively. In line with this is the 'embedded ubiquitous' approach, where learning is embedded with the work, and is provided just at the time of task execution, just enough to accomplish the task at hand. Mobile devices and technology are the first wave in technology that supports this type of learning; it only gets better from here.
Responsive/ Multi-Device Learning	We live in a multi-device world. Almost each one of us today uses a smart phone and a laptop (and that's just the bare minimum requirement!) and other such devices that can support learning. For training administrators, this means developing and delivering learning solutions that work seamlessly across all devices – irrespective of their sizes, shapes, resolutions or OSs. The answer to this is 'responsive E-learning.' Responsive E-learning design, simply put, is used to "provide an optimal viewing experience – easy reading and navigation with a minimum of resizing, panning, and scrolling – across a wide range of devices." But while responsive design provides device/display specific structuring of the content, it is important to ensure the relevance, type and context of the content, and more importantly the 'point of use' and access.

TABLE 4.14 Continued

Trends	Contents
Wearable Computing Technology In Learning	Wearable computing devices and associated technology, though still a novel concept, have been around for a while. Right now, most of these devices seem to be tethered to a phone or other mobile device, but as miniaturization continues this tethering will no longer be required. There are three main reasons it cannot be ignored for learning: • 'Real Sharing' – (life streaming becomes real, searchable, sharable streams of data BIG data becomes real); • The provision of context; • Natural progression from mobile phones.
HTML 5	HTML 5 was hot when it came into focus and till this day remains to be one of the hottest trends to have hit the E-learning industry. So while Flash vs. HTML5 was a point of debate earlier, it's not anymore. For any web-based mLearning, HTML5 is the future of mobile web, even if it's not ready to the extent we believe it to be. Speaking of authoring tools, most tools, today, actually struggle to provide real HTML5 compatibility. Several of them actually just embed non-interactive videos in HTML code. Good quality animations and interactivities are still missing. The tools are expected to become more capable in exploiting the potential of HTML5 in the future. Of the tools available out there, Lectora, Articulate Storyline and Adobe Captivate seems to be the most promising. Some of these also provide options to publish as Flash or HTML5 or as an app. Here's a compilation of 15 Authoring Tools For mEnabling Your E-learning For iPads.
Tin Can API	We had rendered our first impressions of the Tin Can API some time back. That this API promises to address many of the shortcomings associated with the existing SCORM standards, that are now over a decade old, is a given. But what remains to be seen is how it will affect the LMS as we know it and how soon? Very soon, if you ask us; although the actual adoption may take some time. It's true that getting a standard off the ground is a huge challenge, but Tin Can API, with its ability to collect data about the varied experiences an individual has had, both online and offline, capture them in a consistent format and record them in detail, has the power to move more quickly than other standards for a number of reasons.

TABLE 4.14 Continued

Trends	Contents
Gamification In Learning	'Gamification in learning' means attempting to apply the principles that make individuals play games for hours at end. Properly implemented, gamification has the potential to make learning 'stickier,' increase uptake of learning content and also provide a more comprehensive record of learning than is possible using conventional measures in courses. As millennial continue to enter the workplace at an increasing rate, gamification in learning can be used by leveraging the mindset of these young individuals who look for ways to be acknowledged for their accomplishments by their peers by making use of social connections/ media. As in a game, where there are reward points for displaying the player status, in a learning context too, the same mechanism/ strategy can be used.
Informal Learning	Majority of learning in the modern context happens naturally and at most times is embedded in other tasks – contextually and subconsciously, and is always self-initiated. While you cannot 'create and implement' informal learning, you can at best support it by providing an environment, which breeds informal learning. Part of that environment will be the culture of your organization and that's not something any vendor, technology, or tool can do for you. Mobile with its several unique characteristics of being always on, always carried, and a host of sensors in it, could be the ideal enabling technology to begin with and offers a transformational opportunity if pursued properly as discussed in one of our previous posts Mobile Enables Informal Learning.
Videos in E-learning	In early stages of E-learning, video was used by most organizations for training, which later sizzled with the emergence of web and its associated limitation on bandwidth utilization. Fifteen years later, videos are all set to return to E-learning in a big way. And here's why. 1. They provide engagement. 2. They act as on demand Performance Support tools just when your staff needs them. 3. They provide an opportunity to bridge the gap between different screen sizes and multiple platforms. 4. Since they 'show' pictures and can include subtitles, they can act as Cross Language tools for learning.

TABLE 4.14 Continued

Trends	Contents
	5. The cost of video production is low.
	6. They tend to be more viral that other assets.
	7. Employees can record events, processes, problems – just about anything they wish to share creating User Generated Videos.
	In line with videos is a similar asset, which will hold its own place in delivering learning to individuals – Television. In fact, its presence has already been felt with the emergence of products like Apple TV and Google TV, which connect to conventional television sources and the Internet to dish up an interactive experience as opposed to the experience provided by a regular one-way TV. Apart from being made use of for entertainment, TV coupled with video has already been used for learning initiatives in a big manner. Perhaps in a decade from now on, the TV would be completely different from what we know of it today. We had spoken about this in TV In the Future of Learning. In fact, with the APIs and tools for creating applications for TV fast emerging, it now depends on us as to how we, as learning designers, make the most of it.
Algorithmically Generated Content In Learning	Algorithmic generation of content has existed for a while now, most commonly used in games to generate content used to populate the game environment. So how long before we have algorithms that are setup to create 'learning material' by constantly monitoring streams of user generated content, monitoring individual context? Not too far. The first inklings of this are already visible, in the form of search, discovery and sharing services such as Scoop. it and Summify and a whole lot more. The algorithms the services use don't really create the content, but that will change soon. We will have algorithms that actually glean information from various streams and write content.

Source: Thanekar (2013).

4.7.4   Learning Technology Trends in 2014

McIntosh (2014) indicated the E-learning Technology Trends in 2014 as Table 4.15.

4.8   THE CHALLENGE FOR E-LEARNING

Transnational Management Associates (2014) indicated manufacturing models of top–down management have been replaced by knowledge-centric structures that place middle managers at the heart of organizations. As a result, commercial success relies on developing highly skilled, adaptable middle management to execute strategy - by aligning operations to the overall strategy of the organization. But how can we adequately develop our middle managers to execute business strategy when faced with shrinking training budgets and trimmed down HR functions? This has to be one of the biggest challenges we face as training professionals. Is e-learning the answer, as the title of this Viewpoint suggests? The short answer is yes - to a point! Changes to the social, economic and technological environments in which we work support e-learning adoption. The way we work, and the world in which new work, is changing. As Terry Brake points out there are a number of emerging trends that create a positive climate for e-learning success as shown in Table 4.16.

4.9   CONCLUSIONS

The biggest challenge is creating the right environment and culture in which e-learning is valuable to individual career progression as well as organizational success. This requires us to capitalize on the current environmental trends that support e-learning. But it also places the onus on us to build appropriate levels of senior management support to deliver on a vastly under-rated management education promise (Transnational Management Associates, 2014).

According to Christopher (2013), as well put by many field experts, there are four forces that will rock the waters of E-learning in the future: Cloud, Social, Mobile and Information. These forces will drive change and create demand for advanced IT infrastructure that subsequently will profoundly affect the sector's path, trends, initiatives, plans and programs.

TABLE 4.15 Learning Technology Trends in 2014

Trends	Contents
1. Mobile Learning/ HTML5	In both education and the corporate world, there will be further adoption of mobile learning. Teachers and trainers will adopt these tools right in the classroom and well as online and HTML5 will continue to evolve as a programming language to make mobile learning more effective and responsive – automatically adapting to the size and shape of the screen on the learner's device. It will also enable more just-in-time learning.
2. Social/Informal Learning	This trend has been happening over the past five years or so and will continue. The Tin Can API will help enable this.
3. Video	I am concerned that video is often over-used. Talking head video just uses up bandwidth and slows down people like me who learn faster by reading and being able to search. In spite of any concerns I may have, the use of video will continue to grow. Ways will be found to search video content more effectively.
4. Games and simulations	Educators have long known the benefits of games and simulations. The airline industry has used simulators for decades. But formal education institutions and corporations have resisted this for years. It will continue to gain acceptance and increasing numbers of LMS's will become games friendly.
5. E-learning authoring tools	E-learning authoring tools will continue to evolve to include: a. publishing to HTML5 b. cloud-based/collaborative authoring c. responsive (to the screen) and adaptive (to the learner) learning d. just-in-time learning/ performance support e. games, simulations and animations
6. Adaptive e-Learning Design	More design will become adaptive to the learner – responding to the situation and needs of the learner - something classroom teachers have been doing for centuries.
7. Tin Can API adoption	The Tin Can API gives us the ability to track learning outside the LMS – both formal and informal. More LMS's and authoring tools will adopt it. The term Tin Can will stick as opposed to the ADL/SCORM term Experience API.

TABLE 4.15   Continued

Trends	Contents
8. Learning Analytics/Big Data	There will be increasing use of learning data from both inside the LMS and outside of it using data consolidation tools to get learning opportunities to people more efficiently.
9. Industry consolidation and fragmentation	There will continue to be both consolidation at the top level of the business with HR software companies purchasing the larger LMS companies. Smaller companies will continue to come and go. Smaller companies have greater ability to innovate and some of the new ideas will take root.
10. MOOC's	The impact of MOOC's will continue to be debated but also will begin to influence the thinking of corporate training organizations.
11. Personalization	There will be increasing personalization of learning – providing the learner with what is needed when it is needed. This is closely related to adaptive learning and the use of big data tools.
12. The user experience	Most LMS interfaces are less than exciting. There will be more use of graphics in LMS interfaces to make them more engaging for learners.
13. Flipped Classrooms	A fairly recent trend in formal education in which learners are asked to review material before coming to class and the class is used mostly for discussion. E-learning makes this more possible.
14. BYOD – Bring Your Own Device	The trend toward BYOD will continue in which learners are expected to have their own device and systems can adapt to smartphones, tablets, laptops, etc. More and more instructors will use these right in the classroom as well as for remote learning.

Source: McIntosh (2014).

TABLE 4.16   The Challenge for E-Learning

Challenge	Contents
Businesses are evolving into 24/7 work environments	Managers cannot always be pulled off jobs to attend training programmes as and when they come about. They also need to develop their skills quickly and effectively. E-learning is available as and when they need it. The challenge is making it relevant so they are motivated to logon!
Employees are beginning to demand greater work-life balance	Training interventions are disruptive to managers who have to play catch-up when they get back to their jobs. E-learning allows busy managers to learn what they need, when they need it. Online education, designed as a toolkit rather than a linear course, gives managers a just-in-time resource that is always available to them.
Globalization creates a greater reliance on virtual teams	Much of the informal coaching and peer support can no longer be assumed. However, e-learning can provide some of these skills development needs. The real issue is ensuring e-learning content quality and relevance so that it is a valued, meaningful development medium.
Middle managers need to evolve into effective leaders and change agents as well as remain highly skilled workers.	Yet we are all limited as to the types of development programmes we can deliver without encountering unjustifiably huge costs. We are after all in a cost-cutting world. It's an old argument, but e-learning can be cost-effective where targeting large numbers of geographically dispersed people.
Next generation managers are motivated by challenge as well as reward	A new generation is emerging characterized, in part, by people who are young enough to be technologically competent, have a global outlook and who recognize the value of continuous self-development. This is a future face of middle management - a new generation who are comfortable with the online world and understand the need for life-long learning.
Internet technology adoption and deployment spans all aspects of life	There is almost no facet of working life - from email communications, knowledge portals and information databases - that has not been touched by the Internet. The Internet is also a very cost-effective and efficient means to deliver knowledge and skills to the growing bands of knowledge-based workers. These environmental factors create the conditions in which e-learning can help managers develop the skills and knowledge to execute business strategy.

Source: Transnational Management Associates (2014).

The motto "anytime, anywhere and anybody" emphasized character-istics of e-Learning 1.0 – providing easy and convenient access to educa-tional contents (Ebner, 2007). Therefore, e-Learning 1.0 mostly focused on creating and administering content for viewing online.

The factor most lacking in e-learning 1.0 is the social and collabora-tive approach to learning. With the fusion of e-learning 1.0 and Web 2.0, a new set of services has emerged, which makes the learning process more creative and learning experience more enduring. Hart (2008) says that, E-learning 1.0 was all about delivering content, primarily in the form of online courses and produced by experts, that is, teachers or subject mat-ter experts. E-learning 2.0 is about creating and sharing information and knowledge with others using social media tools like blogs, wikis, social bookmarking and social networks within an educational or training con-text to support collaborative approach to learning.

E-learning 2.0 views online learning tools as a platform and not a medium. It aims to create a learning environment, and not just a learning system. By enabling participation and collaboration of students in content creation, organization, and use, e-learning 2.0 creates a "personal learning portfolio." In e-learning 2.0, "everyone is a learner, but also everyone has a potential to be a teacher" (Cobb, 2008). The LMS continues to be the dominant technology for delivery of online courses. E-learning 2.0 com-bines the quality of e-learning 1.0 and the power of Web 2.0. The teaching and learning communities should exploit e-learning 2.0 tools and services to make learning process enjoyable and creative.

E-learning 2.0 is a type of computer-supported collaborative learning (CSCL) system that developed with the emergence of Web 2.0 (Parks, 2013; Karrer, 2007; Downes, 2005). From an e-learning 2.0 perspective, conven-tional e-learning systems were based on instructional packets, which were delivered to students using assignments. Assignments were evaluated by the teacher. In contrast, the new e-learning places increased emphasis on social learning and use of social software such as blogs, wikis, podcasts and virtual worlds such as Second Life (Christine, 2009). This phenomenon has also been referred to as Long Tail Learning (Brown et al., 2008).

E-learning 2.0, in contrast to e-learning systems not based on CSCL, assumes that knowledge (as meaning and understanding) is socially con-structed. Learning takes place through conversations about content and grounded interaction about problems and actions. Advocates of social learning claim that one of the best ways to learn something is to teach it to others (Brown et al., 2008). In addition to virtual classroom environments, social networks have become an important part of E-learning 2.0. Social networks have been used to foster online learning communities around subjects as diverse as test preparation and language education (Manprit, 2012). Mobile Assisted Language Learning (MALL) is the use of handheld computers or cell phones to assist in language learning. Traditional educators may not promote social networking unless they are communicating with their own colleagues (Beverley, 2009).

Virtual worlds for e-learning have been amongst the first applications being deployed in clouds in order to exploit the characteristics of Cloud computing with respect to on-demand provision of resources during runtime (Cucinotta et al., 2010). The Web 2.0 has enabled the generation of a substantial amount of data, both by users and about users. While this data holds substantial value, it is often severely under-utilized, by simply being stored away or even worse discarded. We believe that the Web 3.0 will help users to sift and sort this mass of information by utilizing AI. In the domain of e-learning, AI will likely be used not only for assisting learners, but also for gaining a deeper understanding of the learning process.

Wheeler (2011) considers that the Web 3.0 will be the "Read/Write/ Collaborate" web. E-Learning 3.0 will have at least four key drivers: distributed computing, extended smart mobile technology, collaborative intelligent filtering, 3D visualization and interaction. Halpin (2007) considers that e-Learning 3.0 will be both "collaborative" and "intelligent." Rego et al. (2010) considers that e-Learning concept of "anytime, anywhere and anybody "will be complemented by "anyhow," that is, it should be accessible on all types of devices. The existing forecasts about where e-Learning 3.0 is heading tend to focus on educational aspects, and only briefly mention technical aspects. The study takes a different approach; since from past transformations (from e-Learning 1.0 to e-Learning 3.0), its close relationship with the available technologies of the current generation of the World Wide Web. The challenge then is to correctly identify which technologies will be brought forth by the Web 3.0. Rubens et al. (2011) then hypothesize how these technologies may be utilized in e-Learning 3.0.

Learners need appropriate motivation. As anyone who works in e-learning can testify, e-learner motivation is critical to success. Therefore, the learning effort must demonstrate a tangible association between better execution of business strategy and further career enhancement. This is the ultimate challenge! The result of the learning experience needs to be valuable and meaningful. Even before we tackle the issues of pedagogy and instructional design, the issue of what's-in-it-for-me must be addressed. People learn if they understand the benefits to their job, and are supported and encouraged to learn (Transnational Management Associates, 2014).

The future of E-learning is bright. All we have to do is encourage new E-learning technologies, methods and applications to flourish and older E-learning techniques to evolve. However, to succeed we need to actively address a number of challenges: E-Learning needs to be rigorously aligned with business needs. The value of e-learning can only be realized if we carefully ensure that what is being delivered genuinely helps our managers execute business strategy. Boards and senior management need to exercise leadership to effect this alignment. Business leaders are often not involved with the e-learning initiatives they effectively sponsor. We need their support on three levels (Transnational Management Associates, 2014):

1. Firstly, to encourage training and business people to integrate e-learning into development programmes.
2. Secondly, to encourage line managers to value work-based learning as a fundamental component of working life.
3. Lastly, to validate that e-learning does advance business objectives.

In summary, first of all, this chapter reviews the evolution of e-learning and discusses the literatures from E-learning 1.0 to 3.0 via the perspectives of technology and applications. Secondly, the chapter summarizes content in E-learning, advantages and disadvantages with regards to motivation in e-learning, explores the comparisons of Big Data, Linked Data, Cloud Computing and Data-driven Science, and potential AI utilization by learning methods.

Thirdly, the study discusses E-learning technology and learning management system. E-learning technology includes: media, audio and video, computers, tablets and mobile devices, Open Course Ware (OCW), Social networks, Webcams, Whiteboards, and Screen casting. Learning management systems include Canvas, Blackboard Inc. and Moodle, online distance learning, hybrid learning or blended learning, distributed learning, web enhanced, web supplemented and web dependent, synchronous and asynchronous learning, linear learning, collaborative learning, student-centered learning, learning content management system, computer-aided assessment, electronic performance support systems (EPSS). Also, the study explores future E-learning trends and technologies in the global and analyzes the applications for E-learning which includes: Preschool, K–12, Online schools, Higher education, MOOCs, DOCC, Coursera, Corporate and professional, Public health, ADHD, Classroom 2.0, Virtual classroom, Virtual learning environment (VLE), Augmented reality (AR).

Fourthly, the study summarizes the benefits of E-learning includes: Defray travel costs, More opportunities for extended learning, Wide participation, improved student writing, differentiated instruction, and additional benefits includes: studies completed in "computer intensive" settings found increases in student-centric, cooperative and higher order learning, students writing skills, problem solving, and using technology (An and Reigeluth, 2011). Moreover, positive attitudes toward technology as a learning tool by parents, students and teachers are also improved.

The result of the learning experience needs to be valuable and meaningful. Even before we tackle the issues of pedagogy and instructional design, the issue of what's-in-it-for-me must be addressed. People learn if they understand the benefits to their job, and are supported and encouraged to learn.

Finally, the study analyzes global trends and technologies in the E-learning industry and the challenge for E-Learning:

Global trends and technologies in the E-learning industry:

1. Mobile Learning/HTML5;
2. Social/Informal Learning;
3. Video
4. Games and simulations;
5. E-learning authoring tools;
6. Adaptive e-Learning Design;
7. Tin Can API adoption;
8. Learning Analytics/Big Data;
9. Industry consolidation and fragmentation;
10. MOOC's;

11. Personalization;
12. The user experience;
13. Flipped Classrooms;
14. BYOD – Bring Your Own Device.

The challenge for e-learning:

1. Businesses are evolving into 24/7 work environments;
2. Employees are beginning to demand greater work-life balance;
3. Globalization creates a greater reliance on virtual teams;
4. Middle managers need to evolve into effective leaders and change agents as well as remain highly skilled workers;
5. Next generation managers are motivated by challenge as well as reward;
6. Internet technology adoption and deployment spans all aspects of life.

KEYWORDS

- **Applications**
- **E-learning 1.0**
- **E-learning 2.0**
- **E-learning 3.0**
- **Technology**
- **Trends and challenge**

REFERENCES

1. Alkhateeb F., Al Maghayreh E., Aljawarneh S., Muhsin Z., Nsour A., E-learning Tools and Technologies in Education: A Perspective, E-learning (2010).
2. Allen, I. E., Seaman, J., Staying the Course: Online Education in the United States, 2008 Needham MA: Sloan Consortium (2008).
3. Allen, I. E., Seaman, J., Sizing the Opportunity: The Quality and Extent of Online Education in the United States, 2002 and 2003 Wellesley, MA: The Sloan Consortium (2003).

4. Ambient Insight Research, US Self-paced e-Learning Market Monroe WA: Ambient Insight Research (2009).

5. An, Y. J., & Reigeluth, C. Creating Technology-Enhanced, Learner-Centered Classrooms: K–12 Teachers' Beliefs, Perceptions, Barriers, and Support Needs. Journal of Digital Learning in Teacher Education, 28(2), 54–62. 2011.

6. Anderson P. What is web 2.0. Ideas, technologies and implications for education, 60, 2007, 1–8, Conference Location: Perth.

7. Areskog, N-H., The Tutorial Process – the Roles of Student Teacher and Tutor in a Long Term Perspective and contextual views, TBD. (1995).

8. Aristovnik, Aleksander. The impact of ICT on educational performance and its efficiency in selected EU and OECD countries: a non-parametric analysis. MPRA Paper No. 39805, posted 3, July (2012).

9. Armstrong Scott J. "Natural Learning in Higher Education." Encyclopedia of the Sciences of Learning, (2012).

10. Army, Mil, "DoD gives PTSD help 'second life' in virtual reality | Article | The United States Army." Retrieved 2013–10–22.

11. Awesome Inc., ICT IN EDUCATION, pak123ict.blogspot.com/, Retrieved 22 Oct. 2014.

12. Bååth, J. A. "Distance Students' Learning – Empirical Findings and Theoretical Deliberations," Distance Education, (1982), 3(1), 6–27.

13. Babson Research Study, "Babson Research Study: More Than 6.7 Million Students Learning Online." Retrieved 12 February 2013.

14. Baker R., Yacef K. The state of educational data mining in 2009: A review and future visions. Journal of Educational Data Mining, 1(1):3–17, 2009.

15. Bates, A., Poole, G., Effective Teaching with Technology in Higher Education San Francisco: Jossey-Bass/John Wiley, 2003.

16. Beverley Crane, E., "Using Web 2.0 Tools in the k-12 Classroom" Neal-Shuman Publishers Inc., 2009, p.3.

17. Black, J., McClintock, R. "An Interpretation Construction Approach to Constructivist Design." (1995).

18. Blackboard.com, "Blackboard International | EMEA." Retrieved 2012–10–24.

19. Bloom, B. S., Krathwohl D. R., Taxonomy of Educational Objectives: Handbook 1, (1956).

20. Brown Seely, John, Adler, Richard P. "Minds on Fire: Open Education, the Long Tail, and Learning 2.0." Educause review (January/February, 2008): 16–32.

21. Buckleitner Warren. "So Young, and So Gadgeted." The New York Times. Retrieved 2013–10–22.

22. Calvani, A., et al., Lifelong learning: What role for e-learning 2.0? Journal of e-Learning and Knowledge Society 4 (1), 179–187. Available: http://www.je-lks.it/en/08_01/06Metcalv_en1.pdf (2008).

23. Career FAQs state, "What is Blended Learning?." Career FAQs state. Retrieved 31 March 2013.

24. Cavanaugh, C., Effectiveness of cyber charter schools: A review of research on learning's. Tech Trends, (2009), 53(4), 28–31.

25. Celia Baker, (2014), "Blended learning: Teachers plus computers equal success." Desert News. Retrieved 30 January 2014.

26. Center For Disease Control and Prevention. n.d., "ADHD Data and Statistics." Retrieved June 27, 2014.

27. Challenges and opportunities, Science, (2011) 331(6018), 692–693.

28. Chen C., A constructivist approach to teaching: Implications in teaching computer networking implications in teaching computer networking. Information Technology, Learning, and Performance Journal, (2003) 21(2), 17.

29. Cho, V., Wayman, J. C. Knowledge management and educational data use. Paper presented at the 2009 Annual Meeting of the American Educational Research Association, San Diego, CA. (2009).

30. Christian Dalsgaard, "Social software: E-learning beyond learning management systems." eurodl.org. University of Aarhus. Retrieved 31 March 2013.

31. Christopher Pappas, "Global Trends in the E-learning Industry," International Congress on E-learning 2013.

32. Clark, W. A., R. C., Mayer, R. E. E-learning and the Science of Instruction. San Francisco: Pfeiffer. (2007).

33. Cobb, J. Learning 2.0 for associations: Mission to learn. Available: http://blog.missiontolearn.com/files/Learning_20_for_Associations_eBook_v1.pdf (2008).

34. Comptia.org, "Making the Grade: Technology Helps Boosts Student Performance, Staff Productivity in Nation's Schools, New Comptia Study Finds" (Press release). Retrieved 2014–03–22.

35. Courts, B., Tucker, J. Using Technology To Create A Dynamic Classroom Experience. Journal of College Teaching & Learning (TLC), (2012) 9(2), 121–128.

36. CRTC, "CRTC issues annual report on the state of the Canadian communication system." Retrieved 2014–03–22.

37. Cucinotta T., Checconi F., Scuola Superior Sant'anna, Kousiouris G., Kyriazis D., Varvarigou T., Mazzetti A., Zlatev Z., Papay J., Boniface M., Berger S., Lamp D., Voith T, Stein M., Virtualized e-Learning with real-time guarantees on the IRMOS platform Service-Oriented Computing and Applications (SOCA), 2010 IEEE International Conference on Date of Conference: 13–15 Dec. 2010.

38. Cull, S., Reed, D., Kirk, K. "Student motivation and engagement in online courses." Serc.carlton.edu. Retrieved 2013–10–22.

39. Dennen, V. P., Bonk, C. J. "We'll leave the light on for you: Keeping learners motivated in online courses." Coursesites.com. Retrieved 2013–10–22.

40. Dennis Shiao. "Why Virtual Classrooms Are Excellent Learning Venues." INXPO. Retrieved 18 May 2013.

41. Downes, S. E-Learning 2.0. ACM eLearn Magazine. October 2005. Available: http://www.elearnmag.org/subpage.cfm?section=articlesAmp;article=29–1 (2005).

42. Ebner M., E-learning 2.0 = e-learning 1.0 + web 2.0? Availability, Reliability and Security, International Conference on, 0:1235–1239. (2007).

43. Edweek.org, "Research Center: Charter Schools." Retrieved 2012–10–24.

44. E-learning-companion.com, "The Disadvantages of Online Learning." 12 February 2013. Retrieved 2013–10–22.

45. Electronic-school.com, "Technology's Impact." Retrieved 2014–03–22.

46. eliademy.com, "democratizing education with technology." Retrieved 2014–05–05.

47. Eric Tremblay, "Educating the Mobile Generation – using personal cell phones as audience response systems in post-secondary science teaching." Journal of

Computers in Mathematics and Science Teaching, (2010), 29(2), 217–227. Trentin G. Networked Collaborative Learning: social interaction and active learning, Woodhead/Chandos Publishing Limited, Cambridge, UK, ISBN 978-1-84334-501-5. https://www.researchgate.net/publication/235930117_Networked_Collaborative_Learning_social_interaction_and_active_learning?fulltextDialog=true/, (2010).

48. Fichten, C. S., Ferraro, V., Asuncion, J. V., Chwojka, C., Barile, M., Nguyen, M. N., Klomp, R., Wolforth, J. "Disabilities and e-Learning Problems and Solutions: An Exploratory Study." Technology and Society. (2009).

49. Fischetti. M. The web turns 20: Linked data gives people power. Scientific American, (2010) 12.

50. Forehand, M. Bloom's Taxonomy. From Emerging Perspectives on Learning, Teaching and Technology. Retrieved October 25, 2012, from http://projects.coe.uga.edu/epltt/.

51. Gc.maricopa.edu, "Characteristics of the Successful Online Student." Retrieved 2013–10–22.

52. Gery, G. The quest for electronic performance support. CBT Directions, July, 1989.

53. Glasersfeld E., Constructivism: Theory, perspectives, and practice, chapter Introduction: Aspects of constructivism, (1996) pages 3–7. Teacher College Press.

54. Gonella, L., Panto, E. Didactic architectures and organization models. E-learning Papers. Available: http://www.E-learningeuropa.info/files/media/media15973.pdf (2008).

55. Goroshko O. I., Samoilenko S. A. Twitter as a Conversation through e-Learning Context. Revista de Informatica Sociala, (15), 2011.

56. Government Technology, "Universities Use Second Life to Teach Complex Concepts." Retrieved 2013–10–03.

57. Grabinger, R. S., Aplin, C., Ponnappa-Brenner, G. "Supporting learners with cognitive impairments in online environments." Tech Trends (2008) 52 (1), 63–69.

58. Gu, X., Zhu, Y., Guo, X. Meeting the "Digital Natives": Understanding the Acceptance of Technology in Classrooms. Educational Technology and Society, (2013) 16 (1), 392–402.

59. Guo, Z., Li, Y., Stevens, K. Analyzing Students' Technology Use Motivations: An Interpretive Structural Modeling Approach. Communications of the Association for Information Systems, (2012). 30(14), 199–224.

60. Halpin H., Robu, V., Shepherd, H. The complex dynamics of collaborative tagging. In Proceedings of the 16th international conference on World Wide Web, WWW'07, (2007), pages 211–220, New York, NY, USA, ACM.

61. Hart, J. An introduction to social learning, Available: http://www.c4lpt.co.uk/handbook/sociallearning.html (2008).

62. Hebert, D. G. "Five Challenges and Solutions in Online Music Teacher Education." Research and Issues in Music Education (2007) 5 (1).

63. Hicks, S. D. Technology in today's classroom: Are you a tech-savvy teacher? The Clearing House, (2011) 84, 188–191. http://alliedacademies.org/public/Proceedings/Proceedings24/AEL%20Proceedings.pdf#page=43.

64. Iiyoshi, T., Kumar, M. S. Opening up education: the collective advancement of education through open technology, open content, and open knowledge. Cambridge, Mass.: MIT Press. (2008).

65. Illinois University, "Strengths and Weakness of Online Education." www.ion.uilli-nois.edu/resources/tutorials/overview/. (2 May, 2008).
66. Interconf.fl, COMMUNICATION TECHNOLOGIES USED IN E-LEARNING, http://interconf.fl.kpi.ua/node/1193 (2014).
67. Ipark.hud.ac.uk, "Screen casting Teaching and Learning Innovation Park." Retrieved 2013–10–22.
68. Ite.educacion.es, "Escuela 2.0." Retrieved 2013–10–22.
69. Jaschik Scott, "Feminist Anti-MOOC," Inside Higher Ed, August 19. (2013).
70. John D., Catherine T. MacArthur Foundation, Digital media and learning fact sheet. (2005).
71. Johnson, Henry M. "Dialogue and the construction of knowledge in e-learning: Exploring students' perceptions of their learning while using Blackboard's asynchronous discussion board." Eurodl.org. ISSN 1027–5207. Retrieved 2013–10–22.
72. Jonassen D., Land, S. Theoretical Foundations of Learning Environments. Lawrence Erlbaum Associates (2000).
73. Källander, K., Tibenderana, J. K., Akpogheneta, O. J., et al. "Mobile health (mHealth) approaches and lessons for increased performance and retention of community health workers in low- and middle-income countries: a review." Journal of medical Internet research (2013) 15 (1). doi: 10.2196/jmir.2130.
74. Karl Kurbel, Virtually on the Students' and on the Teachers' sides: A Multimedia and Internet based International Master Program; ICEF Berlin GmbH (Eds.), Proceedings on the 7th International Conference on Technology Supported Learning and Training – Online Education; Berlin, Germany; (November, 2001), pp. 133–136.
75. Karrer, T, Understanding E-learning 2.0. Learning circuit, http://www.astd.org/LC/2007/0707_karrer.htm. (2007).
76. Keengwe, J., Onchwari, G., Onchwari, J. "Technology and Student Learning: Toward a Learner-Centered Teaching Model," AACE Journal. (2009).
77. King G. Ensuring the data-rich future of the social sciences, science, (2011) 331(6018):719.
78. Kogan Page, E-moderating: The Key to Teaching and Learning Online – Gilly Salmon, ISBN 0–7494–4085–6 (2000).
79. Kolowich, Steve. "Conventional Online Higher Education Will Absorb MOOCs, 2 Reports Say." The Chronicle of Higher Education. Retrieved May 15, 2014.
80. KQED, "For Frustrated Gifted Kids, A World of Online Opportunities." KQED. Retrieved 2014–05–24.
81. Lewin, T. Harvard, M. I. T. Team Up to Offer Free Online Courses. New York Times, p. A18 Retrieved November 26, 2012, from http://www.nytimes.com/2012/05/03/education/harvard-and-mit-team-up-to-offer-free-online-courses.html?_r=0.
82. Liesbeth Kester, Kirschner. "Designing support to facilitate learning in powerful electronic learning environments." Computers in Human Behavior 23 (3): 1047. doi: 10.1016/j.chb.2006.10.001. Retrieved 21 January 2014.
83. Lisa Diecker; Lane, Allsopp, O'Brien, Butler, Kyger, Fenty. "Evaluating Video Models of Evidence-Based Instructional Practices to Enhance Teacher Learning." Teacher Education and Special Education 32(2), 180–196. Retrieved 2011–09–17.
84. Loutchko, Iouri; Kurbel, Karl; Pakhomov, Alexei, Production and Delivery of Multimedia Courses for Internet Based Virtual Education; The World Congress

"Networked Learning in a Global Environment: Challenges and Solutions for Virtual Education," Berlin, Germany, May 1–4, (2002).

85. Lukasiewicz T., Straccia U. Managing uncertainty and vagueness in description logics for the semantic web. Web Semantics: Science, Services and Agents on the World Wide Web, (2008). 6(4), 291–308.

86. Luskin Bernard, "Think 'Exciting': E-Learning and the Big 'E'." http://www.educause.edu/ero/article/think-exciting-e-learning-and-big-e (2010).

87. Luskin, B. Media Psychology: A Field whose time is here. The California Psychologist, (1996). 15(1), 14–18.

88. Macs.hw.ac.uk, "Conversational Model." Retrieved 2013–10–22.

89. Madaus, J. W., McKeown, K., Gelbar, N., & Banerjee, M." The Online and Blended Learning Experience: Differences for Students With and Without Learning Disabilities and Attention Deficit/Hyperactivity Disorder." International Journal for Research in Learning Disabilities (2012).

90. Manprit Kaur. "Using Online Forums in Language Learning and Education." StudentPulse.com. Retrieved 2012–08–22.

91. Mansour Iris, "Degreed wants to make online courses count," Fortune, August 15, 2013. (Retrieved August 15, 2013).

92. Marshall C. C., Shipman, F. M., Which semantic web? In Proceedings of the fourteenth ACM conference on Hypertext and hypermedia, HYPERTEXT ' (2003). 03, 57–66, New York, NY, USA, ACM.

93. Martha Beagle and Hudges Don, "Social Networking in Education." http://www.pelinks4u.org/articles/beagle0609.htm, Retrieved 18 Oct. 2014.

94. McCarroll, Niall; Kevin Curran. "Social Networking in Education." International Journal of Innovation in the Digital Economy (2013). 4 (1), 15. doi: 10.4018/jide.2013010101.

95. McIntosh Don, Learning Technology Trends in 2014 Trimeritus E-learning Solutions Inc. (2014).

96. Meetoo-Appavoo A., Constructivist-Based Framework for Teaching Computer Science. International Journal of Computer Science and Information Security (IJCSIS), (2011) 9(8):25–31.

97. Meidlinger K., "Choosing media for children checklist," San. Francisco: Kids Watch Monthly, KQED.org (adapted from Rogow, F.), (2011).

98. Melody Terras, Ramsay, "The five central psychological challenges facing effective mobile learning." British Journal of Educational Technology, (2012) 43 (5): 820. doi: 10.1111/j.1467–8535.2012.01362.x.

99. Michael Biocchi, "Games in the Classroom." Gaming in the Classroom. Retrieved 24 March 2011.

100. Mondahl, M., Rasmussen, J., Razmerita L., Web 2.0 applications, collaboration and cognitive processes in case-based foreign language learning. In Proceedings of the 2nd World Summit on the Knowledge Society: Visioning and Engineering the Knowledge Society. A Web Science Perspective, WSKS (2009) 09, pages 98–107, Berlin, Heidelberg, Springer-Verlag.

101. Moodle.org, "open-source community-based tools for learning." Retrieved 2012–10–24.

102. Moravec J. W., A new paradigm of knowledge production in higher education. On the Horizon, (2008) 16(3), 123–136.

103. Muljadi Paul, E-learning, books.google.com.tw/books?id=XbG4bgcRa8wC (2014) p.6.

104. Murray, Kristine; Rhonda Waller. "Social Networking Goes Abroad." Education Abroad, (2007). 16(3), 56–59.
105. Nagy, A. The Impact of E-Learning, in: Bruck, P.A.; Buchholz, A.; Karssen, Z.; Zerfass, A. (Eds). E-Content: Technologies and Perspectives for the European Market. Berlin: Springer-Verlag, (2005) pp. 79–96.
106. Nelson M. L. Data-Driven Science: A New Paradigm? EDUCAUSE Review, (2009) 44(4):6–7, 23. L. S. Pogue.
107. Nsba.org, "Technology Impact on Learning." Retrieved 2014–03–22.
108. Nsba.org, "Technology Uses in Education." Retrieved 2014–03–22.
109. OECD, E-Learning in Tertiary Education: Where Do We Stand? Paris: OECD (2005).
110. O'Hear, S. E-learning 2.0: How Web technologies are shaping education . Available: http://www.readwriteweb.com/archives/e-learning_20.php (2006).
111. Paolo Pumilia-Gnarini, Didactic Strategies and Technologies for Education: Incorporating Advancements. (2012).
112. Pappano, Laura. "The Year of the MOOC." New York Times. Retrieved 12 February 2013.
113. Parker, D. P., Banerjee, M. "Leveling the digital playing field: Assessing the learning technology needs of college-bound students with LD and/or ADHD." Assessment for Effective Intervention. (2007).
114. Pilgrim, Jodi; Christie Bledsoe." Learning Through Facebook: A Potential Tool for Educators." Delta Kappa Gamma (2011).
115. Rankin, J. How data Systems and reports can either fight or propagate the data analysis error epidemic, and how educator leaders can help. Presentation conducted from Technology Information Center for Administrative Leadership (TICAL) School Leadership Summit. (2013).
116. Ravi Pratap Singh, Top 9 e-Learning Predictions for 2014, http://E-learningindustry. com/top-9-e-learning-predictions-for-2014 (2014).
117. Raybould, B. An EPSS Case Study: Prime Computer. Handout given at the Electronic Performance Support Conference, Atlanta, GA, (1992).
118. Redecker, C. Ala-Mutka, K. Bacigalupo, M. Ferrari, A., Punie. Y., Learning 2.0: the impact of web 2.0 innovations on education and training in Europe. Final Report. European Commission-Joint Research Center-Institute for Prospective Technological Studies, Seville, (2009).
119. Redecker, Christine. "Review of Learning 2.0 Practices: Study on the Impact of Web 2.0 Innovations on Education and Training in Europe." JRC Scientific and technical report. (EUR 23664 EN – 2009).
120. Reeves, Thomas C. "The Impact of Media and Technology in Schools." Retrieved 9 October 2013.
121. Rego, H. Moreira T., Morales E., Garcia, F. J., Metadata and Knowledge Management driven Web-based Learning Information System towards Web/e-Learning 3.0. International Journal of Emerging Technologies in Learning (iJET), (2010) 5(2), 36–44, June.
122. Repetto, M., &Trentin, G. (Eds.). Faculty Training for Web-Enhanced Learning. Nova Science Publishers Inc., Hauppauge, NY. 978-1-61209-335-2. https://www. researchgate.net/publication/235930053_Faculty_Training_for_Web-Enhanced_ Learning/ (2011).
123. Research F. Search engine usage report, (2006).

124. Richardson W. The educator's guide to the read/write web. Educational Leadership, (2005) 63(4), 24.
125. Rideout, V., Vanderwater, E., Wartella, E. Zero to six: Electronic media in the lives of infants, toddlers, and preschoolers. Menlo Park, CA: The Henry J. Kaiser Family Foundation. (2003).
126. Robinson, S., Ritzko, J. Podcasts in education: what, why, and how? Proceedings of 33 the Academy of Educational Leadership, (2009) 14(1).
127. Ross, S., Morrison, G., Lowther, D. Educational technology research past and present: balancing rigor and relevance to impact learning. Contemporary Educational Technology, (2010) 1(1).
128. Rubens N., Kaplan D., Okamoto T. E-Learning 3.0: anyone, anywhere, anytime, and AI. In International Workshop on Social and Personal Computing for Web- Supported Learning Communities, Dec (2011).
129. Rubens N., Louvigne S., and Okamoto T., Learning tweets dataset. Technical report, University of Electro-Communications, (2011).
130. Rupesh Kumar A., "E-Learning 2.0: Learning Redefined." Library Philosophy and Practice (e-journal). (2009) Paper 284. http://digitalcommons.unl.edu/libphilprac/284.
131. Ryan Dan, E - Learning Modules: Dlr Associates Series, books.google.com.tw/books?isbn=1468575201 (2012).
132. Sajadi, S. S., Khan, T. M. An Evaluation Of Constructivism For Learners With ADHD: Development of A Constructivist Pedagogy For Special Need, European, Mediterranean and Middle (2011).
133. Sarasota, Darya; Ali Khalid; Sören Auer; Jörg Unbehauen "Crowd Learn: Crowd sourcing the Creation of Highly-structured E-Learning Content." 5th International Conference on Computer Supported Education CSEDU 2013.
134. Scuola-digitale.it, "Scuola Digitale » Cl@ssi 2.0." Retrieved 2013–10–22.
135. Sendall, P., Ceccucci, W., Peslak, A. "Web 2.0 Matters: An Analysis of Implementing Web 2.0 in the Classroom." Information Systems Education Journal (2008) 6 (64).
136. Sero.co.uk," Curriculum and Pedagogy in Technology Assisted Learning." Retrieved 2013–10–22.
137. Siemens, G. (2010). What are Learning Analytics? ELEARNSPACE. Retrieved Oct. 11, 2014, from http://www.elearnspace.org/blog/2010/08/25/what-are-learning-analytics/.
138. Silva Alves da, N. Morais da Costa G., Prior M., Rogerson S. The Evolution of E-learning Management Systems: An Ethical Approach. International Journal of Cyber Ethics in Education (IJCEE), 1(3):12–24, 2011.
139. Simpson, C. W., Prusak, L. "Troubles with information overload—Moving from quantity to quality in information provision." International Journal of Information Management, (1995) 15 (6).
140. Smith B., Reed P., Jones C., 'Mode Neutral' pedagogy. European Journal of Open, Distance and E-learning." (2008).
141. Smith, Grace and Stephanie Throne, Differentiating Instruction with Technology in the K-5 Classrooms. International Society for Technology in Education, (2004).
142. Sören, Auer, "First Public Beta of SlideWiki.org." Retrieved 22 February 2013.

143. Tamrat T, Kachnowski S. "Special delivery: an analysis of mHealth in maternal and newborn health programs and their outcomes around the world." Maternal and Child Health Journal, (2012) 16(5). doi: 10.1007/s10995-011-0836-3.
144. Tavangarian D., Leypold M., Nölting K., Röser M. Is e-learning the Solution for Individual Learning? Journal of e-learning, (2004).
145. Technologyenhancedlearning.net, Computer-aided assessmenthttp://technologyenhancedlearning.net/tel-initiatives/computer-aided-assessment/, (2014).
146. Thanekar, Pranjalee, Top 10 E-learning Industry Trends For 2013, http://www.upsidE-learning.com/blog/index.php/2013/07/24/top-10-E-learning-industry-trends-for-2013/ (2013).
147. The Learning Curve (2013),"The Size of the Online Education Industry." Retrieved 12 February 2013.
148. The Magazine for Database Professionals, "Open Course Ware: An 'MIT Thing'?" 2006–11, 14(10), 53–58.
149. Transnational Management Associates, The Challenge for E-Learning, http://www.tma-world.com/news-insights/viewpoints-articles/the-challenge-for-e-learning/, (2014).
150. Udaya Sri K., Vamsi Krishna T. V., E-Learning: Technological Development in Teaching for school kids/(IJCSIT) International Journal of Computer Science and Information Technologies, (2014) Vol. 5 (5), 6124–6126.
151. Valerie Strauss, "Three fears about blended learning." The Washington post, Retrieved 31 March 2013.
152. Vangie Beal, CBT, http://www.webopedia.com/TERM/C/CBT.html, (2014).
153. Warner, D., Procaccino, J. D. "Toward wellness: Women seeking health information." Journal of the American Society for Information Science and Technology, (2004) 55. doi: 10.1002/asi.20016.
154. Warschauer, M., Matuchniak, T. New technology and digital worlds: analyzing evidence of equity in access, use and outcomes. Review of Research in Education, (2010) 34, 179–225.
155. Webster, U. "Online education offers a flexible experience." Webster.edu. Retrieved 2013–10–22.
156. Wenglinsky, H. Does it compute? The relationship between educational technology and student achievement in mathematics. Retrieved February 2, 2006, from ftp://ftp.ets.org/pub/res/technolog.pdf.
157. Wheeler S. e-Learning 3.0: Learning through the eXtended Smart Web. In National IT Training Conference, number March, (2011).
158. Whyte, Cassandra B., Lauridsen, Kurt. An Integrated Learning Assistance Center. New Directions Sourcebook, Jossey-Bass, Inc. (1980).
159. Wiki E. Edutech wiki, a resource kit for educational technology teaching, practice and research, (2010).
160. Wiki.laptop.org, "Constructionism – OLPC." Retrieved 2013–10–22.
161. Wikipedia, the free encyclopedia, e-learning 3.0, Retrieved 2014–8-29.
162. William Crain, Theories of Development: Concepts and Applications (6th Edition). Upper Saddle River, NJ: Prentice Hall: Pearson. ISBN 0205810462. (2010).
163. Williams Peter E., Hllman Chan M. "Differences in self-regulation for online learning between first-and second-generation college students." Research in Higher Education, (2004) Vol. 45, No.1. pp. 71–82.

164. Wolf, L. "College Students with ADHD and Other Hidden Disabilities: Outcomes and Interventions." Annals of the New York Academy of Sciences. (2001).

165. Youngberg, David, "Why Online Education Won't Replace College--Yet." The Chronicle of Higher Education. (2012).

166. YouTube, "Nurse education in second life at Glasgow Caledonian University demo." Retrieved 2013–10–22.

167. YouTube, "Second Life Nursing Simulation." 2009–09–16. Retrieved 2013–10–22.

168. Zameer, Ahmad," Virtual Education System (Current Myth & Future Reality in Pakistan)." Ssrn.com. Retrieved 2013–10–22.

169. Zhong N., Liu J., Yao Y., Ohsuga S. Web intelligence (wi). In Computer Software and Applications Conference, 2000. COMPSAC 2000, The 24th Annual International, (2000), pages 469–470. IEEE.

CHAPTER 5

DESIGN OF MOBILE APPLICATION TO MANAGE SENIOR PROJECT

SEIFEDINE KADRY[1] and ABDELKHALAK EL HAMI[2]

[1]*Associate professor, American University of the Middle East, Kuwait*

[2]*Full professor, National Institute if Applied Sciences, Rouen, France*

CONTENTS

5.1   ABSTRACT

The senior project course for engineering students has a vital role in complementing the "learn by doing" education-engineering students in any university. The senior project course is, usually, a two-semester course. This short time frame, poses challenges for both students and faculty. Students have to achieve the outcomes of the project to a certain extent and the faculty has to manage, follow up and evaluates efficiently the work of their students individually and by group. One common problem in senior project is the missing of the control during the visit of the students to a company for project preparation purpose. In this chapter, we propose a new and efficient design of follow up for faculty to manage properly and evaluate effectively the individual progress of the students during the preparation of their senior projects and to avoid "free riders" situation. Our idea is based on the mobile technology to track the students' progress, meeting, and visiting industry and so on. The proposed idea is helpful to avoid free-riding students also provides instructors with solid evidence of the student progress.

5.2   INTRODUCTION

The senior project experience is the conclusion of a student's knowledge and skills in preparation for imminent integration into the workplace. It not only offers students the build up the bridge between the theory and practice but also to increase and practice a maturity towards leadership, team work, communication, client liaison, role playing, ethics, peer assessment and understanding the impact of the integration of their project on the community.

In offering all these benefits there are identifiable challenges of assessing the individual that have been a long-standing issue for instructors and consequently students. In order to evaluate the individual within the group it is essential to provide assessment that is not only fair and just for the individual but also responsible. Individual assessment for the senior project or in general any team project has historically created a situation where subjective assessment is required to differentiate individual student achievements. In this chapter, we propose a new mobile application and desktop application that help students and the instructor as well in the senior project preparation, follow up and fairly grading.

5.3   SENIOR PROJECT REQUIREMENTS AND ORGANIZATION

The regulations implemented in the senior project evolved over a period of several years, through a continuing process of trial and revision by the entire engineering schools (Lekhakul and Higgins, 1994):

1. Students are required to finish their senior design project in one academic year.
2. Students must take the senior design project during their senior year. Because of the breadth and demanding nature of senior design, it is essential to have certain fundamental courses completed before students start their senior design project.
3. Senior design projects are intended to be group projects, so that students learn how to work as an engineering team. Design groups may have two or three students. Groups of four or more students usually are too large to work efficiently or to be easily managed and may not provide sufficient design experience for all students in the group.
4. Projects are supervised by the senior design committee, consisting of the project faculty advisor and two other faculty members. Students are expected to organize their own design group, consisting of their student coworkers. The project faculty advisor is chosen by the students and the remaining faculty committee members are then assigned to the student design group by the professor who is in charge of the senior project.
5. Early in the beginning of semester, the design team meets with the faculty committee for review and approval of the initial design concept, the written project plan and an essential detailed schedule for completion of the project. Throughout the year, the design group must meet with the project advisor weekly and with the committee at the end of each semester to present a design review and progress report.
6. In addition to meeting with the advisor each week, students are required to attend weekly senior seminars. In the first semester, each group makes an oral presentation describing their project to the entire class. At the end of each semester, each design group gives a progress report in which all group members participate. Remaining class periods may be scheduled for guest speakers on topics, such as engineering ethics, occupational safety and health, professional registration, resume writing, interviewing, job search strategies, and technical engineering presentations by practicing engineers from industry.

7. Each group is required to keep a notebook and activity diary throughout the year. The notebook must record daily progress on the project and any schedule revisions. The notebook and activity diary are compatible with industrial practices and are important aids to the faculty committee in measuring project progress, and in making individual grade decisions.

8. A final, cooperatively-written report is required from each design group. These reports are archived in the department library for future reference and may serve as useful aids to future student senior groups.

9. In consultation with the group's faculty committee, the project advisor must give a letter grade to each student at the end of each semester. Incomplete letter grades are not permitted, since this may encourage students to procrastinate on project efforts until the last quarter or semester, when it usually becomes impossible to finish the project on time.

5.4 LITERATURE REVIEW

In Clear (2009), clear have outlined an approach to assessment of student performance in capstone project, which assesses the three dimensions of "product," "process" and "progression in learning." Where product and process must be assessed in group; Progression must be assessed individually, with contributions from reflective reports, evidence of improvement and "observations and opinions of supervisors." Clear is comfortable with this subjective assessment, markers often find it difficult to justify the final grade. This is worsening in senior projects where the student's individual marks can vary from a Distinction to a Fail, a situation that requires justification. Aggarwal and O'Brien (2008) introduced a system that identified the individual student contribution in order to address the issue of "free riders" – "an individual working in a group setting fails to contribute his or her fair share." In Blackshaw and Latu (2005), Latu and Blackshaw interviewed students to identify their preferences when considering the group work grading. They found that 94% of their students wanted to identify "free-riders," 79% felt that individual grades were preferable to group marks. Farell et al. (2012), study the senior projects within the ICT industry. Their study was restricted the student group size from 3 to 5. The aim of their research is to design and develop a system with supporting documentation that will enable a fair allocation of grades to individuals undertaking group work. The system is to ensure that the assessment considers the main objectives of the senior project experience and not just the final project outcome of Kadry and El Hami (2014) and Jaoude et al. (2010) as discussed earlier. In Clear (2010), the authors have developed and deployed an individual contribution assessment method for the students collaborating in Classroom Wiki – a Web-based collaborative writing environment. Lejk and Wyvill (2001) identified six different approaches to the allocation of marks to group work that consider: individual versus group marks, student contribution to assessment distribution, weighing factors for individuals and separating individual and group assessed items. In Vasilevskaya et al. (2014), the authors stated that it is very hard to find a single assessment activity that balances between individual and group assessment, involves both teachers and students, and is summative as well as formative.

5.5   ISSUES ASSOCIATED WITH ASSESSING GROUP WORK

Motivation of participants has been noted to be one of the most serious problems in group work (Kerr and Bruun, 1983; Morganm, 2002). Some group members may be reluctant participants in assessment tasks and be uncommitted to the aims of the group (and the subject for that matter). Motivational issues can arise as a result. Examples of motivational issues associated with group work are social loafing and "free riding." These issues have received considerable attention in the literature (Jones, 1984; Lantane et al., 1979; Ruel et al., 2003). Free-riding has been defined as follows: "the problem of the non-performing group member who reaps the benefits of the accomplishments of the remaining group members with little or no cost to him/herself" (Morris and Hayes, 1997). Free-riding has been distinguished in the literature from "social loafing" (Watkins, 2004). The difference is this: social loafing is a reduction in effort due to not being noticed or lack of identification in a group task. Free-riding is actively obtaining reward for no effort. Thus, social loafing can lead to free-riding.

One way of solving the problem of social loafing and free-riding is to carefully consider the nature of the task given to students and to reward the effort of groups as well as reward the work of individuals. However, this is harder than it sounds. Tasks need to be designed to maximize students' contributions and to recognize and notice their efforts. Some strategies of doing this:

- Work out ways to recognize, monitor and reward the individual effort of group members. Simply tracking the contributions of students' work and requesting that students' names be given on a group assignment might be sufficient. This can either be a matter of negotiation among students themselves or mandated by the instructor.
- As already noted, evaluate the individual's contribution to the group work assignment as well as the work of group.
- Allow group members to notice and evaluate each others' contributions by means such as web-based tools or a peer evaluation procedure.
- An effective assessment procedure that has been trialed in a cross-disciplinary business course such a procedure reduces free riding as measured by a decline of variance between peer evaluation assessments. (It was not clear from this paper whether groups were self-selected or instructor selected. The second variation, given below, involved self selected groups.)

Free-riding has also prompted what is called an "inequity based motivation loss" (sometimes known as the "sucker effect"). The Sucker effect refers to individuals responding to others free-riding upon their efforts by free-riding themselves (Kerr, 1983). It appears that competent students try to avoid being "suckers." They make a calculation of whether or not they are the subject of free-riding from others in the group. If they are, and they feel it unjustifiable, they try to avoid being a "sucker" by reducing their own input to the task. Kerr has shown that students will even choose to fail as a group rather than be a "sucker" (Kerr, 1983). It is suggested that the sucker effect problem is the cause of procrastination in many group work activities. Conscientious students find it hard to get the attention and compliance of free-riders and decide not to proceed alone until a deadline is imminent. But the situation is more complex than it appears. Watkins claims that competent students are less likely to think of themselves as suckers if they genuinely feel that they are covering for a member of the group who is unlikely to succeed by themselves. Thus, one way of minimizing the sucker effect is to allow members of groups to "get to know each other better." If this happens, competent students may be less inclined to feel like "suckers" and are less likely to free-ride (Watkins, 2004). In ad hoc, short term groups—where group members do not socialize as readily—this way of overcoming the problem might be less effective. However, this is only part of a solution, of course. A better solution will reduce free-riding—and maximize the contributions—of all students in group work activities.

In 1996, Strachan and Wilcox announce that the educational literature is littered with ideas and methods of assessing group work, In 2001, Reif and Kruck argue that the issue of assessing individual students' contribution to their group activities is problematic, because group members do not contribute equally to a group's success, and that lecturers normally make few or no observations to assist them in evaluating and determining the students' level of contribution to their group's ultimate success. Consequently if some members of a group are not co-operative and fail to perform their assigned tasks, the workload of other adversely affected. Assessing group work can be extremely difficult because, no matter how teachers derive an individual member's mark, some members will always complain that they have been disadvantaged by the poor efforts of their fellow group members. Group assessment involves the assessment of the product of students' group work by the instructor, or the assessment of the product by fellow students from other groups (inter-peer assessment), or the assessment of the group work by students within a group (intra-peer assessment). For example, in computer courses, the nature of work in the computer industry requires a good deal of team activity. However, Schwalbe (2004) warns that the culture of computer professionals portrays them as nerds who like to hide in dark corners, hacking away on computers, and when they have to communicate with non-computer professionals, they act and talk as if they are talking to someone from another planet. Further computer products are so large and complex, and their development requires a team approach, group work and group assessment should be featured in most courses within computer and ICT related courses, especially in the first and second year of undergraduate engineering study. Some research and studies, argue that group work and peer evaluations may not be appropriate for the first year undergraduate students because they may not possess the necessary prerequisites to handle group dynamics.

5.6   NEW PROPOSED DESIGN

In this chapter, we design a novel mobile application (see Figure 5.1) that helps the instructor to follow their students in the senior project by track their visit to the industry and their work done. Through this application students have the following options (see Figures 5.2 and 5.3):

1.   Sign in using their student IDs.
2.   Scan their fingerprint and send it to the university server along with their location using GPS technology, to proof to the instructor their physically visit to the company. All fingerprint are registered on the university server upon the registration of the students.
3.   Take picture about the system there.
4.   Record sound or video with the employees in the company then upload the file to the university server From the university side, instructor can login locally and see online the location of their

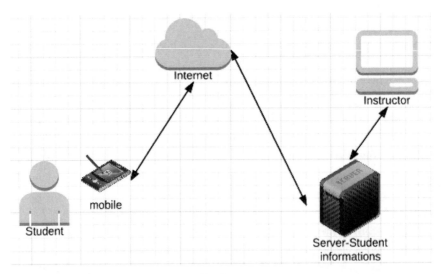

FIGURE 5.1   Design of the mobile application.

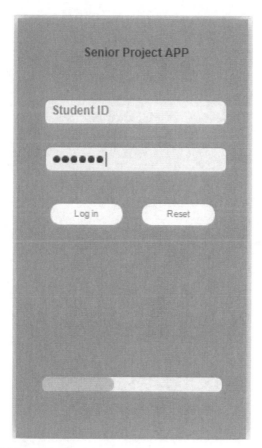

FIGURE 5.2 Mobile application interface.

students using Google map (see Figure 5.4), and can interact with them and track their progress. This will help the instructor to accurately follow up and fairly grade the work of the students during their visit.

5.7 CONCLUSION

In this study, we propose a new design that help instructor to track and follow up their students using students mobile application side and another desktop instructor application. By using the proposed design, instructor

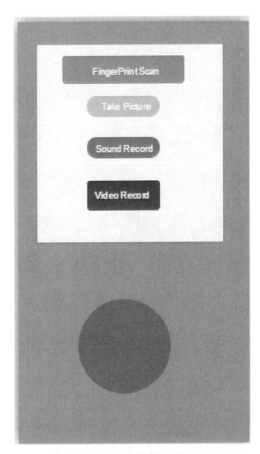

FIGURE 5.3   Mobile application options for students.

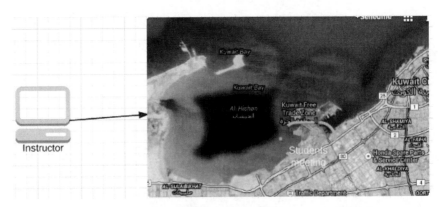

FIGURE 5.4   Instructor is monitoring students' location.

can be sure that their students visit physically the company and grading fairly their work. Future work might include implementation and students survey about the new application.

KEYWORDS

- **levels of evidence**
- **research question**
- **scientific method**

REFERENCES

1. Aggarwal, P., O'Brien, C. L. "Social loafing on group projects: Structural antecedents and effects on student satisfaction," Journal of Marketing Education, vol. 30, no. 3, pp. 255–264, 2008.
2. Blackshaw, R., Latu, S., "Group work and group work assessment for computer courses: a systems analysis and design case study," presented at 16th Australasian Conference on Information Systems, 29 Nov. -2 Dec. 2005, Sydney, Australia.
3. Clear, T., "Thinking issues managing mid-project progress reviews: a model for formative group assessment in capstone projects," ACM Inroads, vol. 1, no. 1, pp. 14–15, March 2010.
4. Clear, T., "Thinking issues: the three p's of capstone project performance," ACM SIGCSE Bulletin, vol. 41, no. 2, pp. 69–70, June 2009.
5. Farrell, V., Ravalli, G., Farrell, G., Kindler, P., and Hall, D., "Capstone project: fair, just and accountable assessment," in Proc. the 17th ACM annual conference on Innovation and technology in computer science education, 2012, pp. 168–173.
6. Jaoude, A. A., El-Tawil, K., Kadry, S., Noura, H., and Ouladsine, M., "Analytic prognostic model for a dynamic system," International Review of Automatic Control, 2010.
7. Jones, G. R. (1984). Task visibility, free riding, and shirking: Explaining the effect of structure and technology on employee behavior. Academy of Management Review, 9(4), 684–695. doi: 10.2307/258490.
8. Kadry, S., El Hami, A. "Flipped classroom model in calculus II," Education, vol. 4, no. 4, pp. 103–107, doi: 10.5923/j.edu.20140404.04. 2014.
9. Kerr, H. L. (1983). Motivation losses in small groups: A social dilemma analysis. Journal of Personality and Social Psychology, 45(4), 819–828. doi: 10.1037/0022–3514.45.4.819.
10. Kerr, N. L., Bruun, S. E. (1983). Dispensability of member effort and group motivation losses; Free Rider effects. Journal of Personality and Social Psychology, 44(1), 78–94. doi: 10.1037/0022–3514.44.1.78.

11. Lantane, B., Williams, K., Harkins, S. (1979). Many hands make light in the work: The causes and consequences of social loafing. Journal of Personality and Social Psychology, 37(6), 822–832. doi: 10.1037/0022–3514.37.6.822.

12. Lejk, M., Wyvill, M., "Peer assessment of contributions to a group project: a comparison of holistic and category-based approaches," Assessment and Evaluation in Higher Education, vol. 26, no. 1, pp. 19–39, 2001.

13. Lekhakul, S., Higgins, R., "Senior design project: Undergraduate thesis," IEEE Trans. Educ., vol. 37, no. 2, pp. 206–230, 1994.

14. Morgan, P. (2002). Support staff to support students: The application of a performance management framework to reduce group working problems. From http://www.heacademy.ac.uk/business/resources/archiverequest

15. Morris, R., Hayes, C. (1997). In Pospisil, R., L. Willcoxson (Eds.), Learning through teaching (pp. 229–233). Perth: Murdoch University. http://lsn.curtin.edu.au/tlf/tlf1997/morris.html

16. Reif, Harry, L., Kruck, S. E. (2001). "Integrating Student Groupwork Ratings Into Student Course Grades." Journal of Information Systems Education, 12(2), 57–65.

17. Ruel, G., Bastiaans, N., Nauta, A. (2003). Free riding and team performance in project education. International Journal of Management Education, 3(1), 26–38.

18. Strachan, I. B., Wilcox, S. (1996). Peer and self-assessment of group work: developing an effective response to increased enrollment in a third year course in microclimatology. Journal of Geography in Higher Education, 20 (3), 343–353.

19. Vasilevskaya, M., Broman, D., and Sandahl, K. (2014). "An assessment model for large project courses," presented at 45th ACM Technical Symposium on Computer Science Education (SIGCSE), Atlanta, USA, March 5–8.

20. Watkins, R. (2004). Groupwork and assessment: The handbook for economics lecturers. Economics Network, from www.economicsnetwork.ac.uk/handbook/printable/groupwork.pdf

PART 2:

E-MAINTENANCE

CHAPTER 6

E-MAINTENANCE MANAGEMENT SYSTEM FOR OPTIMAL FUNCTIONALITY OF MACHINES

MAKINDE OLASUMBO,[1] ADEYERI MICHAEL KANISURU,[2] MPOFU KHUMBULANI,[3] and RAMATSETSE BOITUMELO INNOCENT[4]

[1]*PhD Research Student in Agents System for Reconfigurable Vibrating Screen, Department of Industrial Engineering, Tshwane University of Technology, South Africa, Tel: +27845226300; E-mail: olasumbomakinde@gmail.com*

[2]*Post-doctoral Research Fellow in Agents Technology for Reconfigurable Manufacturing System, Department of Industrial Engineering, Tshwane University of Technology, South Africa, Tel: +27842851620, E-mail: adeyerimichaeltut@gmail.com*

[3]*Researcher in Artificial Intelligent and Research Coordinator, Department of Industrial Engineering, Tshwane University of Technology, South Africa, Tel: +27723614875, E-mail: mpofuk@tut.ac.za*

[4]*PhD Research Student in Reconfigurable Vibrating Screen Development, Department of Industrial Engineering, Tshwane University of Technology, South Africa, Tel: +27721277476, E-mail: ramatsetsebi@tut.ac.za*

CONTENTS

6.1   INTRODUCTION: BACKGROUND AND OVERVIEW

It is worth starting this chapter with the definition of the key words that comprise the topic: E; maintenance; management system; optimal functionality; and machines.

Econnotes electronics way of undergoing a process, activity or task with the use of Internet. It is also an electronic platform provision for carrying out activities in the auspices of digitalized control environment. In the realm of E, many services such as e-government, e-learning, e-banking, e-commerce, e-manufacturing and just to mention but a few are being rendered today through this platform.

Maintenance from layman perspective, maintenance is the process of keeping something in good order. But according to Farnsworth and Tomiyama (2014), it is an efficient and cost effective way to keep all functional key parts of a machine alive and available during its production lifecycle.

In view of this, **e-maintenance** can be defined as a useful tool that is used for carrying out different types of maintenance and associated activities such as technical, administrative, and managerial actions through effective interaction and collaboration of these activities electronically, using network or telecommunications technologies to ensure the optimal functioning of a machine. A simple mathematical equation for e-maintenance to achieve optimal functioning of the machines is expressed in Eq. (1).

$$eM = exM = EMRMS + eMA + eD + eP \Rightarrow SM \Rightarrow$$

$$\textbf{\textit{Optimal Production Control}} \text{ \& } \textit{Lifecycle Management of a Machine} \quad (1)$$

where eM: e-maintenance; exM: excellent maintenance; eMA: e-machine assessment; eD: e-diagnosis; eP: e-prognosis; EMRMS: effective machine routine management system

In view of this formula derived above, maintenance strategy evolution or paradigm can be described as the maintenance strategies evolution and future trend model as depicted in Figure 6.1.

Another school of thought explained that maintenance is an activity, which is applicable to all systems, natural and artificial, which causes such systems to remain unaltered or unimpaired. Okah-Avae (1995) further explained that in natural systems, such as man, animals and plants, nature has perfected an automatic and highly efficient maintenance system for the self-maintenance of these systems. The process of cell regeneration to effect healing, secretion of fluids to continuously lubricate and protect various parts of the body, just to mention a few, are examples of natural maintenance activities which are highly efficient and go on continuously

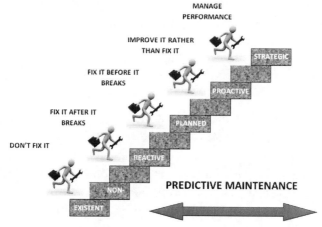

FIGURE 6.1 Maintenance Strategies Evolution (Bangemann, 2005).

to ensure that the systems remain unaltered. In artificial systems, man has had to rely on his knowledge of engineering science to ensure that such systems remain unaltered or unimpaired or at least, if altered, can be repaired.

Marquez and Gupta (2006) viewed the definition of maintenance from another perspective and defined maintenance as "the combination of all technical, administrative and managerial actions during the life cycle of an item intended to retain it in, or restore it to, a state in which it can perform the required function."

Management System is a system that stands to plan, coordinate and control defined activities in order to ensure a smooth running process of the defined task.

Optimal Functionality is all about the peak performance of all elemental parts comprised in a system with 100% efficiency under an ideal situation.

Machines are tools or equipment designed to perform a defined task with the hope of gaining a greater efficiency in executing the defined task.

Thus, this chapter of the book is therefore meant to explain and discuss efficient managerial system that sees to the optimal performance of machines by undergoing maintenance strategies provision through e-platform.

As a matter of fact, machines are broadly one of the cores or acquired elements of significant value that provide services for organization through its effective operations. Nowadays, the management of these machines can become a daunting task and the optimization of their usage is crucial.

Hot Tip

A conglomerate of machines, each; made up of different subsystems that ensures their functionality are been used in producing products required by the customers. In order to ensure its availability, reliability and maintainability when utilized in the production process, what are the measures, techniques vis-à-vis the technologies that should be used by the maintenance managers of these machines to effectively maintain a machine?

Combination of tools to be used and techniques by maintenance managers of machines to effectively maintain a machine in order to obtain a near-zero machine downtime state are:

1. An effective routine maintenance calendar for proper checking of the machine before use daily
2. An application or an e-maintenance system to can be used to determine the health status of a machine
3. An application or an e-maintenance system to troubleshoot and detect faults on a machine
4. An application or an e-maintenance system to predict when any of the subsystem of a machine will fail.

Machines management mainly concerns the optimum maintenance of machines used in an organization in order to ensure its maximum usability, reliability, maintainability and availability at a near-zero machine down time condition. Maintenance engineers of these machines are ceaselessly coming up with new methods of perfectly managing and maintaining equipment used in the factory floor. In order to achieve this phenomenon, relevant historical operating data are needed by maintenance engineer experts to develop real-time intelligent maintenance decision support system. It is a fact that these data are generated by events and actions of humans on these assets and is expected to be useful for decision-making and integrated into the product lifecycle phases.

Machine optimization consists of improving its uptime, throughput, product quality, production costs and is an essential factor to guarantee increased profitability. These days, there is an increasing interest in machines management and optimization, partly due to the economic worldwide recession and to the rapid advances in technology, which force companies and plants to carefully use their assets and resources (Marquez and Gupta, 2006) as well as due to lack of expert and adequate historical machines needed for its optimum management in order to ensure its optimum reliability, maintainability and availability at any particular time in a company. It is common knowledge that nowadays, due to the current economic situation, many companies are still facing problems in achieving greater profitability and better operational efficiencies with fewer human and assets. Therefore, it is necessary to implement new procedures and perform new asset management policies and methodologies. It is a fact that only the companies who learn and adapt will survive. In view of this, maintenance engineers are seeking to maximize their asset usage and hence to demand a comprehensive engineering asset management system. More reliance is now being placed on condition monitoring (CM) systems to extend functional life of the assets and to allow repairs to maximize affordability (Campos, 2009).

The role of maintenance is strategic, and ensures optimum machine management in the industrial environment. Hence, while having this interest in company; industry has been seeing the development of different generations of maintenance systems. Even if maintenance is a necessity, maintenance has a negative image and suffers from a deficiency of understanding and respect. It is usually recognized as a cost, a necessary evil, not as a contributor. Wireman, T. 1990, "World class maintenance management", Industrial Press, New York. conducted a benchmarking exercise and found that the maintenance cost for industrial firms in the USA has increased by 10–15% per year since 1979. High maintenance costs highlight the need to define clearly the maintenance objectives, develop and enhance modern maintenance management methods continuously, integrate maintenance and production activities effectively, and to use intelligent computer-based maintenance systems. In industry, failure-driven and time-based maintenance are two major maintenance management approaches. Failure-driven maintenance (FDM) is also called "run-to-failure" maintenance (Tsang, 2002). It is a reactive management approach, where corrective maintenance is often dominated by unplanned events, and is carried out only after the occurrence of an obvious functional failure, malfunction, or breakdown of equipment. Corrective maintenance action can restore an item of failed equipment by either repairing or replacing the failed component. If the equipment is non-critical or is easily repaired, unplanned stoppages of the equipment will cause minimal disruption to production, and FDM could be an acceptable maintenance approach. However, in the case of purely random breakdown of equipment that would have a serious impact on production, an emergency corrective maintenance action is necessary to avoid the serious consequences of failure (Kans and Ingwald, 2008). The practical implication of emergency corrective maintenance often results in the unpredictable performance in a plant, that is, high equipment downtime, high cost of restoring equipment, extensive repair time, high penalties associated with the loss of production, and a high spare parts inventory level (Iung et al., 2009). Time-based maintenance (TBM) is also called periodic preventive maintenance. In order to slow down the deterioration processes leading to faults, primary preventive maintenance is carried out by periodically lubricating, calibrating, refurbishing, inspecting, and checking of equipment on a regularly scheduled basis. TBM assumes that the estimated failure behavior of the equipment, that is, the mean time between functional failures (MTBF) is statistically or experientially known for equipment and machinery degrading in normal usage (Braaksma et al., 2011). TBM also involves minor or major planned shutdowns of systems for overhaul or predetermined repair activities on still functioning equipment. This can prevent functional failures by replacing critical components at regular intervals just shorter than their expected useful lifetime.

Hot Tip

System overhaul and critical item replacement at fixed intervals are widely adopted by many modern automated plants. Although TBM can reduce the probability of system failure or the frequency of unplanned emergency repairs, it cannot eliminate the occurrence of random catastrophic failure. Some TBM practices may be out of date and do not cope with the actual operating requirement of the modern automated plant. Most of the maintenance decisions are made by experienced planners according to the original equipment manufacturers recommendations, the reported breakdown history or failure data and the operating experience and judgement of the maintenance staff and technicians.

Under a situation of uncertainty, it is very difficult to plan the maintenance activities properly in advance, and so, very often, the maintenance staff is required to work under "firefighting" conditions (Kiritsis, 2010). Most people think that the role of maintenance is "to fix things when they break" but when things break down maintenance has failed (Dale, 2003). Moreover the afore-discussed traditional means or scope of maintenance activities has been limited to the production vs. operation phase. But as the paradigm of manufacturing shift towards realizing a sustainable society, the role of maintenance has to change to take into account a lifecycle management oriented approach (Takata, 2004) for enhancing the eco-efficiency of the product life (DeSimone and Popoff, 1997). In that way, maintenance has to be considered not only in production operation phase but also in product design, product disassembly, product recycling and other production processes (Van Houten et al., 1998). Thus the product can now play a major role in maintenance mainly when the product is "active" (i.e., Intelligent Product, Holon) meaning able to support a part of its knowledge. In order to achieve the aforementioned goal, therefore the integration of e-maintenance management and s-maintenance management into the product lifecycles of the machines used in the industries is vital. In view of this, this book chapter discusses in details the concepts and evolution of e-maintenance management systems, its characteristics, capabilities, tools, existing architectures or platforms, its application, implementation, challenges and contribution and the latest e-functional and prognosis tool and e-temperature machine architecture developed in Tshwane University of South Africa and finally highlight the way forward to have a robust e-maintenance and s-maintenance management systems.

6.2   E-MAINTENANCE TOOLS AND TECHNOLOGIES

The baseline technologies for e-maintenance culture transformation from conventional to e-platform are: web technology, web applications, intelligent agents, multi-agents systems, mobile agents, real time data collection, condition based monitoring, predictive technologies and information pipeline. The aforementioned technologies pool modern information processing and communication tools together in making it feasible through the use of programming languages such as C++, C-sharp, Java scripts, MySQL and etcetera, in order to offer the technical support required to access remote information on machines.

Sensors are the important factors on which basic business process of e-maintenance is being built on. Sensors such as smart sensors (MEMS), which make use of micro-sensor technology are characterized by memory cells, autonomous power, analogue amplification, converter, rectifier, and high response to external and internal stimuli. These sensors find their applications on oil analysis, vibration analysis, thermograph, noise, ultrasonic, ferrographic analysis, spectroscopy analysis, atomic emission and infrared, chromatography, electrical testing (resistance testing, impedance testing, and etcetera) on which condition based maintenance (CBM) principle is built. Hence, these sensors give support for knowing the machine system degradation diagnosis and prognosis analysis (Muller, 2007).

Other hardware tools used for e-maintenance as well as their applicative areas are given in Table 6.1.

Other Mathematical and software tools used for the full implementation of e-maintenance are as stated:

i. **Fourier transform** is one of the important tools used in processing the signals that was obtained from the sensors through data processing into meaningful information that can be used to determine the machine's current functional or health status and its performance prediction. According to Joseph Fourier, Fourier transform deals with the mathematical transformation employed to transform signals between time (or spatial) domain and frequency domain. During signal processing, the domains can be transformed from one form to another.

ii. **Wavelet series** is also another tool used for processing signals obtained from the reading sensors of the machine. It achieves this by representing the signal obtained from the sensors, which is in the form of the square-integral (real- or complex-valued) function by a certain orthonormal series generated by a wavelet.

iii. **Morlet wavelet filtering** is another useful for signal processing. This tool achieve this aim by measuring, analyzing and filtering the signals obtained from a machine in order to detect anomalies during signal processing which will thus be prone to further examination by a maintenance expert to determine the health state of the machine.

iv. **Autoregressive model** is a useful tool used for extracting features on the processed signals obtained from the reading sensors of the machine. Other useful tools that can be used for feature extraction of the processed signals are time-frequency moments and singular value decomposition.

TABLE 6.1   E-Maintenance Tools and Areas of Applications

S/N	E-Maintenance Tools	Areas of Application/Uses
1.	Radio Frequency Identification Device (RFID) tag	It gives support to operator and component identification
		Helps in storage of machine conventional data
		Record tracking of traceable past maintenance actions
		It provides aid for geolocalization of maintenance tools and operators
2.	Global Positioning System (GPS)	It complements the RFID tag technology
		For calculating the location of an operator or maintenance tools
3.	Wireless Technologies: i. Wireless Personal Area Network such as IEEE 802.11, 802.15.4 ZigBee, 802.15.1 ii. Bluetooth iii. Wireless Area Network such as WiFi, WiMax iv. GSM-UMTS	For transmitting and communicating data information at both short and long distance
4.	Web Technology: Internet, Web Service Description Language (WSDL), Universal Description, Discovery and Integration (UDDI)]	Serves as standard communication protocols for data transfer throughout the world UDDI serves to offer directory service for storing information about web services
5.	Diagnostic and prognostic tools	Helps in developing intelligent support for maintenance decision making
6.	Innovative communication equipment (Virtual reality): Graphic Liquid Crystal display (GLCD); Liquid crystal display (LCD); Simple Object Access Protocol (SOAP)	For supporting man/machine or man/man message exchanges in form of text or animated images
7.	PDA, Smart phones, graphic tablets, harden laptops	Provision of support functions to aid operators on site

v. **Fuzzy logic** is also another important tool for predicting the future behavior of a machine. This is done by taking decision on the performance prediction of the machine through experts reasoning results obtained from detailed cognitive mapping of present signatures against the past signatures results on the signatures analyzed. The performance prediction tools highlighted above also consider the level of behavior degradation of the signatures analyzed as well as the closeness of the signatures of the system behavior to any of the previously observed fault, to predict the future performance of a machine.

vi. **Creation of effective database management system** for past signatures serve as a useful tool that can be used to assess the health condition of a machine using Hidden Markov Modeling for analysis.

vii. **A genetic algorithm (GAs) imitates** a biological evolution process and is mostly used to seek optimal solution to a practical problem, expressed by the best-fitted individual string of values (representing parameters of the practical problem). GAs encode the decision variables (or input parameters) of the underlying problem into strings. Each string, called individual, is a candidate solution. To differentiate good candidate solutions from bad candidate solutions, a fitness function is applied as a measure. A fitness function could be a mathematical expression, or a complex computer simulation, or in terms of subjective human evaluation and guide the evolution of solutions to the problem (Vayenas and Nuziale, 2001).

The pseudo-code of a simple GA is as follows (Goldberg, 1989):

Step 1: Initialization: Create initial population I at random or with prior knowledge.

Step 2: Fitness evaluation: Assess the fitness for all individuals in I.

Step 3: Selection: Choose a set of promising candidates C from I.

Step 4: Crossover: Apply crossover to the mating pool C for generating a set of offspring O.

Step 5: Mutation: Apply mutation to the offspring set O for obtaining its perturbed set O.'

Step 6: Replacement: Replace the current population I with the set of offspring O.'

Step 7: Termination: If the termination criteria are not met, go back to Step 2.

The software used for running reliability/maintainability prediction based on GAs using historical data is called GenRel®. The sequential steps and stages in using genetic algorithms as well as the use of GenRel® Software for determining the maximum number of iterations, the convergence limit, and the probability of mutation is depicted in Figure 6.2.

viii. **Elmann neural network** is a performance predicting tool which involves the analysis of different synapses of signatures with the past signatures and results in a network space. This information is being communicated to different users of the information using the Peer-to-Peer (P2P) communication system and the flow of communication between these different users in machine maintenance architecture is depicted in Figure 6.3.

ix. **Augmented virtual reality (AR)** is also a useful tool that can be used to effectively diagnose, detect faults and prognoses a machine. It does this by inserting augmentation tools in a machine to monitored, in which the results is been viewed in a video stream to

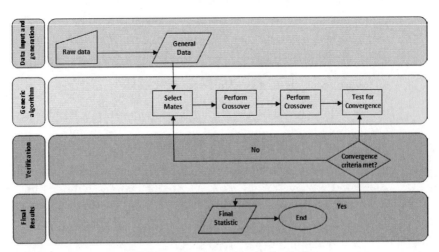

FIGURE 6.2   A schematic illustration of data flow in GenRel®.

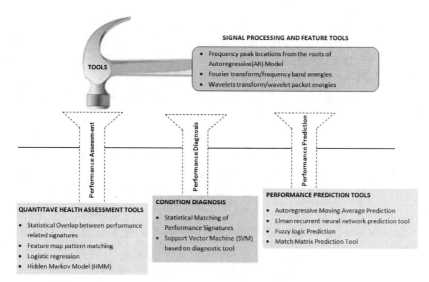

FIGURE 6.3 Quantitative tools for modeling machine performance and prediction.

increase the machine maintenance manager's view in real time, so that he understands the maintenance tasks to be carried out on the machine (Benbelkacem, 2012). The augmentation tool used is a tracking system, which ensures virtual object transfer to the maintenance manager for decision making. Tracking system is a very important research tool used to achieve optimal machine maintenance using augmented reality (Fiala, 2004; Comport et al., 2005), and the vision-based tracking method is usually appropriate. Tracking methods use cameras to capture real life machine subsystems behavior using sensing devices. Therefore, to apply AR to carry out troubleshooting or diagnosis on a machine, the positions and orientations of users in real time must be measured with high accuracy. The marker-based tracking technique is most useful techniques among the other tracking technologies (Kato and Billinghurst, 1999). It uses image-processing technique to measure the relative position and orientation of different subsystems of a machine using a camera and markers (transformation matrix marker/camera). Thus this technique seems to be relevant for optimal industrial machine maintenance. AR Toolkit, which uses square markers is the most popular tracking marker method (Kato and Billinghurst,

1999; Bellarbi et al., 2010). This e-augmented tool is an open-source software library used to develop augmented reality applications for effective machine maintenance. "The Human Interface Technology Laboratory" (HIT Lab) at Washington University developed this useful maintenance tool. The position and orientation of the camera relative to markers is determined using vision techniques embedded in this tool. Thus, the programmer can use this information obtained from the different subsystems of a machine to draw the three dimensional (3D) object and to insert it correctly in the real scene. The virtual object tracking is made feasible by the AR Toolkit when the camera (or user) changes position. The "AR Toolkit" library has several types of markers (Kato and Billinghurst, 1999).

Highlighted below are the major procedures that can be used to carry out maintenance activities on any subsystem that a machine is made up of:

Procedure 1   : Open a video stream

Procedure 2   : For each machine subsystems image that was captured:

Procedure 2.1: Calculate the optimal threshold value from the machine subsystem captured image using Otsu method (Otsu, 1979).

Procedure 2.2: Transform the machine subsystem captured image to a machine subsystem binary image using the calculated threshold value.

Procedure 2.3: Detect black squares markers in the machine subsystem binary image.

Procedure 2.4: Calculate the camera's position and orientation (this calculate the transformation matrix)

Procedure 2.5: Apply the stabilization algorithm (Bellarbi et al., 2010).

Procedure 2.6: Superimpose the virtual subsystems objects upon the captured image (using the calculated transformation matrix).

6.3 THE BENEFITS OF E-MAINTENANCE

The benefits of e-maintenance shall be discussed under two main bodies, the general and specific benefits. The general benefits as it affects managerial settings provided e-maintenance is effectively implemented and carried out are as listed (Alzyouf, 2007):

i. increase in utilization of manufacturing systems;
ii. increase in profit margin as a result of decrease in manufacturing costs, that is fixed cost per unit quantity item due to increase in volume of quantity produced with accepted quality;
iii. less buffer and work-in-progress (WIP) costs;
iv. less product liabilities claims;
v. increase in product price due to improved product quality image and company goodwill as a result of product time delivery;
vi. higher customer satisfaction;
vii. better productivity and profitability;
viii. machinery failure reduction; and
ix. assists in keeping output product constant while reduction in input quantity.

The specific benefits that are attributed to the maintenance structure classifications are as given below:

Remote maintenance: By leveraging information, wireless (e.g., Bluetooth) and Internet technologies, users may log in from anywhere and with any kind of device as soon as they get an Internet connection and a browser. Any operator, manager or expert also has the capability to remotely link to a factory's equipment through Internet, allowing them to take remote actions, such as set-up, control, configuration, diagnosis, debugging/fixing, performance monitoring, and data collection and analysis (Hung et al., 2003). Consequently, the manpower of the machine builder retained at the customer's site is reduced and there are facilities for him to diagnose the problems when an error occurs and, next, to improve the preventive maintenance thanks to the machine performance monitoring (Koc and Lee, 2003). Actually, one of the greatest advantage of e-maintenance is the ability to connect field systems with expertise centers located at distant geographical sites (Ong et al., 2004), allowing notably a remote maintenance decision making (Marquez and Gupta, 2006) that adds value to the top line, trim expenses, and reduce waste. The contribution to the bottom line is significant, making development of an asset information management network a sound investment (Hamel, 2000). Moreover, the web enablement of computerized maintenance management systems (called as e-CMMS) and remote condition monitoring or diagnostic (called as e-CBM) avoid the expense and distraction of software maintenance, security, and hardware upgrade. Computer science experts can add new features and/or migrations without the users even noticing it.

Cooperative/collaborative maintenance: E-maintenance symbolizes the opportunity to implement an information infrastructure connecting geographically dispersed subsystems and actors (e.g., suppliers with clients and machinery with engineers) on the basis of existing Internet networks. The resultant platform allows a strong cooperation between different human actors, different enterprise areas (production, maintenance, purchasing, etc.) and different companies (suppliers, customers, machine manufacturers, etc.). An e-maintenance platform introduces an unprecedented level of transparency and efficiency into the entire industry as shown in Figure 6.4 and it can be an adequate support of business process integration (Tsang, 2002). As a result, there is the chance to radically reduce interfaces, may that be between personnel, departments, or even different IT systems. The integration of business processes significantly contributes to the acceleration

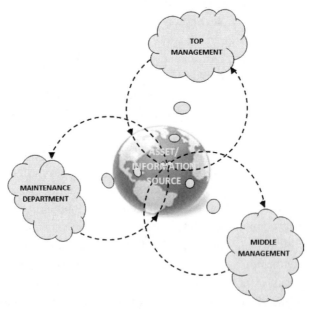

FIGURE 6.4 Implementing e-maintenance.

of total processes, to an easier design (lean processes), and to synchronize maintenance with production, maximizing process throughput, and minimizing downtime costs. In general, this leads to less process errors, improved communication processes, shorter feedback cycles, and hence improved quality. In short, e-maintenance facilitates the bi-directional flow of data and information into the decision-making and planning process at all levels (Goncharenko and Kimura, 1999). By so doing, it should automate the retrieval of the accurate information that decision makers require to determine which maintenance activities to focus resources on, so that return on investment is optimized (Hausladen and Bechheim, 2004).

Immediate/online maintenance: The real-time remote monitoring of equipment status coupled with programmable alerts enable the maintenance operator to respond to any situation swiftly and then to prepare any intervention with optimality. In addition, high-rate communications allow them to quickly obtain several expertise (Ucar and Qiu, 2005) and to accelerate the feedback reaction in the local loop, connecting product, monitoring agent, and maintenance support system. It has almost unlimited potential to reduce the complexity of traditional maintenance guidance through online guidance based on the results of decision-making and analysis of product condition (Moore and Starr, 2006). For example, personal digital assistant (PDA) devices play a key role in bringing mobile maintenance management closer to the daily practice at the shop floor. The PDAs enable the maintenance personnel to directly gain information from monitored machinery. In this context, potential applications of e-maintenance include formulation of decision policies for maintenance scheduling in real time based on up-to date information of machinery operation history, machine status, anticipated usage, functional dependencies, production flow status, and so on.

Predictive maintenance: The e-maintenance platform allows any maintenance strategy, and the improvement of the utilization of plant floor assets using a holistic approach combining the tools of predictive maintenance techniques. The potential applications in this area include equipment failure prognosis based on current condition and projected usage, or remaining life prediction of machinery components. In fact, e-maintenance provides companies with predictive intelligence tools (such as a watchdog agent) to monitor their assets (equipment, products, process, etc.) through Internet wireless communication systems to prevent them from unexpected break-down. In addition, these systems can compare a product's performance through globally networked monitoring systems to allow companies to focus on degradation monitoring and prognostics rather than fault detection and diagnostics (Garcia et al., 2004). Prognostic and health management systems that can effectively implement the capabilities presented herein offer a great opportunity in terms of reducing the overall lifecycle costs (LCC) of operating systems as well as decreasing the operations/maintenance logistics footprint (Wohlwend, et al., 2005).

Fault/failure analysis: The rapid development in sensor technology, signal processing, ICT, and other technologies related to CM and diag-nostics increases the possibilities to utilize data from multiple origin and sources, and of different types (Iung et al., 2003). In addition, by network-ing remote manufacturing plants, e-maintenance provides a multi-source knowledge and data environment. These new capabilities allow the main-tenance area to improve the understanding of causes to failures and system disturbances, better monitoring and signal analysis methods, improved materials, design, and production techniques to move from failures detec-tion to degradation monitoring (Iung et al., 2003).

Maintenance documentation/record: The e-maintenance platform pro-vide a transparent, seamless, and automated information exchange pro-cess to access all the documentation in a unified way, independently of its origin, equipment manufacturer, integrator, and end user (Figure 6.5). Information like task completion form is filled once by user and can be dis-patched to several listeners (software or humans) that registered for such events (Holmberg et al., 2005). At the device level, goods are checked out from stores against a work order or a location and the transaction is recorded in real time. Then, the massive data bottlenecks between the plant floor and business systems can be eliminated by converting the raw machine health data, product quality data, and process capability data into information and knowledge for dynamic decision making (Bangemann et al., 2006). In addition, these intelligent decisions can be harnessed through Web-enabled agents and connect them to e-business tools (such as customer relation management systems, Enterprise Resource Planning (ERP)) to achieve smart e-service solutions (Hung et al., 2003).

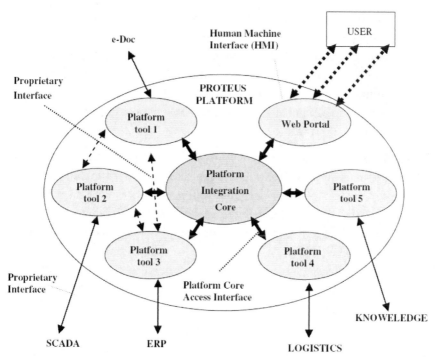

FIGURE 6.5 PROTEUS Platform (Bangemann et al., 2006).

After-sales services: With the use of Internet, Web enabled and wireless communication technology, e-maintenance is transforming manufacturing companies to a service business to support their customers anywhere and anytime (Lee, 2003).

Fault diagnosis/localization: E-diagnosis offers to experts the ability to perform online fault diagnosis, share their valuable experiences with each other, and suggest remedies to the operators if an anomalous condition occurs in the inspected machine (Roemer et al., 2005). In addition, lock-outs and isolation can be performed and recorded on location tanks to wireless technology and palm computing. Consequently, the amount of time it takes to communicate a production problem to the potential expert solution provider can be reduced, the quality of the information shared can be improved and, thereby, the resolution time reduced. All these factors contribute to increase the availability of production and facilities equipment, reduce mean time to repair (MTTR), and significantly reduce field service resources/costs.

Repair/rebuilding: For one, remote operators could via the e-connection tap into specialized expertise rapidly without travel and scheduling delays. Downtimes could conceivably be reduced through direct interaction (trouble shooting) with source designers and specialists (Ong et al., 2004). For another, diagnosis, maintenance work performed, and parts replaced are documented on the spot through structured responses to work steps displayed on the palm top.

Modification/improvement of knowledge capitalization and management: The multi-source knowledge and data environment provided by e-maintenance allows an efficient information sharing and, therefore, important capabilities of knowledge capitalization and management. With the availability of tools for interacting, handling, and analyzing information about product state, the development of maintenance engineering for product lifecycle (PLC) support including maintenance and retirement stages (disassembly, recycling, reuse, and disposal) is becoming feasible (Roemer et al., 2005; Wohlwend et al., 2005).

6.4   GENERAL REVIEW ON E-MAINTENANCE CONCEPTS AND ARCHITECTURE

E-maintenance paradigm emerged in early of 2000 through technological development, and it is based on a platform that utilizes Internet networking, intra-netting and extra-netting based on web technology, sensors application, wireless communications and mobile accessories (Lee, 2001).

Macchi and Garetti (2006) opined that "e-maintenance of an industrial process offers solutions for innovative customer–supplier relationships wherein the customer not only expects high quality products from a manufacturer but also an effective service during product use."

Muller et al. (2008) affirmed that the concept of e-maintenance refers to the integration of the information and communication technologies (ICT) within the maintenance strategy and/or plan to face with new needs emerging from innovate ways for supporting production (e-manufacturing), business (e-business) expected by the Manufacturing Renaissance.

The Open System Architecture in Condition Based Monitoring (OSA-CBM) maintenance architecture has shown in Figure 6.6 is made up of seven functional module software and web interface modules, which are

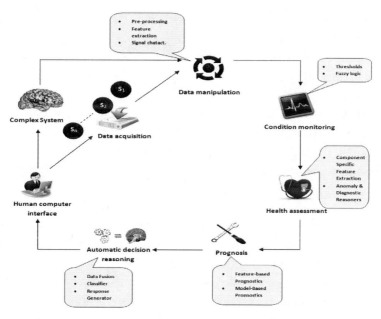

FIGURE 6.6   OSA-CBM architecture (Lebold and Thurston, 2001).

used in assessing the status and properly manage the machine. The first functional module of this architecture is data base management modules, which deals with the acquisition of the real-time and online data using single or multiple sensors with relevant sensor fusion techniques, thus transferring it to the domain or second functional module layer where it is converted or manipulated to obtain useful information about the present status of the machine. More so, the data model for this architecture is made up of the data event, configuration, explanation and extensible class. The data event displays or showcase the real time data to be acquired for ensuring proper management of the machines as describe above. Configuration provides information about the input sources, a description of algorithms used for processing input data, a list of outputs and various output specifics such as engineering units and thresholds for alerts. Explanation provides the data or reference to the data used by architecture to produce the output. In the data base management module, the data are pre-processed, processed and converted into signals that are properly recognized by feature recognition tools, which thus indicate the interpretation obtained from the signals formed through the processed data. The third module provides information about the status of the whole machine using some rule-based decisions obtained from experts tools such as Fuzzy Logic, rule-based expert system and case based reasoning, etc.

The fourth module of this architecture further assesses the component(s) or subsystem(s) of the machine causing the awful working condition of the machine. It performs this using intelligent algorithms obtained from feature extraction and anomaly and diagnostics reasoners. The fifth module then predicts the time the component(s) or subsystem(s) of the machine causing the awful working condition will fail and the need to replace it in order to avoid complete breakdown of the machine and ensure its reliability. This prediction is achieved using intelligent tools and models developed for this architecture such as feature-based prognostics and model based prognostics. The sixth functional module then presents the steps needed to ensure the repair of a predicted subsystem in a machine and also mandate maintenance engineers of these machines of the time to replace irreparable components of a machine while the seventh functional module displays the results of the different functional modules of this maintenance architecture in different web interfaces.

In general, the e-diagnostics architecture is an important subset of the e-maintenance architecture needed for effectively maintaining and managing the machines used in the manufacturing industries. Figure 6.7 presents an architecture which showcases a guide map for selecting communication and data representation technologies to implement an e-diagnostics system for semiconductor manufacturing industry (Roemer et al., 2005). More so, it delineate a set of guidelines, a capability and data taxonomy, security guidelines, industry use case scenarios, and an implementation roadmap needed to effectively diagnose the different subsystems that a machine is made up of. E-diagnostics communication and functional architecture capabilities for any kind of machine to be maintained are divided into four strata or functional layers:

Strata 0—access: collaborative troubleshooting of the different subsystems of the machine with basic remote connectivity.

Strata 1—collection and control: remote performance monitoring of machine during its operation for its effective management and control.

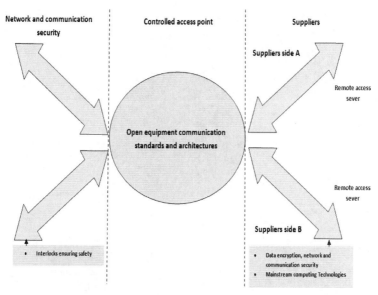

FIGURE 6.7   Overview of the e-diagnostic guidelines (Roemer et al., 2005).

Strata 2—analysis: automated reporting and advanced analysis of the different subsystems of the machines subsystems.

Strata 3—prediction: predictive maintenance, self-diagnostics, and automated notification of the machines used in the manufacturing industries.

Roemer et al. (2005) discussed and presented e-maintenance architecture as depicted in Figure 6.8 for the next generation National Aeronautic and Space Administration (NASA) shuttle vehicles, a road network of a prognostics system architecture that will be used to predict the time that any subsystems or subcomponents of the NASA vehicle will fail or repaired. Real time data are collected at the DSF module for each of subsystem of the vehicle, process and converted into forms where anomalies are detected, which formed the basis of the diagnosis; while the intelligent models called FPSIM (Failure Propagation and Subsystem Interaction Model) is then used to analyze the anomalies detected further and used to predict when to replace or repair any subsystem of the NASA shuttle vehicle to ensure its optimum functionality. The anomaly, diagnostic, and prognostic (A/D/P) tools and technologies of this architecture for the

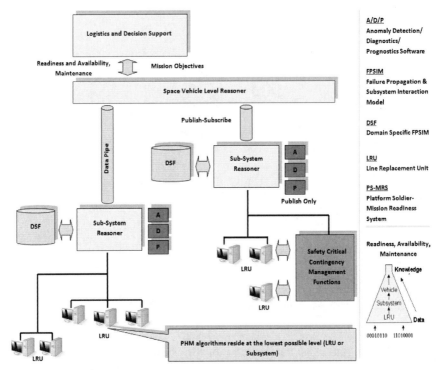

FIGURE 6.8   Distributed Prognostic System Architecture (Roemer et al., 2005).

NASA shuttle vehicles are executed at the lower levels (LRUs) as depicted in the architecture, to detect and foretell off-nominal state of affairs or damage aggravating at an accelerated rate. This information is then scrutinized through the hierarchy of reasoners to make informed decisions on the health of the vehicle subsystems/systems and how they affect the vehicle reliability and availability for its subsequent operations.

6.5   BUILDING OF E-KIT FOR MACHINES

The different tools used for assessing the performance and predicting its subsequent performance has been discussed and showcased in Figure 6.3. The quantitative health assessment of a machine can be determined by matching the present signals obtained from the reading sensors of a machine with failure mode signatures (baseline) obtained in the past through signal processing, feature extraction and sensor fusion modules. If an expert for feature extraction of signal processed in machine is available, then present signals obtained from the multi-sensor systems for the machine and the past failure modes or signatures is then presented in a network space. These signatures in this network space are then compared using CMAC pattern network approach or Logistic regression approach and thus, the current health status of the machine will then be determined. Based on this result, the future performance of the machine is then predicted by an expert, using peer-to-peer functionalities of e-kits as depicted in Figure 6.9. But if the signatures obtained available in the above discussed network space is intricate and complicated, statistical pattern approach, which involves the analysis of the signals using some statistical tools or feature map based approach is employed. Also creation of an effective database management system for past signatures can serve as a useful tool that can be used to assess the health condition of a machine using Hidden Markov Modeling for analysis. In general, a hybrid integrated e-diagnostic and prognostic architecture as depicted in Figure 6.10 has been developed using the relevant quantitative and qualitative tools described above to determine the present and future performance of a machine. The maintenance decision obtained from experts, technical manager and designers of the machines dictate the maintenance activities (i.e., diagnosis, prognosis and repair) that will be carried out on the machine as displayed in the architecture. The information about these activities reaches the maintenance managers of these machines using a feedback loop wireless network system.

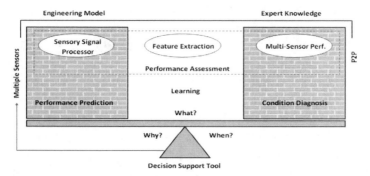

FIGURE 6.9 Peer-to-Peer (P2P) functionalities of the E-Kits for Machine Performance Assessment and Prediction.

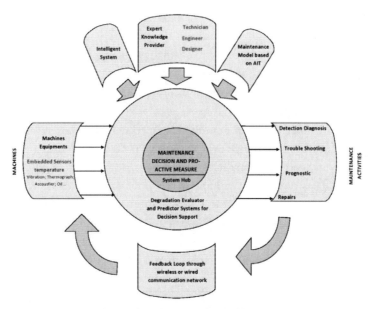

FIGURE 6.10   Hybrid e-diagnosis and Prognosis Machine Maintenance Architecture.

Autoregressive moving average model is a useful tool for predicting the performance of a machine in the future in order to ensure its functioning and availability when needed for operation in the Industry. It does this by generating a parsimonious description of a (weakly) stationary stochastic process in terms of two polynomials, one for the autoregression and the second for the moving average. This model is often called ARMA (p,q) model where p is the order of the autoregressive part and q is the order of the moving average part (as defined below). Another useful tool for predicting the future performance of a machine is matching method, which involves the matching of the previous predicted results obtained past signatures analyzed with the present signatures obtained from the reading sensors of the machine. The diagnosis of any machine can be modeled and achieved using a tool called Support Vector Machine (SVM). SVM are also used to detect faults in machines but are less reliable and more effective than ANNs. SVMs trace their roots back to statistical learning theory, as introduced in the late 1960s. It was not until the early 1990s that the methods used for SVMs began to flourish, and become practical, with the increased computing power available.

Support vector machines (Bishop, 1995; Cherkassky and Mulier, 1998; Vapnik, 1998) have been conceptualized and developed for solving pattern recognition problems. This method involves mapping of the data into a higher dimensional input space and construction of an optimal separating hyper plane in this space. This simply involves solving a quadratic programming problem, while gradient-based training methods for neural network architectures on the other hand are deficient from existence of many local minima (Zurada, 1992; Fauteux et al., 1995; Vapnik, 1995; Burrows, 1996). Kernel functions and parameters are selected in such a way that a bound on the VC dimension is minimized. In addition to the previous work done on this diagnosis technique, the support vector method was extended for solving function estimation problems. In view of this, Vapnik's epsilon insensitive loss function and Huber's loss function have been employed. Besides the linear case, SVMs based on polynomials, splines, radial basis function networks and multilayer perceptions have been applied successfully in the industries. The quality and complexity of the SVM solution does not depend directly on the dimensionality of the input space and this is made evident from the structural risk minimization principle and capacity concept with pure combinatorial definitions (Bishop, 1995; Cherkassky and Mulier, 1998; Vapnik, 1998).

Reliability models of machine are carried out with or without covariates based on the use of rigorous and complex statistical techniques, which include theoretical probability distribution fitting, trend and serial correlation tests, and require assumptions of homogeneous or non-homogeneous Poisson process or assumptions of proportionality of the hazard rate. The assumptions and statistical constraints of probabilistic reliability models limit the ability of these models to accurately represent and fit all real-life mining conditions (Haupt and Haupt, 2004). GAs has several key advantages over conventional mathematical models including: ease of randomized searches while retaining important historical information with the population, computational simplicity, GAs prospects from a population of solutions, not just from a single solution, and they can handle any kind of objective function linear or nonlinear constraints defined in discrete, continuous, or mixed search spaces (Paraszczak and Perreault, 1994; McCormick and Nandi, 1997).

6.5.1 CASE STUDY I ON IEAPMS

IEAPMS (Intelligent E-Assessment and Prognosis Management System) Maintenance Architecture was developed in Tshwane University of Technology with the aim of providing a maintenance engineer of machines in the industry with the necessary strategies and policies for effectively managing their machine fleets in order to increase the availability, reliability and maintainability of these machines, thereby enhancing Just-In-Time (JIT) and Lean production from these machines. This thus ensures that the users of machines in different manufacturing industries moves from the era of "fail, correct and fix" to the era of "prevent and predict," which however serves as an effective means of sustaining these machines. IEAPMS Maintenance Architecture that has been currently developed as shown in Figure 6.11 is made up of eight modules and they are discussed as follows:

Database Machine Management System Module: this is the portfolio where the following operating data are stored for further processing:

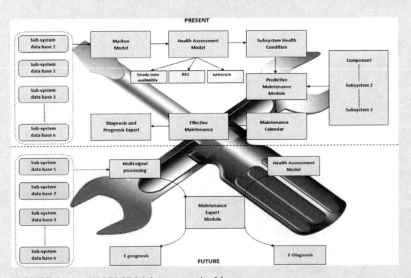

FIGURE 6.11   IEAPMS Maintenance Architecture.

- Machine subsystem date of failures;
- Repair time or downtime experienced by the machine;
- Scheduled hours and stand by hours of operation of the machine for the whole period of investigation, for the first year or month of investigation and for the last year or month of investigation;
- Shelf life of the machine;
- Used life of the machine.

Assessment and Prognosis Tool Module: in this architecture, a model was developed using Markov Model, which thus provides us with the health status of any machine as well as predicting the time that any of the subsystems of this machine will fail and need to be repaired or replaced. This model was mapped out by modeling the present failures and repairs made on the machines from Markov's equations as presented below:

$PM_0(t+dt) =$ [(Probability of being in operating state at time t) AND (Probability of not failing between $t + dt$)] + [(Probability of being in failed states at time t) AND (Probability of being repaired between t and $(t + dt)$)]

If probability of failure between t and dt is $\beta_i dt$ and the probability of not failing is $(1 - \beta_i dt)$. Also, if the probability of repair is $\alpha_i dt$, then using the addition and multiplication rules for probabilities give Eq. (2):

$$PM_0(t+dt) = PM_0(t)[\Sigma(1+\beta_n dt)] + \Sigma\alpha_n dt PM_n(t) \qquad (2)$$

Where n is the number of subsystems that ensures the proper functioning of machine.

Simplifying Eq. (2) further gives Eqs. (3) and (4)

$$PM_0(t+dt) - PM_0(t) = \{PM_0(t)[\Sigma(1+\beta_n dt)] + \Sigma\alpha_n dt PM_n(t)\} - PM_0(t)$$
$$PM_0(t+dt) - PM_0(t) = \Sigma(-\beta_n dt)PM_0(t) + \Sigma\alpha_n dt PM_n(t)$$

$$(3)$$

$$\frac{PM_0(t+dt) - PM_0(t)}{dt} = \sum(-\beta_n)PM_0(t) + \sum(\alpha_n)PM_n(t)$$

$$As\ dt \to 0$$

$$\frac{dPM_0(t)}{dt} = -PM_0(t)\sum\beta_n + \sum(\alpha_n)PM_n(t)$$

$$= \sum(\alpha_n)PM_n(t) - PM_0(t)\sum\beta_n \qquad (4)$$

$$PM_o(t) = P_o$$

$$PM_o(t) = P_n$$

Equating first order derivative of Eq. (3) to zero for a steady state, that is, $\left(\dfrac{dPM_0(t)}{dt} = 0\right)$, the Eq. (4) becomes Eq. (4) as expressed:

$$0 = \Sigma\alpha_n P_n - P_o\Sigma\beta_n$$

$$\Sigma\alpha_n P_n = P_0\Sigma\beta_n$$

$$P_n = \frac{P_0\Sigma\beta_n}{\Sigma\alpha_n} \qquad (5)$$

i.e. $P_1 = \dfrac{P_0\beta_1}{\alpha_1}, P_2 = \dfrac{P_0\beta_2}{\alpha_2} \ldots\ldots\ldots\ldots P_n = \dfrac{P_0\beta_n}{\alpha_n}$

Recall that:

$$P_o + P_1 + P_2 + \ldots P_n = 1 \qquad (6)$$

Substituting the value of $P_1, P_2 \ldots \ldots Pn$ into Eq. (6), then the steady state of any machine (P_o) is found to be:

$$P_1 = \left(\frac{1}{1+\sum\dfrac{\beta_i}{\alpha_i}}\right) \times 100\% = \left(\frac{1}{1+D}\right) \times 100\% \qquad (7)$$

where

$$D = \sum\frac{\beta_i}{\alpha_i} \qquad (8)$$

While the steady state availability of machine subsystems is given as:

$$P_1 = \frac{\lambda_1}{\mu_1}\left(\frac{1}{1+D}\right)$$
$$P_2 = \frac{\lambda_2}{\mu_2}\left(\frac{1}{1+D}\right)$$
$$P_n = \frac{\lambda_n}{\mu_n}\left(\frac{1}{1+D}\right)$$

(9)

Data processing module: this section deals with the rapid processing of data stored in the database, which may take a matter of seconds or perhaps a minute. This process provides the desired machine performance results and an optimized maintenance plan and maintenance calendar that will be use to effectively these machines. The data processing phase comprises the working memory, the inference engine, the agenda and the explanation facility.

The working memory provides a global database of all the useful information used by the rules given in the e-assessment and prognosis architecture algorithm, which involves the step by step coding of the procedures given by the assessment and prognosis module. The inference engine decides which rule is satisfied by the supplied facts during the processing, and then it prioritizes the rule sequentially and executes the said rule accordingly. The agenda, which is a subset of the inference engine, ensures sequential processing of the prioritized rules of the e-assessment and prognosis algorithm while the explanation facility explains application reasoning or results using simple terminologies for the users

Machine Health Condition Assessment Module: this module display the health status result of the machine in three folds as discussed in the assessment and prognosis module. The first fold is Steady State Availability of Machine. As the name implies, it provides relevant information on how often the machine is available to produce products needed by the customers, in order to reduce drastically the downtime experienced this machine to zero and also to promote Just In Time (JIT) production of the desired products in order to meet customers' target. The second fold of result that gives an indication of the health status of any machine is remaining useful life of a machine. As it name implies, this provides a maintenance manager of any machine the remaining time out of the machine's shelf life, that the machine will work effectively to its capacity before it is then subjected to continuous and rapid failures, thereby resulting in increasing downtime experienced by this machine at a rate of exponent progression, high productivity loss and customers target not being met. Thus, the degree of deterioration or degradation experienced by a machine have effect on the machine remaining effectively useful life that is, the higher rate of degradation experienced by a machine due to the mismanagement of this machine or other causes will reduce the remaining effective useful life of this machine over time. In view of this, there is a need to develop an effective maintenance calendar that will be used to maintain the machines in order to maximize the use of the remaining effective useful life (REU) of any machine. This parameter is being determined using steady state availability of the machine formula obtained from the reliability model discussed above and its value is obtained using Eq. (10).

$$L_{remain} = P_0 \times (L_m - L_{used})$$ (10)

where: L_{remain} = remaining effective useful life of the vibrating screen; L_{used} = used life of the vibrating screen; $L_{v.s}$ = life span or shelf life of the vibrating screen.

The third fold is Machine Reliability Decline Growth Rate (MRDGR). This provides maintenance managers of this machine on how effective were their previous machine maintenance management system. This informs them of the need to integrate the effective maintenance plans obtained with their previous maintenance plans in order to ensure the availability of these machines when needed for use in the manufacturing and other related industries. MRDGR of a machine is a function of the failure and repair time of the machine and can be obtained using Eq. (11).

$$MRDGR = \frac{Initial\ MTBF}{Final\ MTBF} \tag{11}$$

where, *Initial MTBF* = Mean time between failures for the first year of investigation; *Final MTBF* = Mean time between failures for final year of investigation.

Also, the health status of each subsystem in the afore-discussed folds can also be determined, which gives a clear indication of which subsystem(s) are mostly and least prone to failure, and thereby as a matter urgency, experts or manufacturers of these machines are consulted to come up with the design of a maintenance strategy for these subsystem(s). This feedback loop mechanism to the machine manufacturer prompts the need for them for sustainable design for reliability and maintainability, which thus evoke the lifecycle management of any machine as 3R'S phases as depicted in Figure 6.12, but ensures these manufacturers view sustainable manufacturing of this products as 6R's as depicted in Figure 6.13, where maintenance formulate a subset or relevant parameter for achieving this.

Predictive Maintenance Module: this module provides an effective maintenance scheduling programme that can be used to effectively manage the subsystems of machines in order to maximize its

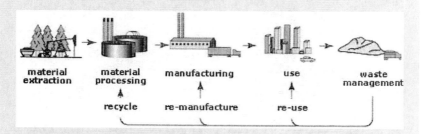

material extraction → material processing → manufacturing → use → waste management

recycle   re-manufacture   re-use

FIGURE 6.12 Cradle to Cradle approach to Life Cycle Management of Manufactured Products (Jawahir, 2008).

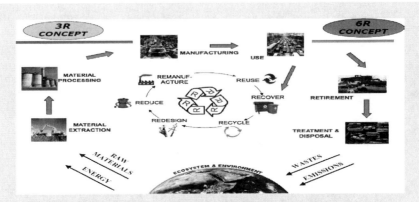

FIGURE 6.13   Sustainable Manufacturing Model (Jawahir, 2008).

remaining effective useful life, thereby enhancing the maintenance paradigm shift from a "fail, correct and fix" strategy to a "predict, prevent and sustain" strategy. This new maintenance scheduling programme combined with the normal routine maintenance system for the machine formulate a strong antidote called an optimized maintenance calendar to effectively manage the machines used in the manufacturing industries.

Maintenance Calendar evaluator module: this is the module that informs the maintenance managers of machines on how effective the maintenance calendar has been beneficial to the sustenance of the machine. If the optimized maintenance calendar contribute immensely to the success of the machine maintenance, then this maintenance calendar should continue to be used and updated from time to time based on different failure scenarios experienced by the machine, but if it is not effective, then an experienced diagnosis, repair and prognosis expert must be consulted to obtain an effective means of detecting when, how and steps needed to avert these failures experienced by a machine. The result obtained from this expert is stored in database, which serve as good tool for achieving big data analytics or algorithms for optimal failure prediction of all the subsystems in a machine. The programming languages and accessories used for developing the IEAPMS Maintenance Architecture are Hypertext

Pre-processor (PHP), JavaScript, Google Graph Application Program Interface (API), JQuery, Structured Query Language (mySQL) database, JavaScript SVG parser and renderer on canvas (canvg) and their functions in the architecture development are:

1. PHP – server side scripting programming language;
2. JavaScript – client side scripting programming language;
3. Google Graph API – for drawing graphs;
4. jQuery – reduces complexity of code in Javascript;
5. MySQL database – saving the results obtained as data;
6. canvg – Javascript SVG parser and renderer on canvas (saving Graphs).

At present, this IEAPMS Maintenance Architecture is limited to offline data for e-assessment and prognosis of machines, but future work has been considered in integrating a real time monitoring of all the subsystems of the machines using sensors as well as collecting this data for optimal e-assessment and prognosis of any machine. The proposed integrated maintenance architecture consists of the following modules as explained below:

Sensor module: this module provides the necessary sensors that will be used to monitor the machine in real time, thereby providing varying signals that need to interpreted, processed and analyzed.

Multi-signal processing and feature extraction module: this module process the signal obtained from the sensor of each of the machine subsystem using signal processing tools such as Fourier transform, wavelet series transformation, time frequency transformation and morlet wavelength filtering techniques. The results of the signatures processed are then analyzed and interpreted using feature extraction tool such as autoregressive model to analyze the signatures processed.

Health assessment module: this module provides the health status of the machine using the analyzed results obtained from the feature extraction tool for each of the subsystems that makes up the machine.

Expert and multi-model tool: this module uses the information obtained from experts of the analyzed signature results as well as set of relevant models such as fuzzy logic, hidden Markov model, support vector machine, artificial neural network and genetic algorithms to diagnose and predict the failure of the subsystems of a machine.

Diagnosis and repair module: this module present the different ways which causes the failures and its effect on the performance of the machine.

Prognosis module: this module uses the results obtained from the signatures analyzed by the expert and multi-model tool explained above to predict the time any of the subsystems will fail. The results obtained from the two precedent modules is then stored in the data base management module discussed above, which will be useful for big data analytics or algorithms for optimal prognosis of the subsystems of a machine when the present and the future IEAPMS Maintenance Architecture is integrated and synchronized for Industry 4.0 smart machine maintenance tool development.

6.5.2   CASE STUDY II ON TEMPERATURE BASED MACHINE MONITORING KIT

Another scenario is to showcase building of e-maintenance kit for extruding machine and cutting machines which major challenge is on temperature control. The processes involved are: modeling of temperature equations, as shown below.

Adeyeri et al. (2012) assumed that if T_i is the initial temperature value, and T_o be the measured temperature before the predicted value of temperature at next planned time of measurement or reading, then the temperature deteriorating factor, U_T is expressed as

$$U_T = \frac{T_i - T_0}{T_m^c} \qquad (12)$$

where, T_i: predicted value of temperature at next planned measuring time T_o; T_o: the current temperature value; U_T: temperature deteriorating factor; T_m^c: critical temperature limit level.

Therefore, T_i^{ta}, which is the predicted value of temperature at next planned time of measurement or reading taken would be

$$T_i^{ta} = [T_i^0 + T_i^0 U_T]_b^{t_n} \tag{13}$$

Simplifying Eq. (13) gives

$$T_i^{ta} = T_i^0 [1 + U_T]_b^{t_n} \tag{14}$$

where t_n is the periodic time numbering of readings and b is a function of speed, environmental condition and product demand.

Therefore, if $T_i^{ta} \geq T_m^c$, then maintenance is required, otherwise do not. The flowchart for implementing the temperature model is as shown in Figure 6.14.

The code for the microcontroller is developed in C language, which is not included in this book chapter. The preliminary exercise carried out in order to get the basis of the conventional machine is to:

i. study the temperature behavioral pattern of the machine to model; and
ii. get the trend of the temperature pattern of the machine system by recording ten to twenty readings of the temperature over a wide period of time and noting the performance rating of the machine against the corresponding temperatures.

While the algorithm developed for the model is as stated below:

Step-1: Setting up: set up display, set up real time clock and set up the internal analog to digital converter
Step-2: Wait for responses
Step-3: Set timer T to zero second and the counter to zero
Step-4: Initialize timer
Step-5: Is timer equals 10 seconds?
Step-6: If no, go back to timer initialization, else get the ADC value

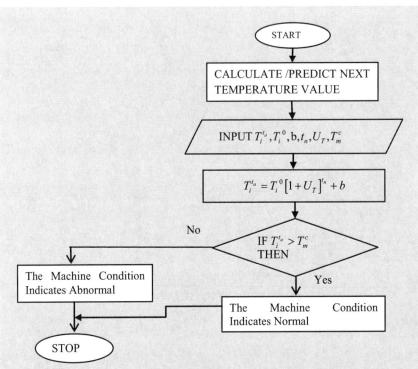

FIGURE 6.14 Flowchart of temperature condition based maintenance monitoring.

Step-7: *Counter stores value to RAM*

Step-8: *Has counter counted to ten values?*

Step-9: *If no, continue with timer initialization, else counter reset to zero*

Step-10: *Calculate average temperature*

Step-11: *Store value in EEPROM, convert temperature value to ASCII and get real time*

Step-12: *Display value and time*

Step-13: *Use inference decision block: for example, if, then maintenance is required, otherwise do not; make inference on machine parts affected and give suggestion*

Step-14: *Is decision made? If yes, display decision*

Step-15: *Delay sets in to cater for decision displayed, else back to reset*

Step-16: *Go to start*

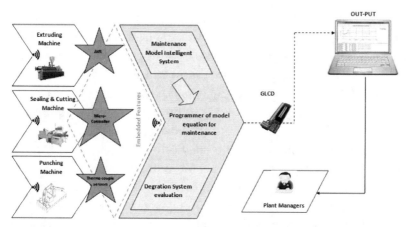

FIGURE 6.15 Temperature E-Maintenance Architecture for Conventional Machines

After this, the e-kit is made possible using the embedded interface. The following components are used in keeping track of the system:

  i. Analog Digital Converter (ADC): The output from the sensor is usually an analog signal that needs to be converted to a discrete signal for proper digitization. In order to achieve this, an analog to digital conversion module is required. For this model, a 10 bit ADC resolution is made used of.
 ii. Microcontroller: The microcontroller used for developing this model is Atmel Atmega 16 MCU, this is an 8 bit MCU but with internal 10 bit resolution ADC module. This makes the MCU a choice for this model.

In summary the whole arrangement is as shown in Figure 6.15.

6.6 CONCLUSION AND FUTURE PROJECTIONS

E-Maintenance is the most current line of maintenance management system used to effectively manage machines and ensure the reliability, availability and maintainability of machines in order to ensure that it attains a near-zero downtime state, thus making them efficient and effective at any time during manufacturing operation. These e-maintenance architectures are been built using relevant e-maintenance tools and accessories as well as quantitative and qualitative tools for the present and future performance analysis and detection. Different e-maintenance architecture for machines has demonstrated the various ways and methods in which the health status as well as their future functioning can be determined and has created a pathway for developing a hybrid e-diagnostics and prognostics maintenance architecture that could be used to optimally manage a machine using quantitative, qualitative and e-tools. Also, IEAPMS Maintenance Architecture and the e-maintenance temperature-based architecture had been developed to ensure optimal functioning of machines used in varying manufacturing industries. The next level of achieving optimal maintenance for proper functioning of machines is the adoption of S-Maintenance Architecture. S-maintenance has been viewed as a useful tool for carrying out maintenance activities on a machine based on the domain expert knowledge, where modules in this network architecture manage this knowledge (formalization, acquisition, discovery, elicitation, reasoning, maintenance use and reuse) and share the semantics to emerge new generation of maintenance services obtained for any machine (as self-X services, Worker self-management, improvement of various types of maintenance, new collaboration methods, etc.) while including e-maintenance characteristics. Thus this new observation obtained on any machine is then been send back using a feed-back loop mechanism to the manufacturers of such machines for optimal improvement and success in future designs of similar machines (Mohamed-Hedi, 2011).

KEYWORDS

- **diagnostic and prognostic tools**
- **e-maintenance**
- **machines**
- **s-maintenance**

REFERENCES

1. Adeyeri, M. K., Kareem, B., Aderoba A. A., Adewale O. S. Temperature Based Embedded Programming Algorithm for Conventional Machines Condition Monitoring. *Proceedings of the Fourth International Conference on Future Computing Technology, International Academy, Research and Industrial Association (IARIA), Nice-Fran*ce, 2012, 51–57.
2. Alzyouf, I. The role of Maintenance in Improving Companies' productivity. *Int. Journal of Production Economic.* 2007, 105, 70–78.
3. Baldwin, R. C. Enabling an E-maintenance infrastructure. 2004. www.mt-online.com.
4. Bangemann,T., Rebeuf, X., Reboul, D., Schulze, A., Jacek Szymanski, J., Jean-Pierre, T., Mario Thron, M., Noureddine Zerhouni, N. PROTEUS – Creating Distributed Maintenance Systems through an Integration Platform. *PROTEUS.* 2005, 1–10.
5. Bangemann, T., Reboul, D., Scymanski, J., Thomesse, J. P., Zerhouni, N. PROTEUS — An integration platform for distributed maintenance systems. *ComputInd [special issue on e-maintenance.* 2006, 57(6), 539–551.
6. Bellarbi, A., Benbelkacem, S., Malek, M., Zenati-Henda, N., Belhocine, M. Amélioration des performances d'ARToolKit pour la réalisationd' applications de réalitéaugmentée. *International Conference on Image and Signal Processing and their Application.* 2010*, ISPA,* Biskra, Algeria.
7. Benbelkacem, S., Zenati-Henda, S., Zerarga, F., AbdelkaderBellarbi, A., Belhocine, M., Malek, S., Mohamed, T. M. Augmented Reality Platform for Collaborative E-Maintenance Systems. 2012, 1–17.
8. Bishop, C. M. Neural networks for pattern recognition, Oxford University Press. 1995.
9. Blann Dale, R. Reliability as a Strategic Initiative: To Improve Manufacturing Capacity, Throughput and Profitability. *Asset Management and Maintenance Journal.* 2003, 16(2). http://www.marshallinstitute.com/inc/eng/justforyou/Body/articles/reliability_str_init.pdf
10. Braaksma, A. J. J., Klingenberg, W., van Exel, P. A review of the use of asset information standards for collaboration in the process industry. *Computers in industry.* 2011, 62, 337–350.
11. Burrows J. H. Predictive and Preventive Maintenance of Mobile Mining Equipment Using Vibration Data. Masters Thesis, Department of Mining Engineering, McGill University, Montreal. 1996.
12. Campos, J. Development in the application of ICT in condition monitoring and maintenance. *Computers in Industry.* 2009, 60(1), 1–20.
13. Cherkassky, V., Mulier, F. Learning from data: concepts, theory and methods, John Wiley and Sons. 1998.
14. Comport, A. I., Kragic, D., Marchand, E., Chaumette, F. Robust real-time visual tracking: comparison, theoretical analysis and performance evaluation. *In IEEE International Conference on Robotics and Automation (ICRA'05).* 2005, 2841–2846, Barcelona, Spain.
15. DeSimone, L. D., Popoff, F., The WBCSD. Eco-Efficiency. MIT Press. 1997.

16. Farnsworth, M., Tomiyama, T. Capturing, Classification and Concept Generation for Automated Maintenance Tasks. *CIRP Annals - Manufacturing Technology.* 2014, 63, 149–152.

17. Fauteux, L., Dasys, A., Gaultier, P. Mechanic's Stethoscope: A technology trader model. 1995, 88(994), 47–49.

18. Fiala, M. A. An improved marker system based on AR Toolkit. *National Research Council*, Canada. 2004.

19. Garcia E, Guyennet H, Lapayre J-C.; Zerhouni, N. A new industrial cooperative tele-maintenance platform. *ComputInd Eng; 46(4)*, 2004, 851–864.

20. Goldberg, D. E. Genetic Algorithms in Search, Optimization, and Machine Learning, Addison-Wesley Pub Co, Boston, MA. 1989.

21. Goncharenko, I., Kimura, F. Remote maintenance for IM [inverse manufacturing]. In: Proceedings of the first international symposium on environmentally conscious design and inverse manufacturing, Tokyo, Japan, 1999, 862–870.

22. Hamel, W. E-maintenance robotics in hazardous environments. In: Proceedings of the 2000 IEEE/RSJ international conference on intelligent robots and systems, Takamatsu, Japan. 2000.

23. Haupt, R. L., Haupt, S. E. Practical Genetic Algorithms, John Wiley and Sons, Hoboken, NJ. 2004.

24. Hausladen, I., Bechheim, C. (2004). E-maintenance platform as a basis for business process integration. In: Proceedings of INDIN04, second IEEE international conference on industrial informatics, Berlin, Germany, 2004, 46–51.

25. Holmberg, K., Helle, A., Halme, J. Prognostics for industrial machinery availability. In: POHTO 2005 International seminar on maintenance, condition monitoring and diagnostics, Oulu, Finland. 2005.

26. Hung, M., Chen K., Ho, R., Cheng, F. Development of an e-diagnostics/ maintenance framework for semiconductor factories with security considerations. Adv Eng Inf 2. 2003, 17(3–4), 165–78.

27. Iung B., Levrat E., Marquez A. C., Erbe H. Conceptual framework for e-Maintenance: Illustration by e-Maintenance technologies and platforms. *Annual Reviews in Control.* 2009, 33, 220–229.

28. Iung, B., Morel, G., Le´ger, J. B. Proactive maintenance strategy for harbor crane operation improvement, *Robotica [special issue on Cost Effective Automation, Erbe H (Ed.)*. 2003, 21(3), 313–324.

29. Jawahir, I. S. Beyond the 3R's: 6R Concepts for Next Generation Manufacturing: Recent Trends and Case Studies. Symposium on Sustainability and Product Development IIT, Chicago, August 7–8 2008, 1–110

30. Kans, M., Ingwald, A. Common database for cost-effective improvement of maintenance performance. *International Journal of production economics.* 2008,113, 734–747.

31. Kato, H., Billinghurst, M. Marker tracking and hmd calibration for a video-based augmented reality conferencing system. *Proceedings of ACM/IEEE Workshop on Augmented Reality (IWAR)*. 1999, 85–92.

32. Kiritsis, D. Closed-loop PLM for intelligent products in the era of the Internet of things. *Computer-Aided Design.* 2011, 43, 479–501.

33. Koc- M., Lee, J. A system framework for next-generation e-maintenance system. In: Proceedings of the second international symposium on environmentally conscious design and inverse manufacturing, Tokyo, Japan. 2001

34. Lebold, M., Thurston, M. Open standards for condition-based maintenance and prognostic systems. In: Proceedings of MARCON —fifth annual maintenance and reliability conference, Gatlinburg, USA. 2001.

35. Lee, J. A framework for next-generation E-maintenance system. In: Proceedings of the second international symposium on environmentally conscious design and inverse manufacturing, Tokyo, Japan. 2001.

36. Lee, J. E-manufacturing: fundamental, tools, and transformation. Robotics Computer-Integr Manuf. 2003,19(6), 501–7.

37. Macchi, M., Garetti, M. Information requirements for e-maintenance strategic planning: A benchmark study in complex production systems. *Computers in Industry.* 2006,57, 581–594.

38. Marquez, A. C., Gupta, J. N. D. Contemporary Maintenance Management: Process, Framework and Supporting Pillars. *Omega.* 2006,*34*, 313–326.

39. McCormick, A. C., Nandi, A. K. Real time classification of rotating shaft loading conditions using artificial neural networks, *IEEE Journal, Transactions on Neural Networks.* 1997, 8, 748–757.

40. Mohamed-Hedi, K., Chebel-Morello, B., Lang, C., Zerhouni, N. A Component Based System for S-maintenance. 9th IEEE International Conference on Industrial Informatics, INDIN'11., Caparica, Lisbon : Portugal. 2011, 1–8.

41. Moore, W. J., Starr, A. G. An intelligent maintenance system for continuous cost-based prioritization of maintenance activities. *ComputInd [special issue on e-maintenance].* 2006, 57(6), 595–606.

42. Muller, A., Marquez, A. C., Iung, B. On the concept of e-maintenance: Review and current research. *Reliability Engineering and System Safety.* 2008, 93, 1165–1187.

43. Muller, A., Suhner, M-C.; Iung, B. Maintenance alternative integration to prognosis process engineering. *J Qual Maint Eng special issue on "Advanced Monitoring of Systems Degradations and Intelligent Maintenance Management.* 2007, 13(2) to be published.

44. Okah-Avae, B. E. The Science of Industrial Machinery and Systems, Maintenance. Spectrum Books Ltd. Ibadan, Nigeria, 1995.

45. Ong, M. H., Lee, S. M., West, A. A., Harrison, R. Evaluating the use of multimedia tool in remote maintenance of production machinery in the automotive sector. In: IEEE conference on robotics, automation and mechatronics. vol. 2 Singapore. 2004, 724–8.

46. Otsu, N. A Threshold Selection Method from Gray-Level Histograms. *IEEE Transactions on Systems, Man, and Cybernetics.* 1979, 9, 62–66.

47. Paraszczak, J., Perreault, J. F. Reliability of diesel powered Load-Haul- Dump machines in an underground Quebec Mine, *CIM Bulletin.* 1994, 87(978), 123.

48. Roemer, M., Dzakowic, J., Orsagh, R., Byington, C., Vachtsevanos, G. An overview of selected prognostic technologies with reference to an integrated PHM architecture. In: Proceedings of the IEEE aerospace conference 2005, Big Sky, United States.

49. Takata, S., Kimura, F., Van Houten, F. J. A. M., Westkamper, E., Shpitalni, M., Ceglarek, D., Lee, J. Maintenance: Changing role in life cycle management. *Annals of the CIRP*. 2004,53(2), 643–656.

50. Tsang, A. Strategic dimensions of maintenance management. *JQME*. 2002,8(1), 7–39.

51. Ucar, M., Qiu, R. G. E-maintenance in support of E-automated manufacturing systems. *J Chin Inst Ind Eng*, 2005, 22(1), 1–10.

52. Van Houten, F. J. A. M., Tomiyama, T., Salomons, O. W. Product modeling for model-based maintenance. *Annals of the CIRP*, 1998, 47(1), 123–129.

53. Vapnik V. The support vector method of function estimation In Nonlinear Modeling: advanced black-box techniques, Suykens J. A. K., Vandewalle J. (Eds.), Kluwer Academic Publishers, Boston. 1998, 55–85.

54. Vapnik, V. The nature of statistical learning theory. 1995 Springer-Verlag, New-York.

55. Vayenas, N., Nuziale, T. Genetic algorithms for reliability assessment of mining equipment. *Journal of Quality in Maintenance Engineering*. 2001, 7 (4), 302–311.

56. Wohlwend, H. et al. E-diagnostics guidebook: revision 2. 1. Technology Transfer #01084153D-ENG, SEMATECH Manufacturing Initiative. 2005, www.sematech org.

57. Zurada, J. M. Introduction to Artificial Neural Systems, West Publishing Company. 1992.

PART 3:

E-PORTFOLIO

CHAPTER 7

THE DEVELOPMENT AND TRANSITION FROM PORTFOLIOS TO E-PORTFOLIO WITHIN EDUCATIONAL CONTEXT

NASIM MATAR[1] and EBTISAM AL-HARITHI[2]

[1]*Assistant Professor at Petra University – Amman , Jordan, Tel: 00962799113593, E-mail: Nasim_matar@yahoo.com*

[2]*PhD Research Student Southampton University, London – UK, E-mail: bfalfahad@gmail.com*

CONTENTS

ABSTRACT

This chapter discusses the importance of portfolios' and the use of e-portfolio within educational context. The uses of portfolios have been incorporated in different educational context, and they proved their educational value for both teachers and students. The current advances in the field of Information and Communication Technologies (ICT) have created a new method for using and incorporating different electronic tools in the educational field. These tools and associated files have been gathered and classified using different ways, and they have been used for showing their competence in the educational field. The previous process have been called electronic portfolio or as abbreviated by e-portfolio. The uses of e-portfolio is not monopolized by the educational field, rather it is used with different disciplines for showing knowledge, achievements and competency. This chapter will present the latest advancement in this field, and will provide the current methods and standards for creating and using e-portfolios. Moreover, it will bring the reader to fully understand the value and use of this tool and how to make the best use of this electronic technology.

7.1   INTRODUCTION

It is important to bring some introduction to the portfolio concept before introducing the uses of e-portfolio, as the e-portfolio is looked at as natural extension for portfolios that came as a result to the current advancement in ICT. Moreover, e-portfolios inherit all the benefits of portfolios, and it adds to them the new benefits that are found by the ICT.

Portfolios are generally defined as a collection of facts and proofs that shows a persons journey through learning and teaching over time and they are used to demonstrate their ability and competences. The use of portfolios can be restricted to specific discipline or they can be used to define a broader concept that wraps a person's lifelong learning. A wide range of different evidences can be included within portfolio such as: writing samples; photographs; videos; research projects; observations and evaluations of supervisors, mentors and peers; teaching and learning artifacts; and reflection of previous evidence.

The most important aspect of portfolios is reflection, as it shows what the portfolio owner has learned, by justifying the reason for the included items within the portfolio (Abrami and Barrett, 2005; Klenowski et al., 2006). According to Kimball (2005), he argues that:

...neither collection nor selections [of pieces to be incorporated into a portfolio] are worthwhile learning tasks without a basis in reflection. Thus it is seen that reflection is the entire pedagogy of portfolios.

Having the previous argument, it is evident that teachers are learning from the reflection that is associated with the process of creating portfolios, through introducing them to different tools and techniques that are used in pedagogical context, and how to use these tools towards teaching, guiding and mentoring students. It is also believed that portfolios have another two important key elements that are:

- measuring learning and development over time (Challis, 2005);
- the process of constructing a portfolio is where the learning takes place, rather than the end product (Smith and Tillema, 2003).

7.2   USES OF PORTFOLIOS

Portfolios are looked at as a main source for providing evidence and they can be employed for different purposes such as, learning, professional development, assessment, job applications, and promotions.

Moreover, they have been used with different audiences such as: lecturers, mentors, employers and for self-use; in addition portfolios provide an alternative form of assessment (Chang, 2001; Smith and Tillema, 2003; Smits et al., 2005; Wade et al., 2005) that can be used with students or teachers in different forms as summative or formative assessment tool (Abrami and Barrett, 2005; Kimball, 2005; Loughran and Corrigan, 1995; Ma and Rada, 2005).

According to the literature, portfolios come in different types and all are based on the use and the audience for the portfolio. According to Zeichner and Wray (2001) portfolios can be of the following types:

- Learning Portfolio: document students learning over time.
- Credential Portfolio: used for registration or certification purposes.
- Showcase Portfolio: used for applying for employment positions.

  Abrami and Barrett (2005), define three different types of portfolio:

- Process portfolio: they represent learning over time.
- Showcase portfolio: they represent achievement in the study or workplace.
- Assessment portfolio: they are used for evaluation purposes.

  Another variation is presented by Smith and Tillema (2003) with the setting of their use in four different types:

- Dossier Portfolio: used for job selection and promotion purposes.
- Training Portfolio: used for learning and development.
- Reflective Portfolio: self-directed portfolio that can be used for many purposes.
- Personal Development Portfolio: self-directed portfolios that reflect learning and development.

From the different variations of portfolios it is obvious that each portfolio has a major aim, so that portfolio developed to show change and progress in learning will not be appropriate for use when applying for a job, just as a portfolio displaying only ideal pieces of work will not be useful for evaluating reflective learning. The option of using portfolios in this study and within Saudi schools is intended to be reflective portfolios, as these types of portfolios are considered more flexible to work with and use. The reflective portfolios can be used with summative and formative assessment procedures and will enable teachers to create many portfolios that are directed to the intended assessment objectives.

7.3 BENEFITS OF PORTFOLIOS

The use of a portfolio has different benefits for its owner, as it helps to focus thinking on a specific issue or matter (Wade and Yarbrough, 1996). Also, a portfolio provides a way to translate theory into practice (Hauge, 2006) and the main benefit is that they document a learner's progress over time (Abrami and Barrett, 2005; Challis, 2005; Smith and Tillema, 2003). In terms of using portfolios with students, a major benefit that has been recorded is students' communication and organizational skills and that they present a way of classifying and recognizing prior learning and also they can lead to new learning outcomes (Brown, 2002). The process of portfolio construction has enabled different students to sense and visualize what they are learning (Young, 2002). Having awareness of their accomplishment creates a better understanding of how their learning takes place (Brown, 2002). So, for students or learners it is a creation process, while evaluators see portfolios as the end product (Darling, 2001).

If portfolios are to be considered in different context, Zeicher and Wray (2001) have posited several important questions for portfolio implementation in educational settings:

- What is the purpose of the portfolio? Is it learning, assessment, professional development or for employment reasons?
- Who decides what should be included in a portfolio? Should it be the student/learner or evaluators?
- How prescriptive should guidelines for creating a portfolio be?
- How should the pieces of evidence in the portfolio be organized? Should they be based on themes chosen by the student/learner or based on programme goals or based on achievement standards?
- What kinds of artifacts are acceptable as pieces of evidence? What should be included or excluded?

- What kind of input should tutors, lecturers and peers have throughout the process of constructing the portfolio?
- How frequently should students/learners be expecting feedback on their progress?
- How should the portfolio be assessed? Should it be by using specific evaluation criteria and grading rubrics or a simple pass-fail system?
- What should happen to the portfolio after it is finished? Should it be publically acknowledgement or a presentation of work?

The previous questions all indicate that there are some problems related to the use of portfolios as an assessment exercises in academic settings. Some of these issues are related to the current shift towards involving electronic environments in educational settings, while others are related to lack of defined guidelines and clear structure towards portfolio construction and use (Smith and Tillema, 2003).

7.4 SUCCESS CRITERIA

Based on the previous considerations of the problems and questions posted, a number of criteria have been set for successful use of portfolios (Loughran and Corrigan, 1995; Smith and Tillema, 2003; Wade and Yarbrough, 1996). The following are a summary of the success criteria:

- Having familiarity with the portfolio concept and understanding the processes and the product of portfolio construction.
- Having clear a framework and guidelines for constructing and using a portfolio.
- Having the portfolio structure tempered with enabling freedom for creativity.
- Providing or gaining feedback during the evidence collection process.
- Have a clear understanding of the value of reflection.
- Understanding the importance and value of portfolios for future use, such as for employment.
- Being motivated (or motivating) to learn and achieve good results.
- Enforcing the concept of portfolio ownership.
- The target audience should be considered.
- Creating a sense of achievement for teachers at overcoming struggles to understand the portfolio concept.

7.5 PORTFOLIOS IN DIFFERENT DISCIPLINES

Portfolios have proved to be beneficial even if employed in different educational disciplines such as teacher education, medical education, nursing, and other disciplines. This field of medical education is shifting towards having assessment based on achievements and maintenance of competencies. In order to have a competency-based evaluation, there should be a way of determining progress against a threshold rather than against other people, thus portfolio assessment shows the greatest promise for meeting these demands (Carraccio and Englander, 2004; Jarvis et al., 2004).

It is acknowledged in medical education that portfolios are tools that can be used to assess performance in authentic contexts (Driessen et al., 2005a), or they can be looked at as collections of evidences that demonstrate education and practice achievements (Davies et al., 2005). Davies et al., (2005), they agree that portfolios compiled by healthcare professionals have the potential to assess outcomes, including attitudes and professionalism, which are both understood as difficult to assess using traditional instruments. Based on different research studies (Davies et al., 2005; Pearson and Heywood, 2004; Duque, 2003; Lynch et al., 2004), there are different benefits of portfolio use within education in medical context such as:

- Reflection on practice and learning from experience, which results in improvements towards patient care.
- The support towards personal and professional development helps clarifying learning goals and monitor progress of medical students.
- Encouraging self-directed learning.
- Progress over time.
- Teach lifelong learning skills.

The previous points are another example of the effective use of portfolios in the context of other professions that seek development of practices and enhancement of educational performance through reflection, support of professional development, measuring progress and lifelong learning skills. To guarantee that portfolios are successfully implemented and used in the medical setting, the following strategies should be addressed:

- Student should be involved in the decision-making process (Driessen et al., 2005b).

- Students should have good supervision (Pearson and Heywood, 2004; Pinsky and Fryer-Edwards, 2004).
- Students should have supportive educational climate where they feel comfortable revealing their weaknesses or mistakes (Pinsky and Fryer-Edwards, 2004).

These benefits of portfolios and successful implementation are not just relevant in the medical education field; they are just as applicable to other fields including education.

7.6   ELECTRONIC PORTFOLIOS

An e-portfolio is simply an electronic version of the paper-based portfolio that is created using computers and it incorporates: graphics, audio, video and interactive elements. Another definition is given by Abrami and Barrett (2005, online) as:

> *a digital container that is capable of storing different elements such as visual and auditory content including text, images, video and sound, that are specifically designed to support a variety of pedagogical processes and assessment purposes.*

Moreover, Challis (2005) provides a more in-depth definition by describing e-portfolio:

> *as selective and structured collections of information that are gathered for precise reasons of showing or evidencing one's accomplishments and growth by storing the evidences digitally and managing them using appropriate software that operates within a web environment and they can be retrieved from a website or delivered by CD-ROM or DVD.*

In terms of data generated and used by portfolios, they can be stored in completely different system as it is shown in the Figure 7.1.

The e-portfolio system can be categorized as three different software systems that are used through the Web or a network. The systems are:

1. E-portfolio Management System: used e-portfolio systems such as Mahara, Epsilen, PebblePad and Taskstream.

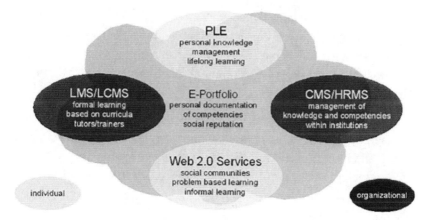

FIGURE 7.1 Different systems used for e-portfolio data management (Source: Himpsl and Baumgartner, 2009).

2. Learning Management System (LMS)/Learning Content and Management system (LCMS) with integrated e-portfolio functions such as Exabis, Fronter and Sakai.
3. Integrated systems with possible portfolio functions such as Drupal ED, Elgg, Joomla, and Wordpress.

All of these systems can be used as e-portfolio systems, and preferring one to another is based on the context and the intended use of the e-portfolio system.

7.7 USES OF E-PORTFOLIOS

E-portfolio uses are classified into three main categories, according to Lorenzo and Ittleson (2005).

Students while studying: portfolio use allows students to demonstrate their competences, develop, demonstrate and reflect on pedagogical practices with enabling them of showing their attitudes, knowledge and skills. Such portfolios are used in educational institutions and colleges (Milman and Kilbane, 2005; Sherry and Bartlett, 2005; Smits et al., 2005; Lorenzo and Ittleson, 2005).

Graduates while moving into or through the workforce: portfolio use in such a context enables the users to apply for licensure, qualification and competencies in job interviews; they can also be used for appraisal, promotion and for critical reflection and learning purposes (Milman and Kilbane, 2005; Pecheone et al., 2005).

Institutions for programme assessment or accreditation purposes: portfolio use in this context is looked as a medium for institution reflection, learning and improvement to express institutional accountability, to make accreditation processes more evident (Lorenzo and Ittleson, 2005).

As an alternative Adamy and Millman (2009) cite two types of e-portfolio:

Accountability portfolios: which involve licensure requirements and evidence of teachers' knowledge of standards.

Formative portfolios: which focus on teachers' acquisition of reflective skills. Figure 7.2 shows the documentation of achievement and documentation of learning using e-portfolio.

The documentation for learning and the documentation of achievements are two widely used methods in e-portfolios. The assessment types used with the previous methods are formative and summative assessments that have been discussed previously in this chapter. From Figure 7.2 it is noticed that the Documentation of Achievement mainly focuses on evaluation towards showcasing or accountability (summative assessment), and it presents the learning objects in thematic organization that consist of predefined, directed, rational pages and screens with a retrospective reflection towards the learned material. On the other hand, the Documentation of Learning focuses on providing feedback towards learning and reflection by having formative assessment. Moreover, the presentation of learning is presented in chronological order that consists of reflective journals, blog entries and discussions that are provided mainly by social networks. The previous two approaches have common shared points which are:

Collection of artifacts: each of the previous approaches has artifacts.

Direction: each of the previous approaches has some form of direction towards the learning objective.

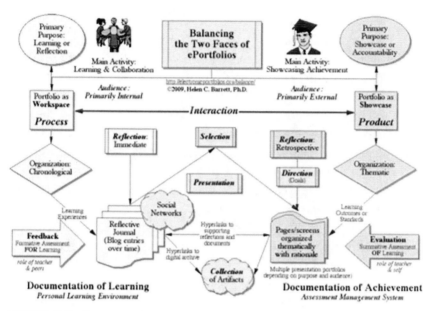

FIGURE 7.2 Documentation of achievement and documentation of learning using e-portfolio (Source: Barrett, 2011).

Presentation: each of the previous approaches has the data presented through an electronic medium.

Collection: each of the previous approaches is working towards collecting the needed materials with respect to the used method either as predefined or reflective.

7.8 BENEFITS OF E-PORTFOLIOS

The use and benefits of e-portfolios are many; the following points address the benefits that have been identified in the literature:

a. E-portfolios enhance skills development: users of e-portfolios are introduced to different skills that include: general literacy, communication and problem solving skills, multimedia and technological skills (Abrami and Barrett, 2005; Barrett, 2000; Heath, 2002, 2005; Wade et al., 2005; Wall et al., 2006).

b. E-portfolios serve as evidence of learning, as they serve as general, flexible and distributed evidence of learning which include different times and places (Love and Cooper, 2004; Wade et al., 2005), also they provide a rich picture of learning showing the past and current learning gains (MacDonald et al., 2004; Wade et al., 2005).

c. E-portfolios show learners competencies and enable them to make connections between their course project and non-academic projects, which facilitates authentic learning (Love and Cooper, 2004). Moreover, e-portfolios help users to manage their professional development, which reflects positively to lifelong learning contribution (Barrett, 2000; Love and Cooper, 2004; Wall et al., 2006) and they provide a significant pedagogical benefit, thought-provoking class discussions and provide student-centered learning (Canada, 2002).

d. E-portfolios provide feedback: the use and construction of e-portfolios has been found to assist in exchanging ideas and providing feedback (Lorenzo and Ittleson, 2005). Users of e-portfolios can have feedback on their work quickly using an electronic media channel which serves in many occasions as a 'feedback loop,' which is an integral process to formative assessment (Cambridge, 2001).

e. E-portfolios provide reflection: the context of using portfolios, either as paper-based or electronic, is to provide the user continuous reflection and justify reasons for choosing certain pieces to be included in their portfolio (Ahn, 2004). It has been found that reflection integrates their learning experience through making meaning out of diverse and unconnected pieces of information (Cambridge, 2001).

f. E-portfolios promote continuous assessment: as users are constructing their portfolios they are constantly revisiting and refining their work and learning resource which provides a better understanding of the assessment criteria and processes they are undertaking which reflects positively on the learning process (Wade et al., 2005, Cambridge, 2001).

Figure 7.3 is based on Kolb's experiential learning cycle that demonstrates the continuous process of learning based around dialogue and collaborative activity with others. E-portfolio use can produce different skills that learners need to effectively navigate

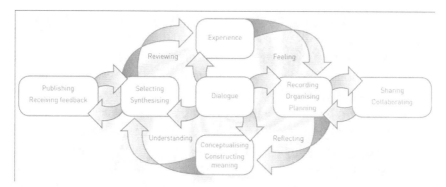

FIGURE 7.3 A model of e-portfolio-based learning, adapted from Kolb (1984).

their way through the complex demands of an information age. Through e-portfolio development, skills of collaboration and selection are acquired. Moreover the most important skills promoted by the use of e-portfolio are those related to reflecting and forward planning in response to an experience or occurrence of learning. These are skills that have relevance across the sectors and in all subject disciplines.

g. E-portfolio use of artifacts: different kinds of artifacts can be used and integrated with e-portfolios such as text, pictures, graphics, audio and video (Abrami and Barrett, 2005; Canada, 2002; Heath, 2005; Love and Cooper, 2004; Milman and Kilbane, 2005; Wade et al., 2005). Moreover, they take advantage of the flexibility that is provided by electronic technology in incorporating different work that is already in an electronic format (Heath, 2002, 2005).

h. E-portfolio maintenance: maintaining and enhancing e-portfolios is considered easy through processing different direct commands such as edit and update which promotes constant revision towards portfolios (Canada, 2002; Heath, 2002, 2005).

i. E-portfolio portability and sharing: e-portfolios are considered portable and easy to share as they provide different ways of sharing either by using CD-ROM, DVD, storage disks or the Web. Also they are easily transferred to another system or working environment (Abrami and Barrett, 2005; Strudler and Wetzel, 2005; Wade et al., 2005).

j. E-portfolio access: e-portfolios are highly accessible if they are used through the Web as they provide the ability for users to work on their portfolios either as students by updating, editing or adding resources, or as evaluators in an any time anywhere purpose (Ahn, 2004; Canada, 2002; Heath, 2005; Wade et al., 2005).

k. E-portfolio audience: as portfolio is based on electronic technologies, it has inherited the high accessibility features that are provide by those technologies, and they can be accessible and viewable by a larger audience which includes students' peers, supervisors, assessors, parents, employers (Ahn, 2004; Strudler and Wetzel, 2005, Wade et al., 2005).

l. E-portfolio organization: the process of organizing e-portfolios is relatively easy due to the inherited nature of electronic resources (Ahn, 2004; Wade et al., 2005; Young, 2002). This enables organizing resources and learning evidences and artifacts in a complex way and presents them in with navigational links that connects ideas with artifacts and resources (Canada, 2002; Heath, 2002, 2005).

m. E-portfolio storage: as portfolios do not rely on large binders full of paper, e-portfolios are easy and efficient to store (Ahn, 2004; Canada, 2002).

n. E-portfolio cost: the use of e-portfolios is inexpensive as they are easy to reproduce. However, there is initial set-up cost for having the software and the equipment that may be quite high (Heath, 2005).

o. E-portfolio standardization: e-portfolios have the flexibility and the ability to be standardized across different regions and countries, if a universal specification can be agreed upon (Abrami and Barrett, 2005).

p. E-portfolio privacy: the use of e-portfolios can include having privacy features in order to provide the needed protection for student work. In many e-portfolio products and tools access policies can be set to those that the students wish to view his/her work.

It is well accepted that using e-portfolios demonstrates what is important about individuals at particular points in time, their achievements, reflection on learning and potentially provides a rich and rounded picture of their abilities, aspirations and ambitions (Figure 7.4).

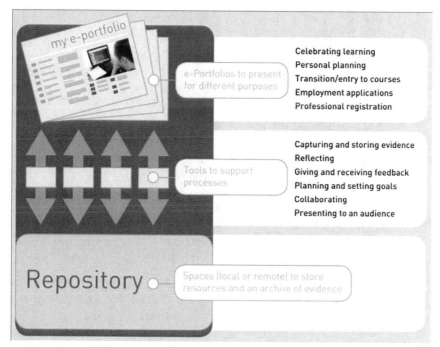

FIGURE 7.4 Impact study of e-portfolios on learning (Source: Hartnell-Young et al. 2007).

7.9 THE USE OF E-PORTFOLIO IN TEACHER EFFECTIVENESS

In their work about e-portfolios, as the pedagogy behind the technology used in Science projects, Chua and Chua (2007) state that e-portfolios can serve as a tool to be utilized in integrating technology to capture the learning process and reflection of individuals. They point out that the pedagogy behind the technology serves as the key to success in using any teaching and learning technology. It is likewise emphasized that in ensuring success in the use of e-portfolios, e-portfolios must be used with consideration towards the use of ICT and teachers ICT competencies Huang (2006).

According to Jafari and Kaufman (2006), there is much promise offered by e-portfolios in the improvement of learning and assessment in teacher education. In order to realize this promise there is:

- A need for new technology to address the need for efficient and effective teacher education. Technology must be considered in the background, and not the focus of this improvement;
- A need for reflection and sharing will be facilitated by this technology, moving teacher education towards a more ideal teaching portfolio that is created across diverse contexts over time, enriched through collaboration, and designed by reflection;
- An ultimate goal being the advancement of student learning through raising teacher effectiveness in interacting with ICT and having appropriate ICT competencies (Jafari and Kaufman, 2006).

By enabling technology to be placed in the background, emphasis would be placed back on theory, reflection, and collaboration. Jafari and Kaufman (2006) encourage teacher educators to use e-portfolios for their advantage that other means cannot achieve by the same totality. The presented ideas by Jafari and Kaufman (2006) are relevant to this research since they provide an insight about the usability of e-portfolios for teachers and ultimately, to the students, for their ultimate learning.

Bartlett (2009) provides an elaboration of a five-step model designed to enhance electronic teaching portfolios, whose conceptualization starts with dissatisfaction with traditional measures of student learning. This dissatisfaction has led to the implementation of performance-based assessment such as portfolios, with an aim to measure teaching effectiveness. Performance assessment, in contrast to standardized tests, deals with collaborative and active learning aiming to ensuring success on tasks that are termed 'real world.' Although being relatively new, there is popular growth in teaching portfolios alongside increased demand by accreditation bodies for tangible evidence of teachers' competencies. In his study Bartlett (2009) implemented e-portfolios with two groups of pre-service teachers. The participants in the programs started with limited technology backgrounds as they created their portfolios. Within the two-year program, the group consisting of graduate students learned how to make web pages, use of PowerPoint, how to create digital movies and how to do Internet research whilst presenting ideas to the class. In addition to leaning technology, the students spent 42 hours creating e-portfolios. Some components comprising portfolios included: welcome page, teaching philosophy, career goals, self-evaluation, and resume. The presentation of teaching development included: used documents, video clips, digital photographs, and scanned work samples of students.

An evaluation model for electronic teaching portfolios includes five steps, which are:

1. A rubric for assessing individual portfolios,
2. Qualitative data on students' assignment perceptions,
3. Qualitative data on e-portfolios advantages and disadvantages,
4. Quantitative data on changes in knowledge and attitudes about technology, and
5. Qualitative data on the use of technology by students as beginning teachers.

The model aims to monitor the effectiveness of the e-portfolio process (Bartlett, 2009). According to Bartlett (2009), rubrics enable students to be informed effectively of what to be expected. The author finds rubrics to lead to better final projects and enable easy grading. A useful framework for assessing e-portfolios is provided by the five-star model. Moreover, Wilson et al. (2003) examined how e-portfolios are implemented in pre-service teacher education programmes, taking into account a three-fold purpose of the process of e-portfolios. These included integrating technology into instruction, reflecting on the uses of technology during instruction and enabling teachers to create a picture of themselves as educators using electronic portfolio.

Two contexts given for the use of portfolios have been identified in the portfolio assessment process these are: student-focused portfolio assessment and teacher-focused portfolio assessment. Using the context of portfolio assessment involving students, the focus is collaboration between a teacher and a student. In terms of portfolio assessment involving teachers, the emphasis is placed on the portfolio being represented as knowledge and competence (Wilson et al., 2003). The work of Wilson and colleagues is relevant to this study although it investigates the implementation of e-portfolio in a pre-service teacher education programme. Furthermore, Huang (2006) focused on investigating the impact of the development of portfolios on self-directed learning (SDL) of pre-service teachers alongside computer technology skills. The study involves three student teachers and two internship students using qualitative methods and some descriptive quantitative analyzes. It is found out that SDL is guided by a natural setting in which problem solving is employed. With SDL the participants are able to be in charge of their own learning, as they create their e-portfolios and determine the purposes and audiences. The study revealed that the e-portfolios used by participants enabled computer technology proficiency and self-directed learning readiness and helped pre-service teachers to become motivated to accomplish the tasks by telling them that their e-portfolio can be used as useful tools for getting a job within their profession. This can however undermine the usefulness of e-portfolios as a means to obtain computer technology literacy and as a self-directed learning tool. It is emphasized that teachers' knowledge can be communicated by e-portfolios in two ways: providing the capacity to integrate various forms of media in one document and communicating ideas across various audiences through the Internet. A study involving electronic teaching portfolios also demonstrated teachers acquired literacy with multimedia technology hence producing rich representations of tasks and knowledge in the classroom (Huang, 2006).

7.10   THE IMPORTANCE OF E-PORTFOLIOS IN EVALUATING PRE-SERVICE TEACHERS

The phases involved in the process of assessment are planning, consensus building, measuring, reflection, analyzing, and improving certain actions based on what is presented in the data about a learning objective or a set of learning objectives. A range of activities is likewise involved in assessment, which includes testing, performances, observations, and project ratings (Orlich et al., 2004, cited in Buzzetto-More and Alade, 2006).

An efficient and viable means of assessing teaching effectiveness can be provided by the use of information technologies (i.e., e-portfolios) as this support traditional and authentic assessment protocol. The measures offered by technology are those that can yield rich sources of data in which educators can understand teaching effectiveness and learning mastery. Those that include the use of information technologies and e-learning, which transform the assessment process are: diagnostic analysis, pre/post testing, the use of rubrics, artifact collection, and data aggregation and analysis, to name a few (Buzzetto-More and Alade, 2006). This article provides a significant contribution to this study in understanding the use of information technologies, particularly e-portfolios, in assessing teacher evaluation.

In order to satisfy the requirements posed by the National Council for Accreditation of Teacher Education (NCATE) standards, e-portfolios have been successfully incorporated into the teachers' curricula by several teacher education programmes (Buzzetto-More and Alade, 2006). The article of Stansberry and Kymes (2007) is relevant to the topic being tackled since it is focused on the design and format of electronic portfolios, specifically on what happens to teachers when they engage in the creation of professional portfolios. It involves the notion of whether teachers are able to transform this experience into a more authentic one in the teaching-learning process. Stansberry and Kymes' work investigates whether teachers find an alteration to their own core beliefs related to these tools apart from acquiring proficiency in the use of portfolio development tools. The study involved 78 teachers who created e-portfolios as they participated in an education course at graduate level. This scenario might well establish the potential capabilities of e-portfolios in evaluating pre-service teachers and developing school improvement, which is the focus of this study. The study used Mezirow's transformational learning theory to explain data which were gathered quantitatively and qualitatively and concluded that albeit not all teachers showed readiness in replicating the process of developing e-portfolios with their students, they were however transformed by the process of e-portfolio creation. Moreover, transformative learning is facilitated by the content that this particular portfolio addresses, alongside the course format that includes collaboration, supportive discussion, and critical reflection, all of which influences this study.

Huang (2006) points out that the multimedia capabilities of web technology may pave the way for the development of a new language of practice in the teaching profession. The study's overall finding indicated:

An improvement in some participants' self-directed learning readiness as well as their computer technology skills, whilst only little improvement was seen in others, as they developed their e-portfolios.

That developing an e-portfolio can be effective for improving the computer technology skills of learners, highlighting the importance of initiative that an individual should take for his or her own learning (Huang, 2006).

This work by Huang (2006) contributes to the development of this study since it points out the benefits that can be gained from utilizing e-portfolios amongst pre-service teachers. Moreover, Huang's work is consistent with the findings made by Bartlett (2002) on the latter's study about the use of e-portfolios in teacher education. Bartlett evaluates the survey responses of pre-service teachers in the creation of instruction units taught to primary pupils through the use of software and multimedia materials. The teacher-respondents rated the assignments positively, indicating opportunities that need to be learned about e-portfolios and educational technology. Time and equipment problems are however two complaints presented in this use of educational technology (Bartlett, 2002).

The above works are also congruent to the efforts made by Kay (2006), which provides a discussion of pre-service education and technology based on 68 refereed journal articles. The review conducted involved ten key strategies, including: the delivery of a single technology course, mini-workshops, integration of technology in courses, the means through which technology may be used (i.e., multimedia, collaboration amongst pre-service teachers and mentor teachers) and access improvement to software, hardware, and/or support. The basis of evaluating these strategies is their effect on computer attitude, use, and ability. Accordingly, most studies tend to emphasize programmes that integrate only one to three strategies. Another is that the use of computers is more pervasive when four or more strategies are utilized. A more important concern is that several students demonstrated severe limitations in methods: small samples, poor data collection tools, unclear sample and programme descriptions, and lack of or absence of statistical analysis. It indicates a need for more rigorous and comprehensive research in order to evaluate the impact of key technology strategies in the education of pre-service teachers.

Kay's work (2006) is relevant to this research as it deals with the strategies based on their effects on computer attitude and use, leading to the effectiveness or otherwise of the use of e-portfolios in evaluating pre-service teacher education. It can thus help in enabling inferences and analyzes on the use of e-portfolios in pre-service teacher evaluation in Saudi Arabia. Figure 7.5 shows how e-portfolios work towards achieving quality standards:

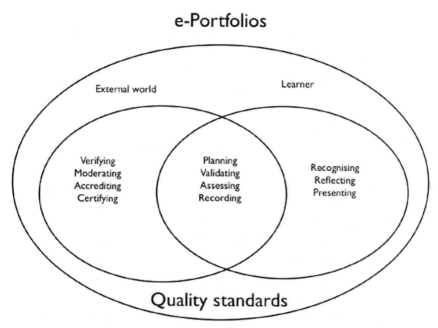

FIGURE 7.5   The use of e-portfolios for raising quality (Source: Attwell, 2010).

Some of the performance indicators posed by e-portfolios include promotion, support, innovativeness, and capacity to engage students in solving authentic problems with the use of electronic tools and resources. Some of the artifacts demonstrating this standard include career objective, inventory of learning styles, and teaching philosophy. In their pursuit to maximize context learning, teachers engage themselves in designing, developing, and evaluating authentic learning experiences and assessment integrating modern tools and resources. Performance indicators leading to this are directed towards designing relevant learning experiences incorporating electronic tools and resources in the promotion of student learning and creativity. Also, developing technology-specific learning environments allowing learners to become active participants in establishing their own academic goals, managing their own learning, and evaluating their own progress. Artifacts in which this standard is demonstrated include: a lesson plan, Web Quest, and software evaluation (Nolan et al., 2011).

It must be emphasized that in the current age of digital learning, teachers represent themselves as innovative professionals in a global and technological society as they demonstrate knowledge, skills, and work processes. This is exhibited by performance indicators, including the demonstration of fluency in technology systems and transfer of current knowledge to new technologies, as well as collaboration with stakeholders with the use of digital tools and resources in an aim to support learner innovation (Nolan et al., 2011). This work is relevant to the topic being investigated as it highlights specific tasks being undertaken in the adoption of digitization in education (i.e., using e-portfolios to enhance planning and learning skills).

Vannatta et al. (2001) emphasize the occurrence of changes in how pre-service teachers think about technology and their corresponding use of technology infusion alongside the role of technology in learning. Findings from an evaluation study involving pre-service teachers are reported, embodying collaboration between teams of teachers and K-12 teachers. Technology is infused in the respective teaching contexts of these teachers whilst creating links between these contexts. Each team in the study pursued constructivist teaching through hands-on experiences with computer technology, two-way interactive videoconference activities, and field experiences in classrooms with technological tools. The findings of the study indicate that a change of view is exhibited by pre-service teachers, from thinking about teaching and learning technology, to thinking about using technology to support students' learning (Vannatta et al., 2001). This study is helpful to the research being undertaken as it highlights the benefits that can be enjoyed by teachers in their use of e-portfolios.

According to Barrett (2007), a culture of evidence must be organized by a learning organization in its pursuit to effectively use portfolios for assessment. The artifacts placed by a learner in an e-portfolio is not the only means that measures evidence in such portfolios, but also the accompanying rationale provided by the individual - such as the notion as to why these artifacts constitute evidence of achievement of specific goals, standards, or outcomes. Moreover, a claim is not substantiated by merely having a learner state that their artifacts are evidence of achievement. The evidence is sometimes validated by some trained reviewers, wherein a well-developed rubric is used alongside identifiable criteria. A simple formula representing this process is stated as: Evidence = Artifacts + Reflection (Barrett, 2007). Barrett provides a fruitful discussion that is useful for this study, since he has highlighted the process in which e-portfolios may be utilized in the learning community.

7.11   SCHOOL IMPROVEMENT THROUGH E-PORTFOLIOS

Related to the discussion of school improvement through e-portfolios is the work of Ring and Foti (2003) in which the authors' assertions are that a problem existing in the educational system is the lack of responsibility on the part of teachers to provide students with a range of communication skills that today's standardized tests have not tested. This is in addition to attending to reading, writing, and other skills that the current standardized tests measure. These communication skills include:

- Communicating effectively with new technologies
- Understanding necessary methodologies to respond to ill-instructed and complex problems.
- Developing awareness where analyzes are required, overlapping cultural, parochial, and disciplinary boundaries.
- Establishing a professional philosophy with a well-thought out rationale.

The rationale behind focusing on these skills is that students will one day compete for jobs and school placements with several students that are adept at modern communication skills and strategies, and it is the teacher's duty to ensure that they compete effectively. In this regard, the development of e-portfolios can potentially bridge the gap between teacher-directed learning and student-centered learning. Moreover, the e-portfolio approach is said to prepare pre-service teachers to utilize technology in order that they may be able to communicate their proficiency in their teacher education programmes (Ring and Foti, 2003).

The work of Wilhelm et al. (2006) indicates the use of e-portfolios in teacher education programmes as a means to provide evidence of the growth and development of pre-service teachers. This electronic version of portfolios is implemented on a widespread basis as programmes are better able to integrate technology into the teacher education curriculum.

The authors describe the process of implementing e-portfolios in three different universities in the United States, describing findings from the first three years of implementation. E-portfolios are considered a natural fit with the reform in teacher education, which is described as standards-based. As schools and implementing bodies define and refine these standards, growth became more appealing with the use of electronics in preparing portfolios (Wilhelm, et al., 2006).

Collaboration with other teachers, students, administrators, and parents is required in a teaching career. In a survey conducted by Jafari and Kaufman (2006), it is pointed out that students were slightly positive about whether the process of portfolio creation is collaborative, and slightly negative about the sufficiency of their opportunities in viewing peers' portfolios. These findings suggest that more peer and faculty reviews would be beneficial to students, with the goal of creating a 'portfolio culture' through the development of a kind of learning environment that promotes care, richness, and intense expectations. In relation to school improvement, it is expected that students would picture themselves as advocates of e-portfolios when they became teachers. In actuality, only few plan to advocate for e-portfolios for student or teacher assessment (Jafari and Kaufman, 2006). Moreover, Jafari and Kaufman (2006) conclude that e-portfolios bear much promise in the improvement of learning and assessment in teacher education. In order to realize this promise, there is a need for new technology that is both simpler and less cumbersome. Technology in this sense must rest in the background and must never be the focus of improving learning and assessment in teacher education.

It is indicated in the study by Maruszczak (2008) that there is a generally favorable perception amongst teachers with regard to the role of school-wide learning expectations and rubrics, as instruction and curriculum are shaped. The school-wide rubrics are used by the teachers in the study as a tool in their classroom assessments. It is likewise emphasized that teachers must be clear about their curriculum and provides an alignment of their curriculum to assist students in mastering essential learning. It is also pointed out that the design and implementation of portfolio tasks have impacted teachers' practice, whereby they are now creating tasks specifically for the purposes of portfolios. It is indicated that teachers use common assessments frequently, but they do not look at students' work frequently to calibrate the school-wide rubrics or utilize the common assessment results to facilitate systematic information of the learning of an individual or group. Moreover, teachers in the study were not utilizing common assessment results in order to help improve the curriculum strategies of the school in a collaborative manner (Maruszczak, 2008). It must be noted that student reflection plays an instrumental role in changing some teachers' practices, as suggested in the study. Student reflection regularly occurs in classrooms by conducting formative reflections for each portfolio. The quality of reflections has improved over time for these students since specific rubrics are provided by the schools on their customized portfolio system (Maruszczak, 2008). This study by Maruszczak is relevant to this research as it highlights how e-portfolios may be used to improve student learning, teacher performance, and the school curriculum in general.

7.12 POINTS OF DIFFERENCE FROM TRADITIONAL PORTFOLIOS

E-portfolios have brought a new concept to the use of portfolios by taking advantage of technological change, and sharing the same conceptual context with traditional portfolios (Barrett and Knezek, 2003; Strudler and Wetzel, 2005). However, they still differ in some major point and characteristics that have been identified through the literature (Challis, 2005; Abrami and Barrett, 2005; Strudler and Wetzel, 2005), the differences are:

- Managing e-portfolios are easier than traditional portfolios through performing tasks such as: search, retrieval, manipulation, refining and reorganizing.
- Using e-portfolios reduces effort and time in comparison with traditional ones.
- E-portfolios are considered more comprehensive and rigorous.
- E-portfolios are able of using more extensive materials due to the electronic features and their advantages.
- E-portfolios can include pictures, sound, animation, graphic design and video.
- E-portfolios are cost effective to distribute and allow fast feedback.
- E-portfolios are directly accessible and easy to share with peers, supervisors, parents, employers and others.
- E-portfolios are capable of having different organizational structures that are not linear or hierarchical.

7.13   ELECTRONIC PORTFOLIO ADOPTION AND IMPLEMENTATION

In order to successfully implement e-portfolios, different factors should be present. Students need to be clearly introduced to the concept of e-portfolios and their purpose and the concept should be linked to the curriculum and programme goals, thus they should be motivated towards using and constructing their own e-portfolios (Chang, 2001; Klenowski et al., 2006; Wetzel and Strudler, 2005). Students need to have clear understanding before and during the use of e-portfolios of what type of evidences are needed and how many pieces they should include, what are the requirements for reflection and self-assessment and how is the portfolio going to be assessed and what mark is going to be addressed to it (Canada, 2002; Chang, 2001; Carliner, 2005). To have a successful implementation, students should be motivated when constructing their own portfolios by enabling student decision-making, also students' need to be assured that they have ownership of their portfolios (Al Kahtani, 1999; Chang, 2001; Tosh et al., 2005). Moreover, they need to have public access to and recognition of their work over the Web. To facilitate this process for students, they can be introduced to past examples of e-portfolios that have been made by their peers and to reveal their efficiency in making learning gains (Abrami and Barrett, 2005). Also, students must be educated in searching, using and exploiting the electronic resources found on the Web or other resources to complete their portfolios (Wetzel and Strudler, 2005).

If the view is to be directed towards portfolio systems, they need to have different criteria to be considered successful (Ahn, 2004; Wetzel and Strudler, 2005; Kimball, 2005) including:

- The need to be flexible towards students' needs.
- They should protect the privacy of those working with portfolios.
- A system needs to 'stand-alone,' without constant interference from academic staff.

Finally, educational institutions using e-portfolios need to be aware that implementing e-portfolio systems is a long-term effort that can present successful outcomes if time is spent in the initial stages before it becomes an available programme or institutionalized (Wetzel and Strudler, 2005; Ahn, 2004).

In order for any institution to have a successful design and creation of e-portfolio systems, they should adopt the following factors according to Yancey (2001) introduced in a series of questions:

- What is/are the purpose/s of having e-portfolios?
- How familiar is the portfolio concept? Is the familiarity considered a plus or a minus?
- Who are the persons that want to create an electronic portfolio, and why?
- Who are the persons that want to read an electronic portfolio, and why?
- Why electronic? And is sufficient infrastructure (resources, knowledge, and commitment) available for the electronic portfolio?
- What processes are needed: What resources are presumed?
- What component does the model assume or include towards faculty development?
- What are the required skills that students need to develop?
- What curricula enhancement does the model assume or include?
- How will the portfolio be introduced?
- How will the portfolio be reviewed?

7.14 BARRIERS TO IMPLEMENTATION

There exist a number of barriers to e-portfolio implementation that have been identified through the literature, the following list summarizes these points (Canada, 2002; Lorenzo and Ittleson, 2005; Sherry and Bartlett, 2005; Tosh et al., 2005; Wetzel and Strudler, 2005).

- E-portfolio systems require the presence of adequate hardware and software.
- Using e-portfolios requires a specific level of computer and technological skills amongst students and staff.
- Technical problems either as hardware or software that are related to e-portfolios.
- The need for technical support when problems are encountered.
- Hardware maintenance.
- Adequate storage space and server reliability.
- Using e-portfolios have demands on staff time.

- Using students' time efficiently.
- Overcoming issues of ownership and intellectual property.
- Problems that are related to security and privacy of data.
- Lack of features or of control over those features.
- Access and permission controls.
- Absence of common standards between different electronic portfolio systems.

7.15   E-PORTFOLIOS: EDUCATIONAL AND PEDAGOGICAL CONSIDERATIONS

The concept of portfolios is based on constructivist philosophy, which is useful for those thinking of implementing portfolio assessment (Abrami and Barrett, 2005; Chang, 2001; Klenowski et al., 2006; Meeus et al., 2006; Strudler and Wetzel, 2005).

According to (Klenowski et al., 2006) constructivism is:

"The knowledge that is constructed through activities such as participatory learning, open-ended questioning, discussion and investigation. Facilitation helps learners construct their own schema for internalizing information and organizing it so that it becomes their own."

Portfolio pedagogy, circles the point of having evidences, reflection that makes learning experiences and making connections between different ideas and actions. Thus it is evident that portfolios seek to encourage and involve students in becoming dynamic participants in their own learning through being the users and the authors of their own portfolios (Kimball, 2005). The technological skills necessary for students to interact with e-portfolios should not be seen as a separate set of skills, but rather as a way of enhancing learning and teaching, also the focus should be made on learning not on the used technology to facilitate learning. Working and interacting with e-portfolios will only be worthwhile if and when they are used to advance important activities in academic life such as reflections by academic staff towards students' e-portfolio construction (Ehrmann, 2006). It has been found that students are capable of making connections between different aspects of their lives that have helped them to form their social identities within their discipline of study through the use of e-portfolios. The use of e-portfolio systems should include consideration of assessment, and purposes of assessment as e-portfolios can be used for formative purposes that help facilitate students' learning and for summative purposes to assess how much a student has learnt over a course of study (Beck et al., 2005; Klenowski et al., 2006).

The Table 7.1 is structured according to Barrett (2004) and it summarizes the differences between using e-portfolios for summative (of learning) and formative (for learning) assessment.

TABLE 7.1 E-Portfolios for Summative/Formative Assessment

Summative Assessment	Formative Assessment
The reasons for having portfolios are prescribed by the institution.	The reason for having portfolios is agreed upon with learner.
Artifacts and activities are overviewed by the institution to determine results of education.	Artifacts are selected by learners in order to tell the story of their learning.
Portfolios are time limited as they are developed at the end of a class, term or programme.	Portfolios are time flexible and they are maintained on an on-going basis throughout the class, term or programme.
Portfolios are scored based on specific and predefined procedures and rules and quantitative data is collected for external audiences.	Portfolios are reviewed with learners and used to provide feedback and reflection to improve learning.
Portfolio structure is based around a set of outcomes, goals or standards.	Portfolio organization is determined by learners or discussed with teachers.
In some cases used for making high stakes decisions.	Rarely used for high stakes decisions.
It includes what has been learned from past to present.	It includes the needs for present to future learning.
It requires external motivation.	It fosters external motivation-engages the learner.
The audiences are external and the learner cannot choose his/her audience.	The audiences could include peers, family, friends; the learner can choose his/her audience.

7.16   E-PORTFOLIO SUPPORT AND TECHNICAL CONSIDERATIONS

In order to have successful implementation of e-portfolio systems, it is important to have a proper plan and pay special consideration to the following points (Barrett, 2000; Heath, 2002; Lorenzo and Ittleson, 2005; McNair and Galanouli, 2002):

- Is the e-portfolio system going to be chosen or designed?
- Is there a clear justification for the chosen technical solution?
- How will the system users be identified?
- How will the e-portfolio audiences be identified?
- What technology skills will be required by the staff and students who will use and interact with the system?
- How will the available financial, hardware and software resources be defined?

When choosing to adopt e-portfolio systems, there are four different options that developers should consider: (Lorenzo and Ittleson, 2005; Strudler and Wetzel, 2005)

1. Adopting in-house designed e-portfolio system, in order to meet the institutions' specific requirements.
2. Adopting open source systems that meet requirements or can be customized to meet more requirements.
3. Commercial systems that the institutions are willing to purchase.
4. Using common tools such as Microsoft Office applications, Internet browsers and so on, to design a portfolio that can then be uploaded to the Web or saved to CD-ROM, DVD, or distributed on memory flash disks.

Different research studies agree that regardless of the chosen e-portfolio system, there should exist some practical and technical requirements that need to be met in order to ensure technical success and minimize any fatigue, contaminants and challenges that might face the users while working and interacting with their chosen solution of e-portfolio system (Abrami and Barrett, 2005; Barrett, 2000; Barrett and Knezek, 2003; Challis, 2005; Lorenzo and Ittleson, 2005; Questier and

Derks, 2006; Siegle, 2002; Tosh et al., 2005). The requirements agreed upon are:

- to provide a structure towards organizing content.
- to be capable of tracking student progress.
- to provide a way of archiving and storing large amounts of data.
- to provide a way of retrieving data.
- to provide a way of linking artifacts to reflective pieces.
- to provide a way of showing assessment results incorporation into the e-portfolio.
- to provide different ways of publishing the portfolio, so it can be produced for different audiences.
- providing flexibility towards organizing and structuring data in e-portfolios.
- to ensure the availability of technical standards necessary so the system can communicate reliably with other systems.
- to enable the system to support and recognize a wide range of file formats used by users.
- to be capable of providing security and access permission for users with respect for their roles.
- the system should be scalable and ensure that a large volume of users can access the system.
- the system should ensure maximum accessibility and usability for users of all levels of skill.
- To define what kinds of technical support are or will be available for users.
- to ensure the protection of privacy and intellectual property of users.
- to specify the time that e-portfolios will exist in the system.
- to ensure portability, so that students can take their e-portfolio to another institution or choose to maintain it on their own.

Having the e-portfolio system bought, developed or customized will definitely affect the educational process and activities in the institution by providing a new pedagogical approach towards educational processes. However, if the previous points and concerns are addressed within e-portfolios systems, they will surely meet the needs of a larger number of students and staff in any institution, which will ensure better adoption and educational results of the educational processes and activities (MacDonald et al., 2004).

7.17   E-PORTFOLIOS QUALITY STANDARD

Back in the year 1997, the Canadian Labor Force Development Board (CLFDB) initiated a research study on the use of "electronic learning record." This study resulted in raising a major concern and setting a defined demand towards e-portfolio standards and quality. The study stated that in order for an e-Portfolio to make a contribution to increasing the effectiveness, efficiency and equitability of a labor force development system, it must consider including an instrument (format, content), and a process (access, development, maintenance) and a utility for all the labor market partners that meet minimum standards for effectiveness, efficiency and equity. This concern was addressed by a specialized educational company under the name of FuturEd Inc. that worked with CLFDB, and created a recommendation for e-portfolio standards to address Human Resources Development policy goals.

The recommendation points for quality standards are:

1. The e-Portfolio should list and describe skills and knowledge in a way that is recognized and respected by all the labor market partners.
2. The e-Portfolio should have the capacity to be a complete inventory of skills and knowledge acquired by the individual regardless of where they were acquired.
3. An individual should develop and own his/her e-Portfolio. Some people may require informed assistance to achieve this. The use of the e-Portfolio and any changes to it should be completely controlled by the individual.
4. The content of the e-Portfolio should be current, accurate and verifiable.
5. The e-Portfolio should allow flexibility to accommodate unique or industry-specific skills.
6. The e-Portfolio should follow a standardized format.
7. The e-Portfolio content and format should link to existing and developing labor market exchange systems.
8. The e-Portfolio and its development process should be relatively simple and straightforward.

9. The development and use of the e-Portfolio for any and all users should be barrier-free; that is to say, social identity, disability and geography should not be barriers to individuals.

10. The development and content of an e-Portfolio should be bias-free.

11. An e-Portfolio should not create barriers; for example, a person who does not have an e-Portfolio is not discriminated against for the lack of one, or for the skills revealed.

7.18 E-PORTFOLIOS DESIGN AND DEVELOPMENT STANDARD

Designing and developing standards towards what to include and how to format the e-portfolio depends heavily on the intended purpose of use and technology. Different studies have defined models and included detailed steps on how to design and develop e-portfolios. However, it is important to consider that an e-portfolio development merges two different processes of Multimedia development and portfolios. Thus when developing an e-portfolio it should be understood that an equal attention should be paid to these complimentary processes as they are both considered essential for effective e-portfolio development. The two most renowned approaches defined by the literature are:

- Helen C. Barett Approach (Barrett, 2000); and
- Kadriye O. Lewis and Raymond C. Baker Approach (Lewis and Baker, 2007).

7.19 HELEN C. BARETT APPROACH

This approach was designed by Helen C. Barett and it depended on the framework designed by Danielson and Abrutyn (1997). The enhanced framework by Helen C. Barett relied on producing two different processes for portfolio and multimedia development. In previous studies related to traditional portfolio, Robin Fogarty, Kay Burke, and Susan Belgrad (1994, 1996) have identified 10 options for portfolio development, further defining the stages and increasing the quality of the portfolio process, and those portfolio development processes are:

1. PROJECT purposes and uses
2. COLLECT and organize
3. SELECT valued artifacts
4. INTERJECT personality
5. REFLECT metacognitively
6. INSPECT and self-assess goals
7. PERFECT, evaluate, and grade (if you must)
8. CONNECT and conference
9. INJECT AND EJECT to update
10. RESPECT accomplishments and show pride

In terms of e-portfolio development Helen C. Barett identifies that the first part of e-portfolio development should include the following processes:

1. Collection: The portfolio's purpose, audience and future use of artifacts will determine what artifacts to collect.
2. Selection: Selection criteria form materials to include should reflect the learning objectives established for the portfolio. These should follow from national, state or local standards and their associated evaluation rubrics or performance indicators.
3. Reflection: Include reflections on every piece in your portfolio and an overall reflection.
4. Projection/Direction: Review your reflections on learning, look ahead, and set goals for the future.
5. Connection: In this phase the creation and publication of e-portfolio hypertext links in order to enable the feedback from others. (This process according to Helen C. Barett, can occur before or after the projection/direction stage).

In terms of multimedia development, the framework suggests the following steps:

1. Assess/Decide: In the first stage there should be focus on needs assessment of the audience, adding presentation goals and defining appropriate tools for e-portfolio presentation. This stage is subdivided into further details as defined by (Helen C. Barett, 2000) that are:

a. Technology Skill Levels

1. Limited experience with desktop computers but able to use mouse and menus and run simple programs
2. Level 1 + proficient with a word processor, basic email and Internet browsing, can enter data into a predesigned database
3. Level 2 + able to build a simple hypertext (nonlinear) document with links using a hypermedia program.
4. Level 3 + able to record sounds, scan images, output computer screens to a VCR and design an original database
5. Level 4 + multimedia programming or HTML authoring, can also create QuickTime movies live or from tape, able to program a relational database

b. Technology Available

1. No Computer
2. Single computer with 16 MB RAM, 500 MB HD, no AV input/output
3. One or two computers with 32 MB RAM, 1 + GB HD, simple AV input (Such as QuickCam)
4. Three or four computer, one of which has 64 + MB RAM, 2+GB HD, AV input and output, Scanner, VCR, Video camera, high-density floppy (Such as Zip Drive)
5. Level 4 + CD-ROM recorder, at least two computer with 128+ MB RAM; digital video editing hardware and software > Extra GB+ storage (Such as Jaz drive)

The technological specification table is considered old with the current technological advancement, thus the levels should be edited to match the current advancement, and the level's can be defined based on the criteria defined by the institution willing to use e-portfolios.

2. Design/Plan: This next stage should focus on the design and organization of the presentation that can include the following important considerations:

- Determining the appropriate content
- Determining Software
- Determining Storage medium

- Determining Presentation sequence
- Constructing flow charts and writing storyboards

3. Develop: This stage includes gathering materials that are going to be included into the presentation, next should be organizing them into sequences using hyperlinks that are produced using appropriate multimedia authoring software.

4. Implement: In this stage the author (teacher/ developer) presents the e-portfolio to the intended audience.

5. Evaluate: This is considered the final stage as the focus should be on evaluating the presentation's effectiveness in light of its purpose and the assessment context.

Based on the merge between Multimedia Development Process and the Portfolio Development Process, the following five stages of e-portfolio Development emerge. The Table 7.2 shows the issues that are going to be address at each stage of this process.

7.20   KADRIYE O. LEWIS AND RAYMOND C. BAKER APPROACH

This approach presents a new framework for developing e-portfolios, it has some common similarities with Barett (2000) approach, and it differs in some major points due to the technological advancement that affected the use and concept of e-portfolios. The current framework consists of the following steps:

1. Define and clarify the scope and purpose of the e-portfolio: In order to achieve this step the framework suggested the Table 7.3 shows the questions to be answered as they are going to drive the process of creating e-portfolios.

2. Creating a flowchart that will give a visual representation and illustration of different e-portfolio aspects such as (sequence, organization, and navigation) of the content of the e-portfolio. This step is considered and important task to be finished as the framework design relates on using flow chars. By creating a flowchart it helps the author to visualize and think ahead about what the e-portfolio will look like when it is completed. Using this steps it has been found that flowchart helps the authors create the direction, structure,

TABLE 7.2 Stages of E-Portfolio Development

Electronic Portfolio Development Stages		
Portfolio Development	**Electronic Portfolio Development**	**Multimedia Development**
Purpose and Audience	1. Defining the Portfolio Context and Goals	*Decide Assess*
Collect Interject	2. The Working Portfolio	*Design Plan*
Select Reflect Direct	3. The Reflective Portfolio	*Develop*
Inspect Perfect Connect	4. The Connected Portfolio	*Implement Evaluate*
Respect	5. The Presentation Portfolio	*Present Publish*

Source: Barrett (2000).

TABLE 7.3 Questions Drive the Process of Creating E-Portfolios (Barett, 2000)

No.	Topic	Questions
1	Purpose	What is the purpose(s) of the portfolio?
2	Audience	Who are the target audiences?
3	Content	What work samples or artifacts will be included?
4	Design	What design processes will be used during the development of the portfolio?
5	Management	How will time and materials be managed in the development of the portfolio?
6	Communication	How and when will the portfolio be shared with pertinent audiences?
7	Evaluation	When and how should the portfolio be evaluated and by whom?
8	Technology	What is the best technology available for this purpose?

and sequence for the portfolio content and define how the different parts of the e-portfolio will relate to each other.

3. This step is concerned with creating or selecting representative artifacts for each item or category defined in the flowchart. In the current context different variety of media types, including text, images, and audio, video, animation, and Internet technologies are available for this step. The proposed framework suggests that the nature of the artifact usually implies the appropriate media type to

be used. The Table 7.4 indicates some examples of content and the appropriate media type to best demonstrate it.

4. In this step users need to position their artifacts and learning objects onto the e-portfolio system. Currently with the wide spread use of Internet applications and services most software are considered easy to use as they also have tutorial in order to show how the system and its functionalities are used. The time and effort spent on learning the effective use of e-portfolios and they way to develop them depends on the previous experience of users with working with similar services found on the Internet.

5. This step is considered the final step as the e-portfolio is published to an appropriate medium that allows viewing. In most e-portfolio systems it is possible for e-portfolios to be published to several different formats as this choice is mainly dependent on the viewing audience as (promotion and tenure committee, instructor, or employer).

The same approach of Kadriye O. Lewis and Raymond C. Baker has adopted an evaluation methodology for evaluating e-portfolios rubrics by setting the following standards that has been used by their research studies. The Table 7.5 shows and defines each criterion.

The previously discussed methods and evaluation standards are widely used in creating e-portfolios for different purposes. The choice and justification for choosing one of these methods is all dependent on the nature, context and use of e-portfolio.

Many other standards are available too, and many will be defined in the future as the information and communication tools are in continuous development and use. But it is important to drive the use of these tools by

TABLE 7.4   Media Types Artifacts

No.	Content	Media Type
1	Teaching Philosophy/Syllabus	Text
2	Didactic Lecture	PowerPoint Presentation with voiceover, Streaming video
3	Evaluation Forms, Certificates of recognition	Scanned documents (e.g., pdf format or images)
4	Images, Photographs, Graphics	Image files (e.g., Jpeg, gif, png)

TABLE 7.5 Criteria to Evaluate the E-Portfolio Rubrics

Criterion	Definition/Standard
Design	Well organized; unique/imaginative approach to design; highly visual; excellent use of design, audio, and text elements.
Navigation	All of the portfolio navigation links and all sections (standards, artifacts, and reflections) connect back to the main table of contents. All external links to all connecting websites connect.
Technical	Links work; content includes audio/video, digital images, slide show, or PDF docs; documents are error free; portfolio has been converted to CD or posted to a website.
Layout and Text Elements	The e-portfolio is easy to read with visual organization of information using fonts, point size, bullets, italics, bold, and indentations for headings and subheadings. The layout uses horizontal and vertical white space appropriately. The background and colors enhance the readability of text.
Artifacts	Artifacts are related to reflections; categories are complete; good variety of artifacts included; excellent quality digital or video images and sound.
Caption	Each artifact is accompanied by a caption that clearly and accurately explains the importance of that particular work including title, author, date, standard addressed, and description of the importance of the artifact.
Reflection	Reflections are clearly related to artifacts, demonstrate growth over time, are well written, and reveal depth and breadth of experiences.
Multimedia	All audio and video files effectively enhance reflective statements, create interest, and are appropriate examples for one or more standards. Background audio does not overpower the primary audio. Creativities and original ideas enhance the content of the e-portfolio.
Writing Mechanics	The text has no errors in grammar, capitalization, punctuation, and spelling.

Source: Kadriye O. Lewis and Raymond C. Baker (2007).

educational standards and theories and not to be driven by the technological features only.

7.21 CONCLUSION

This chapter presented portfolio's and discussed some important issues related to their use and benefits. Later the chapter presented the use of e-portfolios, In addition the benefits and uses have been outlined with different implementation of e-portfolio in educational context. The use of e-portfolios in teacher effectiveness and the evaluation of pre-service teachers have been presented and outlined. The presented information on e-portfolio, presented the effect of e-portfolio on raising the quality of teachers performance and the improvements it has on schools. The chapter presented e-portfolio's adoption and implementation and the expected barriers of implementation. Moreover, e-portfolios educational and pedagogical considerations have been presented too. E-portfolio's support and technical considerations have been outlined and the used quality standards have been defined. In terms of standards for quality, two popular standards that are Hellen C. Barett and Kadriye O. Lewis, and Helen C. Baker have been presented and briefly discussed as they are used for constructing e-portfolios.

KEYWORDS

- barriers
- criteria
- development standards
- e-portfolio
- education
- educational technologies
- improvement
- learning objects
- pedagogies

REFERENCES

1. Abrami, P. C., Barrett, H. (2005). Directions for research and development on electronic portfolios. *Canadian Journal of Learning and Technology,* 31(3), online version.
2. Adamy, P., Millman, N. B. (2009). Evaluating electronic portfolios in teacher education. US: Information Age Publishing Inc.
3. Ahn, J. (2004). Electronic portfolios: Blending technology, accountability and assessment. Retrieved 5 June 2012 from: http://thejournal.com/articles/16706.

4. Al Kahtani, S. (1999). Electronic portfolios in ESL writing: An alternative approach. Computer Assisted Language Learning, 12(3), 261–268.
5. Attwell, G. (2010). Rethinking e-portfolios. Retrieved 19 February 2012 from: http://www.pontydysgu.org/projects/taccle/.
6. Barrett, H., Knezek, D. (2003). *E-portfolios: Issues in assessment, accountability, and preservice teacher preparation.* ERIC# ED476185, Education Resources Information Center.
7. Barrett, H., Knezek, D. (2003, April 22) E-portfolios: Issues in assessment, accountability and preservice teacher preparation. Paper presented at the American Educational Research Association Conference, Chicago, IL.
8. Barrett, H., Wilkerson, J. (2004). Conflicting paradigms in electronic portfolio approaches. Retrieved June 24 2012 from: http://electronicportfolios.org/systems/paradigms.html.
9. Barrett, H. (2000). Electronic teaching portfolios: Multimedia skills + portfolio development = powerful professional development. Retrieved 22 April 2012 from: http://www.electronicportfolios.com/portfolios/site2000.html
10. Barrett, H. (2004). Electronic Portfolios as Digital Stories of Deep Learning: Emerging Digital Tools to Support Reflection in Learner-Centered Portfolios. [Retrieved January 21, 2005 from: http://electronicportfolios.org/digistory/epstory.html]
11. Barrett, H. C. (2007). Researching electronic portfolios and learner engagement: The REFLECT initiative. *Journal of Adolescent and Adult Literacy 50(6), 436–449.*
12. Barrett, H. C. (2011). *Balancing the two faces of e-portfolios.* Retrieved 20 Feb 2012 from: http://electronicportfolios.com/balance/Balancing2.htm,
13. Barrett, Helen (2000, April). Create Your Own Electronic Portfolio. Learning & Leading with Technology Vol. 27, No. 7, pp. 14–21
14. Bartlett, A. (2002). Preparing pre-service teachers to implement performance assessment and technology through electronic portfolios. *Action in Teacher Education,* 24(1), 90–97.
15. Bartlett, A. (2009). A Five-Step Model for Enhancing Electronic Teaching Portfolios. In P. Adamy and N. Milman (ed.) *Evaluating electronic portfolios in teacher education.* USA: Information Age Publishing Inc.
16. Beck, R. J., Livne, N. L., Bear, S. L. (2005). Teachers' self-assessment of the effects of formative and summative electronic portfolios on professional development. *European Journal of Teacher Education,* 28(3), 221–244.
17. Brown, J. O. (2002). Know thyself: The impact of portfolio development on adult learning. *Adult Education Quarterly,* 52(3), 228–245.
18. Burke, Kay; Fogarty, Robin; Belgrad, Susan (1994). The Mindful School: The Portfolio Connection. Palatine: IRI/Skylight Training & Publishing
19. Buzzetto-More, N. A., Alade, A. J. (2006). Best practices in e-Assessment. *Journal of Information Technology Education,* Vol.5.
20. Cambridge, B. L. (2001). Electronic portfolios as knowledge builders. In B. L. Cambridge, S. Kahn, D. P. Tompkins and K. B. Yancey (eds.) *Electronic portfolios: Emerging practices in student, faculty, and institutional learning* (pp. 1–11). Washington, DC: American Association for Higher Education.
21. Canada, M. (2002). Assessing E-Folios in the Online Class. *New Directions for Teaching and Learning,* 91, 69–75.

22. Carraccio, C., Englander, R. (2004). Evaluating competence using a portfolio: A literature review and web-based application to the ACGME competencies. *Teaching and Learning in Medicine*, 16(4), 381–387.
23. Challis, D. (2005). Towards the mature ePortfolio: Some implications for higher education. *Canadian Journal of Learning and Technology*, 31(3), online version.
24. Chang, C. (2001). Construction and evaluation of a web-based learning portfolio system: An electronic assessment tool. *Innovations in Education and Teaching International*, 38(2), 144–155.
25. Chua, G., Chua, G. (2007). Electronic portfolio: The pedagogy behind the technology in Science projects. In T. Hirashima (ed.) *Supporting learning flow through integrative technologies*. Amsterdam: IOS Press.
26. Danielson, C., Abrutyn, L. (1997). *An Introduction to Using Portfolios in the Classroom*. Alexandria: Association for Supervision and Curriculum Development.
27. Darling, L. F. (2001). Portfolio as practice: The narratives of emerging teachers. *Teaching and Teacher Education*, 17(1), 107–121.
28. Davies, H., Khera, N., Stroobant, J. (2005). Portfolios, appraisal, revalidation, and all that: A user's guide for consultants. *Archives of Disease in Childhood*, 90(2), 165–170.
29. Driessen, E., van der Vleuten, C., Schuwirth, L., van Tartwijk, J., Vermunt, J. (2005 a). The use of qualitative research criteria for portfolio assessment as an alternative to reliability evaluation: *A case study. Medical Education*, 39(2), 214–220.
30. Driessen, E., van Tartwijk, J., Overeem, K., Vermunt, J., van der Vleuten, C. (2005b). Conditions for successful reflective use of portfolios in undergraduate medical education. *Medical Education*, 39(12), 1230–1235.
31. Duque, G. (2003). Web-based evaluation of medical clerkships: A new approach to immediacy and efficacy of feedback and assessment. *Medical Teacher*, 25(5), 510–514.
32. Ehrmann, S. C. (2006). Electronic portfolio initiatives: A flashlight guide to planning and formative evaluation. In A. Jafari and C. Kaufman (Eds.), Handbook of research on ePortfolios (pp. 180–193). Hershey, PA: Idea Group Reference.
33. Hartnell-Young et al. (2007). The Impact of e-Portfolios on Learning. Coventry: Becta [Online]. Available at http://www.jiscinfonet.ac.uk/infokits/e-portfolios/becta-2007.pdf (Accessed: 13 August, 2012).
34. Hauge, T. E. (2006). Portfolios and ICT as means of professional learning in teacher education. Studies in Educational Evaluation, 32(1), 23–36.
35. Hawkins, R. J. (2002). Ten Lessons for ICT and Education in the Developing World.
36. Heath, M. (2002). Electronic portfolios for reflective self-assessment. *Teacher Librarian*, 30(1), 19–23.
37. Heath, M. (2005). Are you ready to go digital? The pros and cons of electronic portfolio development. *Library Media Connection*, 23(7), 66–70.
38. Himpsl, K., Baumgartner, P. (2009). Evaluation of E-Portfolio Software. *International Journal of Emerging Technologies in Learning*, 4(1), 16–20.
39. Huang, Y. C. (2006). *E-portfolios: Their impact of self-directed learning and computer technology skills on pre-service teachers*. Ann Arbour, MI: ProQuest Information and Learning Company. *International Education Journal*, 5(3), 367–380.

40. Jafari, A., Kaufman, C. (Eds.) (2006). *Handbook of research on ePortfolios.* Hershey. PA: Idea Group Reference.
41. Jarvis, R. M., O'Sullivan, P. S., McClain, T., Clardy, J. A. (2004). Can one portfolio measure the six ACGME general competencies? *Academic Psychiatry,* 28(3), 190–196.
42. Kadriye O. Lewis, Raymond C. Baker, 2007, "The Development of an Electronic Educational Portfolio: An Outline for Medical Education Professionals," Journal of Teaching and Learning in Medicine, Volume *19,* issue (2), 139–147, 2007 by Lawrence Erlbaum Associates, Inc.
43. Kay, R. H. (2006). Evaluating strategies used to incorporate technology into preservice education: A review of the literature. *Journal of Research and Technology in Education,* 38(4), 383–408.
44. Kimball, M. (2005). Database e-portfolio systems: A critical appraisal. *Computers and Composition,* 22(4), 434–458.
45. Klenowski, V., Askew, S., & Carnell, E. (2006). Portfolios for learning, assessment and professional development in higher education. *Assessment and Evaluation in Higher Education,* 31(3), 267–286.
46. Lorenzo, G., Ittleson, J. (2005). *An overview of e-portfolios.* Retrieved 14 April 2012 from: http://www.educause.edu/LibraryDetailPage/666?ID=ELI3001
47. Loughran, J., Corrigan, D. (1995). Teaching portfolios: A strategy for developing learning and teaching in preservice education. *Teaching and Teacher Education,* 11(6), 565–577.
48. Love, T., Cooper, T. (2004). Designing online information systems for portfolio-based assessment: Design criteria and heuristics. *Journal of Information Technology Education,* 3, 65–81.
49. Lynch, D. C., Swing, S. R., Horowitz, S. D., Holt, K., Messer, J. V. (2004). Assessing practice based learning and improvement. *Teaching and Learning in Medicine,* 16(1), 85–92.
50. MacDonald, L., Liu, P., Lowell, K., Tsai, H., Lohr, L. (2004). Graduate student perspectives on the development of electronic portfolios. *Tech Trends,* 48(3), 52–55.
51. Maruszczak, J. P. (2008). *The design and implementation of an electronic portfolio assessment system as a high school graduation requirement: A multiple case study of three Rhode Island high schools.* Ann Arbour, MI: ProQuest Information and Learning Company.
52. McNair, V., Galanouli, D. (2002). Information and communications technology in teacher education: Can a reflective portfolio enhance reflective practice? *Journal of Information Technology for Teacher Education,* 11(2), 181–196.
53. Meeus, W., Questier, F., Derks, T. (2006). Open source e-portfolio: Development and implementation of an institution-wide electronic portfolio platform for students. *Educational Media International,* 43(2), 133–145.
54. Milman, N. B., & Kilbane, C. R. (2005). Digital teaching portfolios: Catalysts for fostering authentic professional development. *Canadian Journal of Learning and Technology,* 31(3), online version.
55. Nolan, Jr., J., Nolan J., Hoover, L. A. (2011). *Teacher supervision and evaluation.* NJ: John Wiley & Sons, Inc.

56. Orlich, H., Callahan & Gibson (2004). *Teaching strategies: A guide to better instruction.* New York: Houghton Mufflin.

57. Pearson, D. J., Heywood, P. (2004). Portfolio use in general practice vocational training: A survey of GP registrars. *Medical Education,* 38(1), 87–95.

58. Pecheone, R. L., Pigg, M. J., Chung, R. R., Souviney, R. J. (2005). Performance assessment and electronic portfolios: Their effect on teacher learning and education. *The Clearing House,* 78(4), 164–176.

59. Pinsky, L. E., Fryer-Edwards, K. (2004). Diving for PERLs: Working and performance portfolios for evaluation and reflection on learning. *Journal of General Internal Medicine,* 19(5), 582–587.

60. Questier, F., Derks, T. (2006). Open source eportfolio: Development and implementation of an institution-wide electronic portfolio platform for students. *Educational Media International,* 43(2), 133–145.

61. Ring, G., Foti, S. (2003). Addressing standards at the program level with electronic portfolios. *Techtrends,* 47(2), 28–32.

62. Sherry, A. C., Bartlett, A. (2005). Worth of electronic portfolios to education majors: A 'two by four' perspective. *Journal of Educational Technology Systems,* 33(4), 399–419.

63. Siegle, D. (2002). Creating a living portfolio: Documenting student growth with electronic portfolios. *Gifted Child Today Magazine,* 25(3), 60–65.

64. Smith, K., Tillema, H. (2003). Clarifying different types of portfolio use. *Assessment and Evaluation in Higher Education,* 28(6), 625–648.

65. Smits, H., Wang, H., Towers, J., Crichton, S., Field, J., Tarr, P. (2005). Deepening understanding of inquiry teaching and learning with e-portfolios in a teacher preparation program. *Canadian Journal of Learning and Technology,* 31(3), online version.

66. Stansberry, S. L., Kymes, A. D. (2007). Transformative learning through 'teaching with technology' electronic portfolios. *Journal of Adolescent & Adult Literacy,* 50(6), 488–496.

67. Strudler, N., Wetzel, K. (2005). The diffusion of electronic portfolios in teacher education: Issues of initiation and implementation. *Journal of Research on Technology in Education,* 37(4), 411–433.

68. Tosh, D., Light, T. P., Fleming, K., Haywood, J. (2005). Engagement with electronic portfolios: Challenges from the student perspective. *Canadian Journal of Learning and Technology,* 31(3), online version.

69. Vannatta, R., Beyerbach, B., Walsh, C. (2001). From teaching technology to using technology to enhance students learning: Preservice teachers' changing perceptions of technology infusion. *Journal of Technology and Teacher Education,* 9(1), 105–127.

70. Wade, A., Abrami, P. C., Sclater, J. (2005). An electronic portfolio to support learning. *Canadian Journal of Learning and Technology,* 31(3), online version.

71. Wade, R. C., Yarbrough, D. B. (1996). Portfolios: A tool for reflective thinking in teacher education? *Teaching and Teacher Education,* 12(1), 63–79.

72. Wetzel, K., Strudler, N. (2005). The diffusion of electronic portfolios in teacher education: Next steps and recommendations from accomplished users. *Journal of Research on Technology in Education,* 38(2), 231–243.

73. Wilhelm, L., Puckett, K., Beisser, S., Meredith, E., Sivakumaran, T. (2006). Lessons learned from the implementation of electronic portfolios at three universities. *Techtrends,* 50(4), 62–71.
74. Wilson, E. K., Wright, V. H., Stallworth, B. J. (2003). Secondary preservice teacher's development of electronic portfolios: An examination of perceptions. *Journal of Technology and Teacher Education,* 11(1), 515–527.
75. Yancey, K. B. (2001). General patterns and the future. In B. L. Cambridge, S. Kahn, D. P. Tompkins & K. B. Yancey (Eds.) *Electronic portfolios: Emerging practices in student, faculty, and institutional learning* (pp. 83–87). Washington, DC: American Association for Higher Education.
76. Young, J. R. (2002). 'E-portfolios' could give students a new sense of their accomplishments. *Chronicle of Higher Education,* 48(26), A31–A32.
77. Zeichner, K., Wray, S. (2001). The teaching portfolio in US teacher education programs: What we know and what we need to know. *Teaching and Teacher Education,* 17(5), 613–621.

PART 4:

E-SYSTEM

CHAPTER 8

PARALLEL PROCESSING FOR VISUALIZING REMOTE SENSING IMAGERY DATA SETS

SAFA AMIR NAJIM[1] and ALAA AMIR NAJIM[2]

[1]*Assistant Professor in Software Engineering, Computer Science Department, Science College, Basrah University, Basrah, Iraq, Tel: +9647712043595, E-mail: safanajim@gmail.com*

[2]*Lecturer in Graph Theory, Mathematical Department, Science College, Basrah University, Basrah, Iraq, E-mail: alaanajim2014@gmail.com*

CONTENTS

In this chapter, we present a parallel method to visualize remote sensing imagery data sets and measure their efficiency on the graphics processing unit (GPU). Visualization of remote sensing imagery data sets is a common challenge task in the dimensionality reduction (DR). The requirement to accelerate the projection process and efficiency measurement of the visualization comes from the large size of the data sets. We have implemented the trustworthy stochastic proximity embedding (TSPE) method on GPU to speed up its projection process. To measure the efficiency of the visualization, the parallel codes of the two well-known metrics in this field namely, correlation and residual variance are introduced. The results showed that the high computational efficiency of the GPU helped to reduce the time spent on processing the results and computing their efficiency.

8.1   PARALLEL PROCESSING BY GRAPHICS PROCESSING UNIT

Nowadays, the implementation time has become a key factor for the success of the software. Not too long ago, the focus was on the central processing unit (CPU) in order to speed the computation, where it was required to wait at least 12 months for minor changes to the current speed. Although multiple CPUs have been introduced instead of a single CPU to implement the tasks in parallel, the efficiency of the speed has been limited to simple applications because of the vast amount of data.

Over the past few years, the processing power and memory bandwidth of the new generation of graphics cards have become significantly better than the CPU. In addition to their capability of displaying graphics, the GPU can accelerate general-purpose computations. It is useful for information extraction, in visualization, in telecommunications, and in many scientific fields, including biology and chemistry, and because its computational power goes far when compared with the CPU (Sanders and Kandrot, 2011). Figure 8.1 shows that the exponential increase in the performance of the GPU in the last nine years is better than that of the CPU.

8.1.1 GPU AND CPU

The tendency to use the GPU is due to the following reasons: Firstly, the technology of the CPU is not capable of managing a large scale of computations, and we can say that the improvement in the CPU technology has reached a stable scale. That means that the CPU alone in a computer system cannot provide what people demand. Secondly, the GPU's speed is 10–100 times faster than CPU, which makes it suitable to implement the challenge computational problems. Thirdly, complex computation

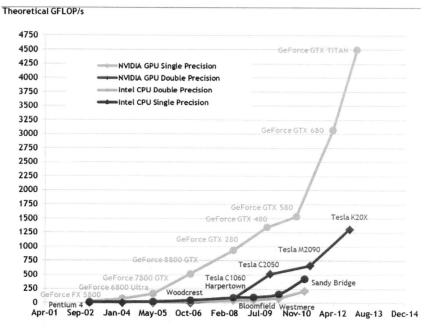

FIGURE 8.1   The performance of the GPU has increased exponentially in recent years, which makes it deliver superior performance in parallel computation compared with the CPU (NVIDIA, 2013).

tasks can be sent to the GPU by dividing them into several smaller subtasks that process at the same time. Finally, the GPU provides a chance to solve the impractical tasks that the CPU cannot because of its limited technology.

The GPU has been developed independently to have hundreds of cores, as in Figure 8.2. The GPU depends on the phenomenon "single instruction multiple data (SIMD)" to execute the same instruction on different shared processors by using different data. The GPU requires a parallel structure of a method in order to implement it; otherwise, the execution would be very poor.

8.1.2 COMPUTE UNIFIED DEVICE ARCHITECTURE

Compute Unified Device Architecture (CUDA) is the hardware and software NVIDIA parallel computing architecture that integrates with the environment, such as Microsoft Visual Studio C++, to provide a way to write the CUDA C++ (or CUDA C) program. It has two types of functions: host and kernel. Host functions are responsible for the execution of the sequential codes on the CPU, and the computing control is

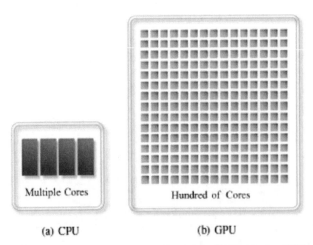

(a) CPU   (b) GPU

FIGURE 8.2 The general structure of the CPU and the GPU. (a), the CPU contains very few cores that can work in parallel. (b), the GPU, in the right image, contains hundreds of small cores that can work in parallel.

given to the GPU by calling the kernel function in order to execute the parallel codes. The GPU consists of a large set of threads that are grouped into blocks, and many blocks can be grouped into grids. Each thread has a small private memory, and each block has memory to share their threads together (NVIDIA, 2009). Threads in a block synchronize their cooperation in accessing shared memory. Figure 8.3 shows the architecture of CUDA.

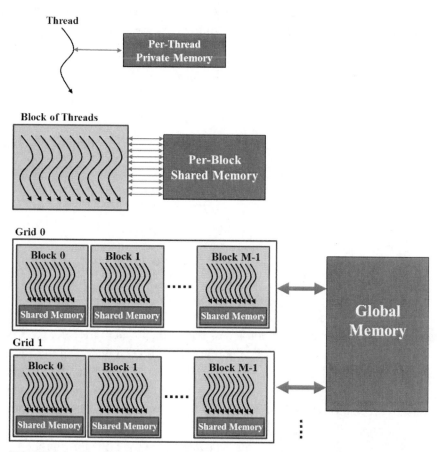

FIGURE 8.3   The GPU consists of many threads that are grouped into blocks, which are grouped into grids of blocks. Three types of memory (private memory for a thread, shared memory for a block and global memory for all threads) with more cores lead the GPU to execute the parallel program independently, and in less time.

According to size of the data sets (N), the number of threads in a block can be specified (Sanders and Kandrot, 2011). For example, in the following CUDA C++ codes, the block is declared as one dimension vector has (M = 256) threads:

```
int ThreadsPerBlock = 256;
int BlocksNeeded = (N + ThreadsPerBlock – 1)/ThreadsPerBlock;
dim3 dimBlock(ThreadsPerBlock);
dim3 dimGrid(BlocksNeeded);
```

Another possibility is the block can be declared as two-dimensional matrix has 16 x 16 threads, as in the following codes:

```
int Bx = 16;
int By = 16;
dim3 block(Bx, By);
dim3 grid(N/Bx; N/By);
```

Each thread in a block has unique index, *threadIdx*, and index of each block in a grid is *blockIdx*. Thus, thread index in a grid is done by:

Index x = blockIdx.x * blockDim.x + threadIdx.x

or

Index y = blockIdx.y * blockDim.y + threadIdx.y

where *blockDim.x* or *blockDim.y* is used to provide number of blocks in a grid.

The GPU and the CPU are working together to achieve higher performance (Kirk and Hwu, 2010). However, the relation between them is not direct because they cannot access each other's memories directly, as in Figure 8.4. To call the GPU, that data should be copied from the CPU memory to the GPU memory by using the *cudaMemcpy* function under *cudaMemcpyHostToDevice* mode. The same thing is happening when results are returned back from the GPU to the CPU: the data are copied from the GPU to the CPU by using the *cudaMemcpy* function under *cudaMemcpyDeviceToHost* mode. The number of the GPU threads should be greater or equal to the size of the moved data sets to it. The GPU parallelism benefits more by a large-scale data sets, because the CPU computation speed is enough to execute simple computations.

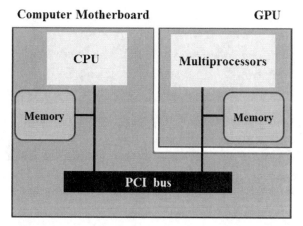

FIGURE 8.4 CPU and GPU are cooperative in executing CUDA program. Data should be copied from the CPU memory to the GPU memory for parallel execution because the GPU cannot access directly the CPU memory, and vice versa.

8.1.3 APPLICATION OF USING THE GPU

The main strong point of the GPU is its highly parallel computations, which solve the common problem encountered by many different applications. The GPU is used in different applications to achieve high performance. Some of the applications are (Bachoo, 2010):

1. Visualization: The GPU is a significant advancement for the visualization field because enormous amounts of computations are often required. It can be used to speed up processing large-scale data sets to find their color image.
2. Medical image: Using the GPU in this field is important in order to speed up results to do extra processing or treatments according to the analysis.
3. Video enhancement: The techniques used in video enhancement are always slow. Thus, the GPU has a more powerful computation engine that can be applied to these techniques rather than the CPU.
4. Bioinformation and life science: Analyzing protein and DNA sequences are more general subjects that use the GPU because of its better performance.
5. Data mining and neural networks: The GPU is used to find useful information among large size of databases and train the artificial

neural network on a lot of information in order to reduce computation time.

6. Software development: The GPU is very advantageous in developing software. For example, using the GPU with Matlab and Labview increases the computation speed.

In this chapter, we will visualize remote sensing imagery data sets by the DR method on the GPU to speed up the projection process. In addition, the measurement of the efficiency by correlation and residual variance metrics of the visualization are also done on the GPU. The comparison with the CPU implementation will be done in order to see the efficiency of the parallel codes.

8.2 INFORMATION VISUALIZATION BY DIMENSIONALITY REDUCTION

Visualization of high-dimensional data sets is widely used to analyze data in many fields of study, including remote sensing imagery, biology, computer vision, and computer graphics. Its purpose is to provide rich information to assist with data analysis (Zhang, 2008). Dimensionality reduction (DR) is an important step for data pre-processing in visualization and knowledge discovery, and it is used for different purposes, such as information visualization, noise reduction, and imaging applications (Borg and Groenen, 2008). Formally, for a set of n input points $X \subset R^D, \varnothing(X)$, is used to project the D-dimensional data points $x_i \in X$ to d-dimensional data points $y_i \in Y$, where d « D.

$$\varnothing : R^D \rightarrow R^d \quad (1)$$

$$x_i \mapsto y_i, \forall\ 1 \leq i \leq n \quad (2)$$

The high-dimensional data sets have several features; however, some might not be relevant to specific data analysis. DR is used to discover the main and important features by which to make analysis and visualization possible. The fundamental information in the original data sets is reflected in the distances between pairs of data points, and this information should be preserved by using a gradient step and fitting the input distances r_{ij} to output distances d_{ij}. Thus, the goal of preserving the distance is to represent the original data sets in a projected space (Nishisato, 2006; Lee and Verleysen, 2007). In reality, reducing dimensionality of large data sets to a low-dimensional space without losing information might be impossible. In general, DR attempts to minimize as in the following equation:

$$\varnothing = \sqrt{\sum_{i,j=1}^{n}\left(r_{ij} - d_{ij}\right)^{2}} \tag{3}$$

The cost function in Eq. (3) measures the difference between the distances in the input space and the corresponding distances in the projected space, and the final values should be minimized according to the data in a projected space.

Unfolding complex high-dimensional data sets into low-dimensional representation should focus on preserving the nearby neighborhood relationship between points rather than on creating additional points. There are two ways by which to de ne neighboring points for a point. The first supposes that all points are neighbors for a point, but the nearest k points are strong neighbors (Yang, 2011). Each point has a fixed number of neighbors and this number will not change through the projection process. The second method uses a fixed circular radius (r_{c}), where the neighboring points are inside this domain for a point (Agrafotis, 2003). Thus, the number of neighbors is not the same for all points in the space.

8.2.1 TYPE OF DIMENSIONALITY REDUCTION

A variety of strategies has resulted in the development of many different DR methods. Linear and nonlinear DR methods are the best examples to describe them. Linear projection: principle components analysis (PCA) uses orthogonal linear combination to find linear transformation space of data set. Because of its simplicity, it is used for data visualization (Jolliffe, 2002). Other visualization methods aim to preserve distances, such as multidimensional scaling (MDS). It computes distance matrix among points by computing pairwise Euclidean distances. PCA and MDS fail to find satisfactory low dimension representation of nonlinear data. Many DR methods use nonlinear projections to project the data into low dimensions. Kernel PCA (Taylor and Christianini, 2004) is a nonlinear version of PCA, and isometric feature mapping (Isomap) (Tenenbaum et al., 2000) uses geodesic distance rather than Euclidean distance in MDS. Other methods use local linear relationships to measure the local structure, as in local linear embedding (LLE) (Roweis and Saul, 2000), maximum variance unfolding (MVU) preserves direct neighbors while unfolding data (Weinberger and Saul, 2006), and Laplacian Eigen maps (Belkin and Niyogi, 2002) take a more principled technique by referring to the spectral properties of the resulting dissimilarity matrix. Some methods, like stochastic neighbor embedding (SNE) (Kerstin Bunte et al., 2012), t-distributed stochastic neighbor embedding (t-SNE) (Maaten and Hinton, 2008) and neighborhood retrieval visualizer (NeRV) (Venna et al., 2010), attempt to match probability distributions induced by the pairwise data dissimilarities in the high dimensional space and low dimension space, respectively. Stochastic proximity embedding (SPE) proceeds by calculating Euclidean distance for global neighborhood points within fixed radius (Agra Otis et al., 2010). It is an enormous step in computational efficiency over MDS, and faster than Isomap.

Recently, we proposed a new trustworthy dimension reduction method, called trustworthy stochastic proximity embedding (TSPE), to visualize different data sets (Najim and Lim, 2014). The benefit of visualizing by TSPE is that it is able to recognize the features by preserving their point distances between the projected space and the original data sets. TSPE can overcome many of the problems introduced by false neighborhood by deriving higher quality point relationships in its low-dimensional representation. The TSPE is better than many DR methods because the TSPE prevents false neighborhood errors to occur in the results (Najim and Lim, 2014).

8.2.2 VISUALIZATION BY TRUSTWORTHY STOCHASTIC PROXIMITY EMBEDDING (TSPE)

DR is an important step for data pre-processing in visualization and knowledge discovery, and it is used for different purposes, such as information visualization (Saeed et al., 2013; Than et al., 2014; Bunte et al., 2010; Atyabi et al., 2013; Prasartvit et al., 2013; Wang, 2014). TSPE has the ability to visualize difficult data sets, and, in terms of visualization, it gives satisfactory results (Najim and Lim, 2014). This ability comes from dealing adequately with the projected space. In general, a projected space is improved through projection process, consequently, TSPE focus on this state by using decreasing neighborhood size in order to continue this improvement. In each step of the projection process, the neighborhood size is reduced in order to keep pace with improvements in the projected space. In addition, the space optimization is reduced gradually, with TSPE focusing, to begin with, on maintaining a distant relationship between the points and then maintaining the nearby relationships. The points in projected space are updated depending on their relation, as follows:

$$y_j \leftarrow y_j + \lambda(t)T(d_{ij})\frac{r_{ij} - d_{ij}}{d_{ij} + \varepsilon}(y_j - y_i) \tag{4}$$

$$T(r_{ij}) = \begin{cases} 1 & if\,(d_{ij} \le d_c(t)) \vee ((d_{ij} > d_c(t)) \wedge (d_{ij} < r_{ij})) \\ 0 & \text{Otherwise} \end{cases} \tag{5}$$

where $d_c(t)$ is a decreased neighborhood radius over time. TSPE starts iteratively, on the projected space, with selecting a random point in time t, which updates all the local neighborhood points in a sufficient region within local neighborhood radius $d_c(t)$, so the coherent structure will be constructed by sending false neighbors away; according to Equation 4. The local neighborhood radius d_c starts with large value at t_0 to include all points, and then gradually this value is decreased through times by $d_c(t_0) = (t + 1)$ to keep neighborhood points with the improvement of projected space. TSPE uses projected space and a decreasing neighborhood radius $d_c(t)$ in the definition of $T(d_{ij})$, in Eq. (4), which makes the proposed method overcomes the DR problems. Figure 8.5 shows the general idea of the TSPE.

8.2.3 QUALITY OF VISUALIZATION

When measuring the quality of visualization for a given data sets, it is important to know which DR method is suitable for the task at hand. Furthermore, the user cannot compare the quality of a given visualization with the original data by visual inspection due to its high-dimensionality. Thus, the formal measurements should evaluate the amount of the preserving neighborhood color distances in the visualization with their

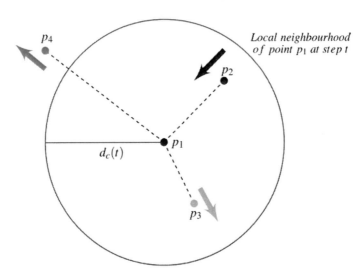

FIGURE 8.5   The main idea of the TSPE. At step t of the iteration process, the point $p_1 \in Y$ is selected. The radius of local neighborhood of this point at this step is $d_c(t)$. p_1 preserves its distance with true neighbors, as with p_2, when $(d_{ij} < d_c(t)) \wedge (d_{ij} \geq r_{ij})$. It pushes away the false neighbor points, as with p_3, when $(d_{ij} < d_c(t)) \wedge (d_{ij} < r_{ij})$. The points which are outside the local neighborhood, which their $(d_{ij} > d_c(t)) \wedge (d_{ij} < r_{ij})$, as with p_4, are pushed further away.

corresponding distance in original data. Correlation (γ) and residual variance (Stress) and are the good metrics can be used in this matter. If we suppose X is a vector of all points of the data sets in original space and Y is a vector of all the corresponding points in projected space. A and B are the vectors of all pairwise distance of X and Y, respectively, then:

Correlation function (γ): this metric computes the linear correlation between original input distances and color distances in visualization (Mignotte, 2012). The value of correlation is equal to 1 when all distances are perfect preserved, where positive slope between two vectors with perfect linear. In the other hand, the value equal to -1 if the two vectors have prefect linear relationship with negative slope. The correlation metric is defined as the follow:

$$\gamma = \frac{\dfrac{A^T B}{|A|} - \overline{AB}}{\sigma_A \sigma_B} \tag{6}$$

where $|A|$ is the number of components in A, and $\overline{A}$ and σ_A are the mean and standard deviation of A, respectively.

***Residual Variance* (*Stress*)** is a metric used to compute the standard error of difference between visualization and original space [14]. It calculates the sum of squares of differences between original data point distances and projected color distances, as in the following equation:

$$Stress = \sqrt{\frac{\sum_{i=1}^{N}(a_i - b_i)^2}{N-2}} \qquad (7)$$

Small stress value indicates that visualization has very little error and higher efficiency in preserving the original information.

8.3 REMOTE SENSING IMAGERY

Remote sensing imagery is a well-known technique to observe the earth and urban scenes by producing a large number of spectral bands (Smith, 2012). However, the challenge is how to display the abundant information contained in these images in a way that is more interactive and easy to analyze, in a 3-D image cube, for example, by a user (Tyo et al., 2013). Due to the difficulty of using these bands that are greater than 100, several DR algorithms are produced to overcome this problem by finding the better relationships among color values in three color channels after applying complex formulas to shrink the dimensionality of the original space. The precision of the results depends on the type of algorithm used, which should preserve and gather the relation among neighbors in the original space (Onclinx et al., 2009). DR provides a good way to visualize remote sensing imagery by generating its color image (Bachmann, 2006; Kotwal and Chaudhuri, 2010).

The relationship between the amount of reduction and efficiency is inverse, where the efficiency is reduced when the dimensionality of low-dimensional space is very low, and vice versa. Inefficiency indicates the inability of the DR method to preserve the original information in low-dimensional space. Although, in general, visualization of a remote sensing imagery data sets requires that the dimensionality of low-dimensional space is equal to three (Cui et al., 2009), the efficiency of visualization is reduced because the relationship between the neighbors are loose.

Thus, some of the colors in the visualization of remote sensing imagery data sets do not represent the correct relations derived from the original system, and, therefore, those colors are false colors. On the other hand, the preservation of the original information is increased when the dimensionality of low-dimensional space is increased.

8.4   OUR CONTRIBUTIONS IN GPU APPLICATION

Evaluating the efficiency of each point with the remaining points in the data sets in a general way has been used by most methods, where the final efficiency of the data sets represents the average of efficiencies from all those points. According to this premise, the principle of divide and conquer is the best way to speed up processing. The evaluation of each point is computed by computing the value of its neighborhood relationship with all other points in parallel. If we suppose that there are N points in the data sets, the processing requires the sending of that point with all other points (N) to the GPU, as in Figure 8.6. The N points in the GPU are processed in parallel to measure the efficiency of that point. Parallel implementation helps a lot in reducing the execution time, especially if the size of the data sets is very large. We will use this idea to implement TSPE, correlation and residual variance metrics on the GPU.

8.4.1   PARALLEL TSPE

The TSPE can be implemented directly on the GPU because the method equation is a parallel equation. Figure 8.7 showed that the kernel TSPE

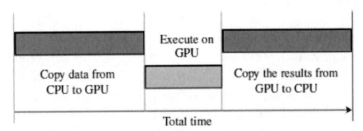

FIGURE 8.6   Evaluation of one point requires to move N points from CPU to GPU. These points are processing in parallel to evaluate their relation with point in the hand.

```
__global__ void TSPE_GPU(float i, float *Y, float *X, float lambda, float dc, int
Dimension, int LDimension, int N) {
    int j = blockIdx.x * blockDim.x + threadIdx.x;
    if (j < N)
    {   float sum = 0.0;
        for (int k=0; k<LDimension; k++)
            sum += (float) ((Y[i+ N*k] - Y[j+ N*k]) * (Y[i+N*k] - Y[j+N*k]));
        float dij = sqrt(sum);
        sum = 0.0;
        for (int k=0; k<Dimension; k++)
            sum += (float) ((X[i+ N*k] - X[j+ N*k]) * (X[i+N*k] - X[j+N*k]));
        float rij = sqrt(sum);
        if ((dij<=dc) || ((dij > dc) && (dij<rij)))
        {   float T = (float) (lambda * (rij - dij) / (dij + 1e-8));
            for (int k=0; k<LDimension; k++)
                Y[j+N*k] += (float) (T * (Y[j+N*k] - Y[i+N*k]));
        }
    }
    __syncthreads();
}
```

FIGURE 8.7 TSPE is implemented on the GPU by using CUDA C++ codes. This kernel attempts to update the coordinates of all points in the data sets to their relationship of the point i. The variable dimension is the dimension of the original space X, and dimension is the dimension of the projected space Y. N is the size of the data sets.

GPU updates the coordinates of all points in the data sets in parallel depending on the selected point i. In the host, a point is selected each time and sent to the GPU through kernel TSPE GPU to do a parallel task of updating the other points according to the TSPE equation. The result of applying TSPE, which is projected space and has three dimensions, is projected into CIE Lab color space.

8.4.2 PARALLEL CORRELATION METRIC

The visualization of remote sensing imagery data sets by TSPE should be measured to see how it preserves the original information. Although the implementation time of the correlation metric needs to be long in the CPU because of the large volume of remote sensing imagery data sets, parallel implementation of it can overcome this limitation. To implement Equation 6 on the GPU, one kernel is not enough because the standard deviations (σ_A and σ_B) depend on the mean values ($\overline{A}$ and $\overline{B}$, respectively).

To implement it, we will use two stages to construct CUDA C++ codes. In the first stage, we compute the values of the variables $\bar{A}$, $\bar{B}$ and of each data point in the data sets. Figure 8.8 shows that the kernel (*step_one*) computes the vectors of these values in parallel on the GPU. The final values for these variables represent the average of the total sum of their values of all points, which have been computed on the CPU. In the second stage, we compute the variables for each point in the data sets, σ_A and σ_B. The kernel (*step_two*), in Figure 8.9, is responsible for implementing this task on the GPU. In the CPU, the overall average is computed to get the sum of all their values and to know the final value of the data sets. Therefore, all variables of Eq. (6) are ready for application directly to get the final correlation value of the visualization by TSPE.

8.4.3 PARALLEL RESIDUAL VARIANCE METRIC

Residual variance is also important to measure the stress error of the visualization by TSPE, but it needs to be implemented on the CPU for a long time.

```
__global__ void step_one(float i, float *B, float *A, int Dimension, int LDimension,
float *S_a, float *S_b, float *S_ab, int N) {
    int j = blockIdx.x * blockDim.x + threadIdx.x;
    if (j < N)
    {    float sum = 0.0;
        for (int k=0; k<LDimension; k++)
            sum += (float) ((B[i+ N*k] - B[j+ N*k]) * (B[i+N*k] - B[j+N*k]));
        float dij = sqrt(sum);
        sum = 0.0;
        for (int k=0; k<Dimension; k++)
            sum += (float) ((A[i+ N*k] - A[j+ N*k]) * (A[i+N*k] - A[j+N*k]));
        float rij = sqrt(sum);
        S_a[j] += rij / N;
        S_b[j] += dij / N;
        S_ab[j] += (dij * rij) / N;
    }
    __syncthreads();
}
```

FIGURE 8.8   First step to compute correlation metric on the GPU (CUDA C++). This kernel attempts to compute the $\bar{A}, \bar{B}$ and for the point i. The results are assigned to the variable S_a, S_b and S_ab, respectively. The variable Dimension is the dimension of the vector A. Dimension is the dimension of the vector B. The projected and original spaces have the same size of data points, which are N.

Equation (7) is much simpler than Equation 6, where implementation of the residual variance on the GPU requires just one kernel, as in Figure 8.10. This kernel attempts to compute the stress error (*Stress_i*) for a point *i*.

```
__global__ void step_two(float i, float *B, float *A, int Dimension, int LDimension,
int N, float Mean_a, float Mean_b, float *Std_a, float *Std_b) {
    int j = blockIdx.x * blockDim.x + threadIdx.x;
    if (j < N)
    {    float sum = 0.0;
        for (int k=0; k<LDimension; k++)
            sum += (float) ((B[i+ N*k] - B[j+ N*k]) * (B[i+N*k] - B[j+N*k]));
        float dij = sqrt(sum);
        sum = 0.0;
        for (int k=0; k<Dimension; k++)
            sum += (float) ((A[i+ N*k] - A[j+ N*k]) * (A[i+N*k] - A[j+N*k]));
        float rij = sqrt(sum);
        Std a[j] += (rij – Mean_a) * (rij – Mean_a) / N;
        Std b[j] += (dij – Mean_b) * (dij – Mean_b) / N;
    }
    __syncthreads();
}
```

FIGURE 8.9   Second step to compute correlation metric on the GPU (CUDA C++). This kernel attempts to compute the σ_A and σ_B for the point i. The results are assigned to the variable Std_a and Std_b, respectively. The variable Dimension is the dimension of the vector A. L Dimension is the dimension of the vector B. The projected and original spaces have the same size of data points, which are N. Mean_a and Mean_b represent the $\overline{A}, \overline{B}$, respectively, of all data sets.

```
__global__ void Residual_variance(float a, float *Y, float *X, int Dimension, int
LDimension,
int N, float *Stress_i) {
    int b = blockIdx.x * blockDim.x + threadIdx.x;
    if (b < N)
    {    float sum = 0.0;
        for (int k=0; k<LDimension; k++)
            sum += (float) ((Y[a+ N*k] - Y[b+ N*k]) * (Y[a+N*k] - Y[b+N*k]));
        float dab = sqrt(sum);
        sum = 0.0;
        for (int k=0; k<Dimension; k++)
            sum += (float) ((X[a+ N*k] - X[b+ N*k]) * (X[a+N*k] - X[b+N*k]));
        float rab = sqrt(sum);
        Stress_i[b] += (rab - dab) * (rab - dab) / (N-2);
    }
    __syncthreads();
}
```

FIGURE 8.10   CUDA C++ codes of residual variance. This kernel computes the stress error (Stress_i) for a point a.

8.5   EXPERIMENTAL RESULTS

In this section, the implementation on Intel(R) i7–930 2.80 GHz CPU with 12 GB memory on Windows 7. We ran our proposed method in Microsoft Visual Studio C++ 2008 with CUDA 4.2 and NVIDIA GeForce GTX 480 graphics card with a butter size of 1 GB. We used the AVIRIS Moffet Field data sets from the southern end of San Francisco Bay, California, done in 1997. The data sets contain 224 bands, where each band has 1800 x 600 points.

Figure 8.11 shows the ability of TSPE in visualizing remote sensing imagery data sets. Although the results of parallel implementation on GPU have exactly the same results as sequential results on the CPU, execution time by the GPU achieves a higher speed-up than the CPU. Table 8.1 shows the comparison between CUDA C++ (GPU) and the CPU codes in time execution for different data sizes. The computation time of TSPE is further reduced because the CPU and the GPU are working together, where the CPU uses one (or very few) core(s) to execute host function, and the GPU is called to execute kernel function in parallel.

In order to measure the efficiency of the visualization, we have to use correlation and residual variance measurements metrics. These metrics are implemented on the GPU by using CUDA C++. Table 8.2 shows the comparison between CUDA C++ and the CPU codes in time execution for

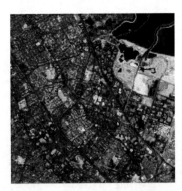

FIGURE 8.11   The TSPE method is used to visualize small part of remote sensing imagery data sets, which has dimension 300 x 300 x 224. TSPE reduced the 224 dimensions of original data set into 3 dimensions, which is then projected to CIE Lab color space to generate the final visualization.

TABLE 8.1   TSPE Speed Performance Comparisons Between the CPU Codes and the CUDA C++ Codes With Different Size of Remote Sensing Imagery Data Sets

Data Size	CPU	GPU	Times GPU faster than CPU
1 x 300 x 300	26.80	0.021	1276
2 x 300 x 300	104.044	0.040	2601
3 x 300 x 300	238.06	0.060	3967
4 x 300 x 300	468.751	0.076	6167
5 x 300 x 300	716.511	0.096	7463
6 x 300 x 300	1205.302	0.113	10666
7 x 300 x 300	1961.032	1.300	1508
8 x 300 x 300	2090.713	3.389	616
9 x 300 x 300	2497.094	6.054	412
10 x 300 x 300	3755.885	9.037	415
11 x 300 x 300	5633.828	12.864	437
12 x 300 x 300	8450.741	15.034	562

Time unit is minutes.

TABLE 8.2   Performance Comparisons of the CPU Codes and the CUDA C++ Codes in Small Data Sets (300 x 300 x 224)

	GPU	CPU	Times GPU faster than CPU
TSPE	0.021	26.8	1276
Correlation	7.698	111.126	14
Stress	3.771	47.374	13

data sets have N=300 x 300 points. The CPU and the GPU are working together, where the CPU uses very few cores to execute the host function, and the GPU is called to execute the kernel function in parallel. It should be noted that the moving of information from host to device, and from device to host, takes a lot of time, as in Figure 8.12. Therefore, we move long the vector of high-dimensional data sets and the vector of low-dimensional space just one time from host to device through the projection before starting the TSPE in order to avoid losing more time.

Correlation or Stress computation has about 10 times speed-up whereas TSPE has over 1000 times speed-up because TSPE is not required to run

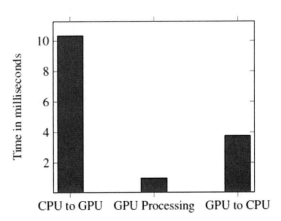

FIGURE 8.12   Times of moving data from the CPU to the GPU is greater than pure execution in the GPU. Moving the results back to the CPU is also greater than execution in the GPU. In this experiment, the time of moving data from the CPU to the GPU is greater than moving them from the GPU to the CPU because of the large size of input data.

on all point, where random selected points are enough to update the other points. Times unit is in minutes.

When the GPU receives the computation control, N threads work parallel. Thus, the GPU parallelism is more beneficial with large remote sensing imagery data sets because the CPU computation speed is enough to execute simple computation.

8.6   CONCLUSION

In this chapter, DR was implemented on the GPU to visualize remote sensing imagery data sets. The benefit of a parallel implementation is to obtain the results in as short a time as possible. The results showed that CUDA implementation of TSPE is faster than their sequential codes on the CPU in calculating floating-point operations, especially for large data sets, such as remote sensing imagery. The GPU is more suitable to the implementation of the correlation and residual variance measurement methods because they do a large computation. We illustrated that this massive speed-up requires a parallel structure to be suitable for running on the GPU. Large data sets, such as remote sensing imagery, is a better candidate data sets to be implemented on the GPU to accelerate the computation time, which will be one hundred times speed-up over the CPU.

KEYWORDS

- **dimensionality reduction**
- **graphics processing unit**
- **remote sensing imagery**
- **visualization**

REFERENCES

1. Agrafotis, D. K., Xu, H., Zhu, F., D. Bandyopadhyay, Liu, P., "Stochastic proximity embedding: Methods and applications." Molecular Informatics, vol. 29, pp. 758–770, 2010.
2. Agrafotis, D. K., "Stochastic proximity embedding." Computational Chemistry, vol. 24, 2003.
3. Atyabi, A., M. H. Luerssen, Powers, D. M., "PSO-based dimension reduction of EEG recordings: Implications for subject transfer in BCI." Neuocomputing, vol. 119, pp. 319–331, 2013.
4. Bachmann, C. M., T. L. Ainsworth, R. A. Fusina (2006). "Improved manifold coordinate representations of large-scale hyperspectral scenes." IEEE Transactions on Geoscience and Remote Sensing, 44:2786–2803.
5. Bachoo, A., "Using the CPU and GPU for real-time video enhancement on a mobile computer," in IEEE 10th International Conference on Signal Processing (ICSP), 2010.
6. Belkin, M., Niyogi, P., "Laplacian eigenmaps and spectral techniques for embedding and clustering." Advances in Neural Information Processing Systems, vol. 14, p. 585–591, 2002.
7. Borg, I., Groenen, P., "Modern Multidimensional Scaling: Theory and Applications." Springer Verlag, 2005.
8. Bunte, K., Hammer, B., A. Wismller, Biehl, M., "Adaptive local dissimilarity measures for discriminative dimension reduction of labeled data." Neuocomputing, vol. 73, pp. 1074–1092, 2010.
9. Cui, M., Razdan, A., Hu, J., Wonka, P., (2009), "Interactive hyperspectral image visualization using convex optimization." IEEE Transaction on Geoscience and Remote Sensing, 47:1673–1684.
10. Jolliffe, I. T., "Principal Component Analysis." New York: Springer-Verlag, 2002.
11. Kerstin Bunte, M. B., Sven Haase, Villmann, T., "Stochastic neighbor embedding (SNE) for dimension reduction and visualization using arbitrary divergences." Neurocomputing, vol. 90, pp. 23–45, 2012.
12. Kirk, D. B., Hwu, W. M. W., "Programming Massively Parallel Processors." Morgan Kaufmann Publishers Inc. San Francisco, CA, USA, 2010.
13. Kotwal, K., Chaudhuri, S., (2010), "Visualization of hyperspectral images using bilateral filtering." IEEE Transaction on Geoscience and Remote Sensing, 48: 2308–2316.

14. Lee, J. A., Verleysen, M., "Nonlinear Dimensionality Reduction." Springer, 2007.
15. Maaten, L., J. P. V., Hinton, G., "Visualizing high-dimensional data using t-SNE." Machine Learning Research, vol. 9, pp. 2579–2605, 2008.
16. Mignotte, M., "A bicriteria optimization approach based dimensionality reduction model for the color display of hyperspectral images." IEEE Transactions on Geoscience and Remote Sensing, vol. 50, pp. 501–513, 2012.
17. Najim, S. A., Lim, I. K., "Trustworthy dimension reduction for visualization different data sets." Information Science, vol. 278, pp. 206–220, 2014.
18. Nishisato, S., "Multidimensional Nonlinear Descriptive Analysis." Boca Raton, FL: Chapman & Hall, 2006.
19. NVIDIA, "GPU computing applications," NVIDIA corporation, 2013.
20. NVIDIA, "NVIDIA next generation CUDA compute architecture: Fermi," in http://www.nvidia.com/object/fermi-architecture.html, NVIDIA Corporation, 2009.
21. Onclinx, V., V. Wertz, M. Verleysen (2009)." Nonlinear data projection on noneuclidean manifolds with controlled trade-off between trustworthiness and continuity." Neurocomputing, 72:1444–1454.
22. Prasartvit, T., Banharnsakun, A., B. Kaewkamnerdpong, Acha-lakul, T., "Reducing bioinformatics data dimension with ABC-kNN." Neuocomputing, vol. 116, pp. 367–381, 2013.
23. Roweis, S. T., Saul, L. K., "Nonlinear dimensionality reduction by locally linear embedding." Science, vol. 290, pp. 2323–2326, 2000.
24. Saeed, M., K. Javed, Babri, H. A., "Machine learning using bernoulli mixture models: Clustering, rule extraction and dimensionality reduction." Neurocomputing, vol. 119, pp. 366–374, 2013.
25. Sanders, J., Kandrot, E., "CUDA By Example." Addison-Wesley Professional, 2011.
26. Smith, R. B. (2012). "Introduction to Hyperspectral Imaging." MicroImages, Inc.
27. Taylor, J. S., Christianini, N., "Kernel methods for pattern analysis." Cambridge University Press, 2004.
28. Tenenbaum, J. B., V. de Silva, Langford, J. C., "A global geometric framework for nonlinear dimensionality reduction." Science, vol. 290, pp. 2319–2323, 2000.
29. Than, K., T. B. Ho, Nguyen, D. K., "An effective framework for supervised dimension reduction." Neurocomputing. http://dx.doi.org/10.1016/j.neucom.2014.02.017, 2014.
30. Tyo, J. S., Konsolakis, A., D. I. Diersen, R. C. Olsen (2003). "Principal components based display strategy for spectral imagery." IEEE Transaction on Geoscience and Remote Sensing, 41:708–718.
31. Venna, J., Peltonen, J., Nybo, K., H. Aidos, Kaski, S., "Information retrieval perspective to nonlinear dimensionality reduction for data visualization." Journal of Mach. Learn. Res, vol. 11, pp. 451–490, 2010.
32. Wang, J., "Real local-linearity preserving embedding." Neuocomputing, vol. 136, pp. 7–13, 2014.
33. Weinberger, K., Saul, L. K., "An introduction to nonlinear dimensionality reduction by maximum variance unfolding," in Proceedings of the National Conference on Artificial Intelligence, Boston, MA., 2006.
34. Yang, L., "Distance-preserving dimensionality reduction." Wiley Inter-disc. Rew.: Data Mining And Knowledge Discovery, vol. 1, pp. 369–380, 2011.
35. Zhang, J., "Visualization For Information Retrieval." Springer-Verlag Berlin Heidelberg, 2008.

CHAPTER 9

FRAUD DETECTION IN PROCESS-AWARE INFORMATION SYSTEMS USING PROCESS MINING

SHAHLA MARDANI[1] and HAMID REZA SHAHRIARI[2]

[1]Computer Engineering and Information Technology Department, Amirkabir University of Technology, Tehran, Iran, Tel: +982164542716, E-mail: Shahla.Mardani@aut.ac.ir

[2]Assistant Professor, Computer Engineering and Information Technology Department, Amirkabir University of Technology, Tehran, Iran, Tel: +982164542716, E-mail: Shahriari@aut.ac.ir

CONTENTS

Process aware information system (PAIS) is a software system that helps to organizations to implement their processes in a flexible way. It is vulnerable to occupational frauds due to the flexibility. Occupational fraud means the use of one's occupation for personal enrichment. Flexibility in these systems gives the opportunity for fraudsters to commit illegal activities. Strict security controls on these systems at runtime reduce their flexibility. Thus, use of fraud detection method is an appropriate solution to keep both flexibility and security. Most related work in this domain use process mining to detect frauds. In this chapter, at first we propose a brief review of fraud detection, process aware information system and process mining and finally we investigate fraud detection methods that are presented in process aware information system's domain.

9.1 INTRODUCTION

Today, Organizations should be changed quickly to remain competitive. Process Aware Information Systems (PAISs) are used to help the organizations to adopt quickly. They are flexible about implementation of the business processes. In PAISs, business processes are defined in a loose way to users can change them according with requirements in the real world. The flexibility of PAIS makes it vulnerable against occupational fraud. Occupational fraud is illegal acts, which are committed by the employees in an organization to achieve personal gain. Due to the flexibility, PAISs provide fraudsters with opportunity for illegal activities. Strict access control could not avoid it, because it also would reduce the flexibility. Thus researchers are looking for a solution to ensure security along with the flexibility. An appropriate solution is fraud detection that uses system's records to detect illegal activities after execution. In this chapter, fraud detection methods and their performance are surveyed.

There is not enough information about fraudulent behaviors in PAISs because the process models are not fixed. While the normal behavior is not known, the abnormal behavior is also unknown. Therefore, it is not possible to use the misuse detection methods to detect frauds. Thus, the best approach is use of the anomaly detection methods. In this way, we need a profile of the normal behavior. According to the frequent changes, it is not actually available. Thus, we should obtain it from the PAIS log. The log consists of both normal behavior and abnormal behavior. Most works in this domain use process-mining techniques to discover some information about the executed processes. In these works, one or more process model is mined from the log and is used as a base for detecting frauds (Van der Aalst and De Medeiros, 2005; Bezerra and Wainer, 2007, 2008, 2011, 2013; Rozinat and van der Aalst, 2008; Bezerra et al., 2009; Jalali and Baraani, 2010). In general, the most important challenge in this domain is to obtain the normal behavior of the system. Researchers are looking for methods that can overcome the challenge and show better performance.

This chapter aims to answer some basic questions:

- Why it is important to detect frauds in information systems?
- Why do organizations use PAIS while it may cause occupational frauds?
- How do organizations use fraud detection methods to ensure security in PAISs?
- How do process mining help to fraud detection in PAISs?
- What is the performance of current approaches and what are the open problems?

9.2   FRAUD

Today, fraud is an important subject in many companies and organizations, because it causes great financial losses and spends a portion of the revenues each year. According to the 2012 report of Association of Certified Fraud Examiners (ACFE), companies lose an estimated 5% of their revenues due to fraud each year (ACFE, 2012). Nevertheless, many of the frauds are never detected due to concealed commitment of them. Also new technologies provide the fraudsters with many new opportunities for committing the fraud. Thus organizations should be aware of the new technologies and the new methods of fraud detection. In this section, we explain the fraud definition, types of the fraud and fraud detection methods.

9.2.1   FRAUD DEFINITION

There are many different definitions of fraud in various domains. In the general sense, fraud is deception and trickery and a fraudster is someone who makes losses to others and benefit from them through tricking. ACFE defines fraud as follows (Wells, 2011):

"Fraud can encompass any crime for gain that uses deception as its principal modus operandi. Of the three ways to illegally relieve a victim of money – force, trickery, or larceny- all offenses that employ trickery are frauds ... although all frauds involve some form of deception, not all deceptions are necessarily frauds."

The term of fraud includes many crimes and is defined in different ways for each domain. What is common between them is that someone deceives other to gain something. Our focus in this chapter is occupational fraud. Thus we propose a definition that covers this type of fraud. Occupational fraud is (ACFE, 2012):

"The use of one's occupation for personal enrichment trough the deliberate misuse or misapplication of the employing organization's resources or assets".

Thus, from here, whenever we use word "fraud," we mean recent definition.

9.2.2 TYPE OF FRAUD

Until now, researchers present several classifications of the frauds. For example, ACFE classifies the occupational frauds to three classes (ACFE, 2012): corruption, asset miss-appropriation and financial statement fraud. Nevertheless, there is another classification that provides the more general overview. In this classification, frauds are classified to two classes: internal fraud and external fraud. Internal fraud includes two categories: financial statement fraud and transaction fraud. Figure 9.1 shows this classification of the frauds.

External fraud is committed by someone out of the organization. In contrast, internal fraud is committed by the individuals within the organization. In the financial statement fraud, fraudster misstates financial values to show the profitable appearance and deceive shareholders. It is obvious that no assets are stolen, but fraudster changes the view of organization performance. This type of the fraud is usually committed by the managers that have more power. Unlike the financial statements fraud, in the transaction fraud, fraudster intends to steal or embezzle organizational assets. It would include crimes such as bribery, using company resources for personal gain or payroll abuses. Transaction fraud can be committed by managers or non-managers.

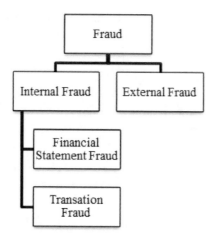

FIGURE 9.1 Fraud classification.

9.2.3   FRAUD DETECTION

Fraud detection means finding the frauds as soon as possible. The fraudsters misuse new technologies for committing illegal activities. They use many different ways. Thus we need new fraud detection methods based on the new technologies. Researchers propose many fraud detection methods in different domains and applications. In general, these methods are classified in three classes (Flegel et al., 2010):

- Misuse detection;
- Anomaly detection;
- Specification-based detection.

The base of detection in the three classes is different. In the first class, there is the information about all of the frauds that might be committed. Therefore, the frauds are modeled in an appropriate format such as state machine or petri-net. If a sequence of interrelated events consistent with these models is seen, a fraud is detected and an alert is issued. Since the frauds are known; the alerts could present some information about the frauds. In this way, the false positive rate is equal to zero, but the false negative rate might be increased due to the unknown frauds. Thus the knowledgebase and the models should be updated to reduce false negative rate. It might be costly.

In contrast of the misuse detection, in anomaly detection approach, there is no information about the frauds. In this approach, the normal behavior of a user, a process or a protocol is modeled and each deviation from this model is detected as an anomaly or a fraud. When the normal behavior is training to the system, we should be confident that there is no fraud in the training data. In this approach, the alerts could not be accurate because the information about the frauds is not explicit. There is two points: first, there is no a legitimated reason for why a behavior is a fraud. It might be caused to a high false positive rate. Second, if the deviation is not noticeable, it is difficult to detect it. It might be addressed to a high false negative rate.

In the third approach, a specification of the safe behavior is defined and each deviation from this specification is labeled as a fraud. In the limited domain, the specification is defined in an automatic way. In this approach, the alerts are not informational due to lack of knowledge about the frauds.

As you seen, each approach has some strengths and weaknesses. Therefore, we cannot specify the best approach. The appropriate approach for fraud detection is selected based on the domain and the situations.

9.3 PROCESS AWARE INFORMATION SYSTEM

From the past to the present, business processes flow through people and systems implicitly in the organizations. Business challenges direct the organizations to pay more attention to these processes. The organizations become aware that business processes are the main source of the competitive advantage. Therefore, they need a business process management to meet their requirements. In this regard, Information Technology (IT) is proposed as a key enabler for managing business processes. In this way, some challenges are emerged. The organizations should change their processes depending on the business environment changes to remain competitive. The Process Aware Information Systems (PAISs) that use fix process models cannot change their business processes easily and quickly. Thus researchers are looking for a flexible system to meet this need. PAISs are formed to achieve this purpose. In this section, we present a brief review of PAISs and their features.

9.3.1 HISTORY

Information systems are placed in four layers that are shown in Figure 9.2. Central layer is related to the infrastructure including hardware and operating system. Second layer contains generic applications such as database management system and word processor. Third layer includes domain-specific applications that are used just in organizations or specific domain (such as accounting package). Finally forth layer includes tailor-made applications that are generated specially for organizations.

In the 1960s, there were not the second and third layers and all applications were custom-built. Gradually, these layers emerged and their applications were developed. In the 1970s and 1980s, the information systems were developed in data-driven approaches. The focus of IT was on storing and retrieving data. With web development and growth of these applications that take many users and tasks, a need for overall view of the

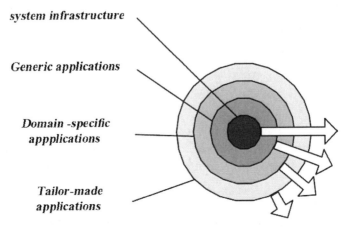

FIGURE 9.2   Some of the ongoing trends in information systems (Dumas, 2005).

information systems arose. In order to meet this requirement, a move from programming to integrating applications has begun. In this way, instead of programming, software components are integrated with each other. In this regard, the view of information systems was changed: move from data oriented to process oriented along with a trend from programming to assembling [14]. In early 1990s, the emphasis on the processes was increased and the process-driven approaches were considered more than before.

In this trend, the need of response to the organization's environment caused system engineers reused existing applications in the new system rather than built from scratch. Thus, approaches of system development focus on the rapidly adapting existing software in the response to changes. Process Aware Information system (PAIS) is a software system that follows these approaches.

PAIS is made for responding to environment changes quickly. They are based on the applications of second layer (such as Workflow Management System) and third layer. These systems can change quickly because of separating process layer from underlying applications. In this way, the process logic is pulled out from the applications and the programs and redesign is possible.

9.3.2   DEFINITION

Process Aware Information System is "a software system that manages and executes operational processes involving people, applications, and/or information sources on the basis of process models" (Dumas, 2005).

PAISs should change their process models, implement new processes and adapt to organization's environment. Thus, they should support the flexibility and the uncertainty.

9.3.3 PROPERTIES OF PAIS

PAISs should have some properties (Reichert et al., 2009) to handle their adaptability in a quick way:

- Process implementation is based on the loose-defined models instead of the fixed models. It allows users to complete the processes in the different ways.
- Implemented process should adapt with exceptional instances without effect on the other instances.
- The process models should change and the other models are replaced.

These properties show the change in the PAISs as a fundamental requirement. In this section we explain how change is done in the PAISs.

9.3.3.1 Process Change

The process layer is separate from the application layer in the PAIS. Process logic is defined in the models at design time and PAIS orchestrates applications based on these models at runtime. Thus each business process should be defined using process model. There are several models for each process type. For example, consider a product ordering process: This process would change over time and could be completed in different ways. For each way, we should define a new model. The model is a directed graph in which the nodes are activities and the edges are relations between them. Process instances are made based on these models and are orchestrated by PAIS. Figure 9.3 shows main activities of a PAIS.

First, one model is made from a business process at design time (1). The model is defined manually or using process mining. Process instances are made based on the model at runtime (2) and performers perform tasks of the model (3). When an exception is occurred in an instance, the performers could change the model for that (4). The changes are done according the access level of the performer. If the changes are repeated more, one alert is

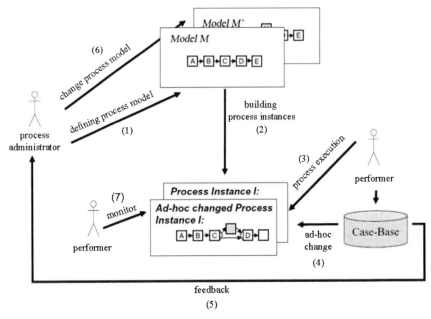

FIGURE 9.3   The main activities in a PAIS (Weber et al., 2005).

sent to process administrator to change the model (5). Process administrator change the model based on the feedback (6). Performers could monitor each other's actions (7) (Weber et al., 2005).

Therefore, the changes are performed at two levels: the process type level and the process instance level (Weber et al., 2005; Reichert et al., 2005; Rinderle-ma, 2008, 2008). When a process instance is changed in real world, the process model should be changed. This change is in the process type level. In this situation, some process instances might be running and should be decided for migration or non-migration. Ad-hoc changes are occurred in the process instance level. These changes are used in the exceptional or unexpected situations. The changes are performed in such ways that do not change the process model or other process instances. For example, consider a treatment process with a test for patient. If performing the test is impossible or dangerous for one patient, an exception is occurred. In this situation, the process model is changed for this patient (process instance) through deleting or replacing the test task.

9.3.3.2   How to Change

The process model can be changed using primitives (such as add or remove an edge or a node) or using high-level operators (a set of primitives). Use of primitives seems simple but it might cause confusion or error. In high-level operators, a set of primitives is used. For example operator *replace* include primitives *delete a node* and *add a node*. Use of high-level operators has following advantages:

- easy to use;
- error reduction;
- ensuring the soundness (lack of deadlock and endless loop);
- easy to understand change log;
- change optimization.

In general, the use of high-level operators has less complexity. Thus PAISs prefer to use these operators instead of primitives. In 2008, Weber et al., specify 18 change patterns of control flow [19] in high-level that are divided into two major categories: Adaptation patterns and Patterns for changes in predefined regions. Adaptation Patterns are used at runtime by changing the process model. Second category needs some predictions at design time to provide an opportunity for performing desired tasks at runtime. Figure 9.4 shows 18 change patterns.

There are four patterns in second category: late selection, late modeling, late composition of process fragments and multi-instance activity. In late selection, some different implementations of a region are present at design time. At runtime, performer selects an implementation depending on the situation. Late modeling provides user with more discretion. In this pattern, some different implementations of activities are provided and performer decides which activities and which order of activities should be used. In late composition, some different orders of activities are available for performers to complete process. In multi-instance activity, some different instance of one activity is created and is used to complete process at runtime. PAIS provides loose-defined models through these patterns.

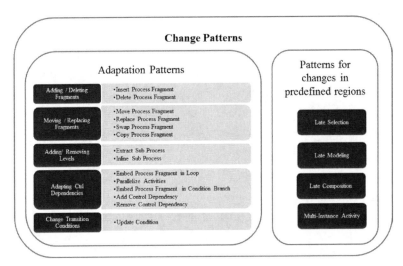

FIGURE 9.4   Change patterns in PAIS (Weber et al., 2008).

9.3.3.3   Impact of Change on the Process Instance

When a process model is changed, process instances are running. What will happen to the process instances? To answer this question, consider the following example:

One process has four activities: A, B, C and D. Activity E is added between B and C. Figure 9.5 shows this change.

Consider four process instances represented in Figure 9.6 that are running. If execution of the process instance has not reached to the change position (task C in this example), the process instance migrates to new model. Otherwise the process instance is followed earlier model.

In instance (a), activity B is running. In instance (b) activity B is finished and activity C is active but is not running. These instances can migrate to new model. In contrast, instances (c) and (d) cannot migrate because activity C has started.

9.3.3.4   Change Pattern Support

PAISs support specified primitives and adaptation patterns. Table 9.1 shows the change patterns that are supported at the process type level and Table 9.2 presents the change patterns that are supported at the process instance level.

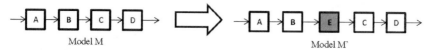

FIGURE 9.5 Change in a process model using add a new task.

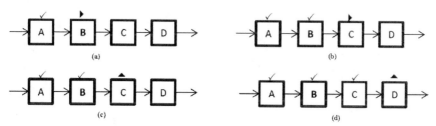

FIGURE 9.6 Running process instances of the model M.

9.3.4 SECURITY IN PAIS

In PAISs, process are performed and controlled by different users. They can change process model and add or remove some activities. In this situation, security becomes more important. Current commercial systems such as Sttafware, Websphere MQ or AristaFlaw and current scientific systems such as YAWL use role-based access control (RBAC) mechanisms to ensure security. These mechanisms connect business activities and organizational structure through access rules. Moreover, authorization constraints and separation of duties are other security mechanisms that are used in PAISs. Access control mechanism and authorization constraints can be enriched using context information (such as time and position) (Leitner et al., 2011). Nevertheless, these policies cannot be rigid and fix because they restrict the performance of PAIS and reduce the flexibility. Thus researchers balance the flexibility against the security.

9.3.5 FRAUD IN PAIS

Although many organizations use many security policies to prevent abuse and fraud but it is inevitable. Today, new technologies provide fraudsters with new ways to commit illegal activities. Thus fraud prevention or fraud detection is difficult and many frauds remain hidden.

TABLE 9.1 Change Patterns Support at Process Type Level (Weber et al., 2008)

| Commercial | | Academic | | | | | | | | Change |
Staffware	Flower	YAWL+ Worklets/ Exlets	WIDE	WASA2	PoF	MOVE	HOON	CAKE2	ADEPT2/ CBRFlow	
Primitives										
+	+	+	+	+	+	+	+	+	-	Add Node
+	+	+	+	+	+	+	+	+	-	Remove Node
+	+	+	+	+	+	+	+	+	-	Add Edge
+	+	+	+	+	+	+	+	+	-	Remove Edge
-	-	+	-	-	-	-	-	+	-	Move Edge
Change Patterns										
-	-	-	+	-	-	-	-	-	+	Insert Fragment
-	-	-	+	-	-	-	-	-	+	Delete Fragment
-	-	-	-	-	-	-	-	-	+	Move Fragment
-	-	-	+	-	-	-	-	-	-	Replace Fragment
-	-	-	-	-	-	-	-	-	-	Swap Fragment
-	-	-	-	-	-	-	-	-	+	Extract Fragment

TABLE 9.1 Continued

Commercial		Academic									Change
Staffware	Flower	YAWL+ Worklets/ Exlets	WIDE	WASA2	PoF	MOVE	HOON	CAKE2	ADEPT2/ CBRFlow		
-	-	-	-	-	-	-	-	-	+		Inline Fragment
-	-	-	-	-	-	-	-	-	+		Embed Fragment in Loop
-	-	-	-	-	-	-	-	-	+		Prallelize Activities
-	-	-	+	-	-	-	-	-	-		Embed Fragment in Conditional Branch
-	-	-	-	-	-	-	-	-	+		Add Control Dependency
-	-	-	-	-	-	-	-	-	+		Remove Control Dependency
-	-	-	+	-	-	-	-	-	+		Update Condition
-	-	-	-	-	-	-	-	-	-		Copy Fragment

TABLE 9.2 Change Patterns Support at Process Instance Level (Weber et al., 2008)

Commercial				Academic						Change
Staffware	Flower	YAWL+ Worklets/ Exlets	WIDE	WASA2	PoF	MOVE	HOON	CAKE2	ADEPT2/ CBRFlow	
Primitives										
-	-	-	-	+	-	-	-	+	-	Add Node
-	-	-	-	+	-	-	-	+	-	Remove Node
-	-	-	-	+	-	-	-	+	-	Add Edge
-	-	-	-	+	-	-	-	+	-	Remove Edge
-	-	-	-	-	-	-	-	+	-	Move Edge
Change Patterns										
-	-	-	-	-	-	-	-	-	+	Insert Fragment
-	+	-	-	-	-	-	-	-	+	Delete Fragment
-	-	-	-	-	-	-	-	-	+	Move Fragment
-	-	+	-	-	-	-	-	-	-	Replace Fragment
-	-	-	-	-	-	-	-	-	-	Swap Fragment
-	-	-	-	-	-	-	-	-	+	Extract Fragment
-	-	-	-	-	-	-	-	-	+	Inline Fragment
-	-	-	-	-	-	-	-	-	+	Embed Fragment in Loop

TABLE 9.2 Continued

| Commercial | | Academic | | | | | | | | Change |
Staffware	Flower	YAWL+ Worklets/ Exlets	WIDE	WASA2	PoF	MOVE	HOON	CAKE2	ADEPT2/ CBRFlow	
-	-	-	-	-	-	-	-	-	+	Prallelize Activities
-	-	-	-	-	-	-	-	-	-	Embed Fragment in Conditional Branch
-	-	-	-	-	-	-	-	-	+	Add Control Dependency
-	-	-	-	-	-	-	-	-	+	Remove Control Dependency
-	-	-	-	-	-	-	-	-	+	Update Condition
-	-	-	-	-	-	-	-	-	-	Copy Fragment

Meanwhile, PAISs that are used in commercial environments are more vulnerable due to their flexibility and frequent changes in processes, guidelines, rules and policies.

Thus many organizations are seeking ways to detect frauds and avoid financial losses. Today, designing and implementing fraud detection systems is an appropriate solution. PAISs also use these systems to ensure security along with flexibility.

If a benefit-cost analysis for implementation or non-implementation of these mechanisms is done, the cost of non-implementation is more. Thus implementation of a fraud detection system is not only conservative but is also necessary. It is both a part of risk management strategy and a part of business strategy (Khan et al., 2010).

Detecting fraud in PAIS needs special methods due to its nature. It is not possible to use a fix and specified model to detect fraud in PAISs. In these systems, the process model is not specified and it changes depending on the environment changes. Thus we do not know the normal behavior of the system. Also we cannot find the abnormal behavior of the system. Even though, there is just one model, we cannot be certain about what is happen. Therefore, most approaches for detecting fraud follow the log processing that is called process mining.

9.4   PROCESS MINING

In this section we present a brief review on the process mining. Process mining show precise information about what happen in the system.

9.4.1   DEFINITION

At first, Process mining was considered process model discovery. It was defined:

"Discovering process model from the log generated by the information system" (van der Aalst et al., 2003, 2004).

Today, the concept of process mining is more than process model discovery. Now, every log is seen in a form of process can be mine using process mining. Also not only process model can be discovered but some information about organizational structure and cases can be extracted.

Thus, process mining is knowledge mining from the log in a form of process log and use of it for detailed analysis.

The log that is used for process mining should include process instances and activity's events (a well-defined step of process). Sometime we use more information such as performer, timestamp and case data. Figure 9.7 shows a sample of a log.

9.4.2 PROCESS MINING APPROACHES

The important subject in process mining is that: "what is happen is not what we think." It is a good reason to use process mining for fraud detection. Process mining can be performed in three perspectives:

- process perspective;
- organizational perspective; and
- case perspective.

Process perspective tries to reply the question "How?" It focuses on the activities' order. Organizational perspective replies the question "who?" and focuses on the performers. Case perspective replies the question "what?" and focuses on the process instances. There are many algorithms for process mining in the each perspective but process perspective is more considered. We review some important approaches in this section.

In 2003, Van Der Aalst et al. have presented a detailed review of process mining approaches (van der Aalst, 2004). Here we have a brief look at their work.

9.4.2.1 Approach Based on Petri-net Theory

In this theoretical approach, it is assumed that there is no noise in the log and enough information is provided. Alpha algorithm is followed this approach. The algorithm assumes that the log is complete, that is, if a task can be followed directly by another task; an instance of this behavior is seen in the log. Alpha algorithm is based on the four sequential relations that are obtained from the log. These relations are represented with specific signs: $>_w$, $\rightarrow_w$, $\#_w$, and $_w\|$ where W is a log and a set of tasks $(T \ (W \in P(T))$. The relations are defined as follow:

Directive description	Event	User	yyyy/mm/dd hh:mm
Case 10			
	Start	bvdongen@staffw_e	2002/06/19 12:58
Register	Processed to	bvdongen@staffw_e	2002/06/19 12:58
Register	Released by	bvdongen@staffw_e	2002/06/19 12:58
Send questionnaire	Processed to	bvdongen@staffw_e	2002/06/19 12:58
Evaluate	Processed to	bvdongen@staffw_e	2002/06/19 12:58
Send questionnaire	Released by	bvdongen@staffw_e	2002/06/19 13:00
Receive questionnaire	Processed to	bvdongen@staffw_e	2002/06/19 13:00
Receive questionnaire	Released by	bvdongen@staffw_e	2002/06/19 13:00
Evaluate	Released by	bvdongen@staffw_e	2002/06/19 13:00
Archive	Processed to	bvdongen@staffw_e	2002/06/19 13:00
Archive	Released by	bvdongen@staffw_e	2002/06/19 13:00
	Terminated		2002/06/19 13:00
Case 9			
	Start	bvdongen@staffw_e	2002/06/19 12:36
Register	Processed to	bvdongen@staffw_e	2002/06/19 12:36
Register	Released by	bvdongen@staffw_e	2002/06/19 12:35
Send questionnaire	Processed to	bvdongen@staffw_e	2002/06/19 12:36
Evaluate	Processed to	bvdongen@staffw_e	2002/06/19 12:36
Send questionnaire	Released by	bvdongen@staffw_e	2002/06/19 12:36
Receive questionnaire	Processed to	bvdongen@staffw_e	2002/06/19 12:36
Receive questionnaire	Released by	bvdongen@staffw_e	2002/06/19 12:36
Evaluate	Released by	bvdongen@staffw_e	2002/06/19 12:37
Process complaint	Processed to	bvdongen@staffw_e	2002/06/19 12:37
Process complaint	Released by	bvdongen@staffw_e	2002/06/19 12:37
Check processing	Processed to	bvdongen@staffw_e	2002/06/19 12:37
Check processing	Released by	bvdongen@staffw_e	2002/06/19 12:38
Archive	Processed to	bvdongen@staffw_e	2002/06/19 12:38
Archive	Released by	bvdongen@staffw_e	2002/06/19 12:38
	Terminated		2002/06/19 12:38

FIGURE 9.7   A sample of a log (van der Aalst, 2004).

If $a, b \in T$:

1. $a>_w b$: if and only if there is a sequence like $\sigma = t_1 t_2 t_3 ... t_{n-1}$ such that:
   $i \in \{1, ..., n-2\}$ ‹ $\sigma \in w$ ‹ $t_i = a$ ‹ $t_{i+1} = b$.
2. $a \rightarrow_w b$: if and only if $a >_w b$ and $b \not>_w a$.
3. $a \#_w b$: if and only if $a \not>_w b$ and $b \not>_w a$.
4. $a \parallel_w b$: if and only if $a >_w b$ and $b >_w a$.

$a >_w b$ is when task a is followed by task b at least once. It does not indicate a causal relationship between a and b, because a and b can be performed in parallel. $\rightarrow_w$ is a causal relation and $\parallel_w$ and $\#_w$ are respectively parallel relation and selection relation. Since all relations can be made using $>_w$, it is assumed that the log is complete with respect $>_w$.

Alpha algorithm cannot support duplicate tasks and special loops but it can mine the logs with timestamps and support all performance metrics.

9.4.2.2 Heuristic Approach

In the previous approach, we assume that the log is complete. But in fact, logs are rarely complete or without noise. Heuristic techniques are developed to overcome this problem. They are not noise-sensitive and support special loops and non-free choice structures. This approach includes three steps:

- First step: construction of dependency/frequency table;
- Second step: determining relations table from dependency/frequency table;
- Third step: reconstruction process model from relation table.
- At first, a dependency/frequency table is constructed. For each task a, following information is achieved:
- The overall frequency of task a (#A);
- The frequency of task a is directly proceed by task b (#B < A);
- The frequency of task a is directly followed by task b (#A < B);
- The frequency of task a is directly or indirectly proceed by task b (#B<<<A);
- The frequency of task a is directly or indirectly followed by task b (#B>>>A);
- The metric that shows the strength of causal relation between task a and task b (#A→B).

If task a is often performed and then task b is performed with a short interval, it is possible that the task a causes the task b. On the other hand, if the task b is performed a little before task a, it is unlikely that the task a causes the task b.

For calculating the strength of causal relation between task a and task b, #A→B-causality counter is changed with a factor δ^n. δ is a fall factor and $\delta \in \{0.0, \ldots, 1.0\}$. In an instance, if task a is performed before task b and n task is placed between them; #A→B-causality counter is increased with a factor δ^n. If in an instance, task b is performed before task a and n task is placed between them; #A→B-causality counter is decreased with a factor δ^n. After processing all instances, #A→B-causality counter is divided by the minimum overall frequency of task a and b (min (#A, #B)).

After construction a dependency/frequency table, a relation table is made. In second step, main relations ($>_w$, $\to_w$, $\#_w$, and $_w\|$) is constructed using simple heuristic techniques and based on the dependency/ frequency table.

In third step, process model is constructed using relation table. in this step alpha algorithm and any algorithm that work based on the relation table, can be used.

9.4.2.3   Inductive Approach

Inductive approach consists of two steps: induction and transformation. In induction step, an algorithm is used to map the instances into a Stochastic Task (Activity) Graph. In transformation step, the Stochastic Task Graph is transformed to a block-structured process model in the form of Adonis (a business process management system). This transformation is necessary because Stochastic Task Graph cannot represent parallel and selection structures.

The difference between inductive approach and first approach is the way that detects dependencies. Graph producer algorithm considers each pair of tasks that is seen in the similar process instances to detect relations, while the first approach considers direct successors.

9.4.2.4   Data Mining Approach

Data mining approach has two notable differences with previous approaches. First, only block-structured models are considered. Second, the mining algorithm is based on the rewriting techniques instead of graph-based techniques. Moreover, this approach aims to mine the minimum and complete models. Completeness means all cases in the log is covered by mined model and minimum means only the cases that are in the log are covered.

Block-structured models are built using nested blocks. These blocks are different in operators and constants. Operators make process flow while constants are tasks that are embed into the process flow. For example S is an operator *sequence*, P is an operator *parallel* and a, b and c are different tasks. In this way, term $S(a,P(b,c))$ shows task a is performed before task b and task c that are performed in parallel. Each term in this definition is always well-defined.

This approach consists of five steps: First, a process instance is made based on the event data of a process. Second, a time-forward algorithm builds a process model from all instances. The model is in the Disjunctive

Normal Form (DNF). It starts with an operator *choice* and places all paths inside this block. The algorithm makes one block for each group of instances and adds it into operator *choice* that is root.

The next step considers the relations between tasks that are obtained from the random order of tasks without real precedence relation between them. These pseudo precedence relations must be identified and removed from the model. To identify pseudo precedence relations, the model is transformed by a term rewriting system that calculates all possible orders in operator *parallel*. Then a search algorithm determines which orders are pseudo precedence relations. Next, the smallest subset of the orders that describes corresponding blocks is determined and other orders are removed as pseudo precedence relations.

At the end of this step, the first transformation of term rewriting systems is reversed because the process model is built on DNF. The last step is a decision mining that is based on a decision tree induction. In this step, an induction is done for each decision point in the model. Performing this step needs the data about the form of process for each trace. A tree induction algorithm makes the decision tree using these data. The tree is transformed and in the end of all steps, a blocked-structure model is obtained that is complete and optimized.

9.5 FRAUD DETECTION IN PAIS USING PROCESS MINING

In the last decade, several works have been done in the domain of fraud detection. Most of these works are based on the process mining. This chapter will provide a brief review of some works at two perspectives: process perspective and organizational perspective. The case perspective is ignored because this perspective uses the business intelligence and data mining techniques. Although each knowledge discovery from process log is considered as process mining but surveying the case perspective is out of time and patience.

9.5.1 PROCESS VIEW

As mentioned earlier, the process perspective tries to reply the question "how?" And focuses on the order of activities. Most efforts have been done in the domain of fraud detection using process mining have emphasized on this perspective. In general, fraud detection in this perspective is based on the anomalies that occur in the control flow.

We know that business processes in PAISs are not fully known, but each event that occurs will be recorded. We can discover much information from these records. For example, you can specify the sequence of performed tasks in the system.

When an instance of a process is performed, its activities are executed respectively and the executions are recorded in the log of system. For example, when a sequence of activities is recorded as *abc* in the log, it means that task *a* was performed before task *b* and task *b* was performed before task *c*. Such sequence is called a trace.

Considering the rare sequences as anomalies is common in the process perspective and most methods use it. However, an abnormal sequence is necessarily a rare sequence but a rare sequence is not necessarily anomaly. Thus, we face a fundamental challenge: Which rare sequence is abnormal?

Van Der Aalst et al. (2005) have proposed a method in 2005 that has considered the rare sequences as anomalies. They claimed that their method can be used for all level of security from low level intrusion detection to high level fraud detection. In their method, they discover a process model from the log and investigate new traces in the model using token game. No abnormal trace would run in the model and stocked in a region of it. It was difference between normal behavior and abnormal behavior. In this way anomalies are detected. However, it is a limited solution because it needs a log of the normal traces to discover process model.

Subsequently, researchers continued the work of Van Der Aalst et al. Fabio Bezerra et al. presented some works in 2007 and 2008. They proposed three anomaly detection algorithms in Bezerra and Wainer (2007, 2008). The base of the algorithms was the method of Van Der Aalst et al. (2005) but they added the notion "Inclusion Cost." The algorithms did not assume that there is a normal log without anomaly. They considered when a trace is not an instance of model; the model needs a structural change. Thus an abnormal trace needs probably more structural changes (Bezerra and Wainer, 2008). It is the concept of inclusion cost. Inclusion cost is the measure of the amount of changes that is done in process model to cover the new trace (Bezerra and Wainer, 2008).

Bezerra et al. used different process mining algorithms and some different measures to optimize their algorithms and presented new works in Bezerra and Wainer (2007, 2011). They have used three measures: Fitness, Structural Appropriateness, and Behavioral Appropriateness in Bezerra and Wainer (2007) and also they have used another measure, size in Bezerra and Wainer (2011).

The works in Bezerra and Wainer (2007, 2011) state that the conformance level between the traces of the log and the mined model of them is less when the log includes abnormal traces. In this way, each trace has conformance variance more than a specified threshold is known as anomaly or fraud.

In order to provide the mentioned base, it is necessary to obtain the conformance variance of the normal log. Analyzing the value of conformance variance in the normal logs helps to specify a threshold for it. The threshold is used to separate abnormal traces from normal traces. Abnormal trace causes more conformance variance than threshold. The conformance variance in Bezerra and Wainer (2007, 2011) has calculated using the measures presented in Rozinat and van der Aalst (2005, 2008). Measures are based on the two dimensions: fitness and appropriateness. The dimensions are defined as follow:

- Fitness: "the extent to which the log traces can be associated with valid execution paths specified by the process model (Rozinat and van der Aalst, 2005)."
- Appropriateness: "The degree of accuracy which the process model describes the observed behavior combined with the degree of clarity in which it is represented (Rozinat and van der Aalst, 2008)."

Appropriateness includes two states: structural appropriateness and behavioral appropriateness. Structural appropriateness shows how much behavior which never seen in process log is proposed by the model (Rozinat and van der Aalst, 2008). Structural appropriateness is related to control flow. There are several grammatical methods to provide similar behavior in the process model. A simple way for measuring the structural appropriateness is calculating the number of different tasks. In order to measure the fitness between the log and the process model, the traces of the log are replayed in the model and the mismatches are counted. Measuring the fitness and appropriateness has been described in detail in Rozinat and van der Aalst (2008).

Bezerra et al. have added size as a new measure in Bezerra and Wainer (2011) and calculated conformance variance using it. Size is similar to structural appropriateness and focuses on the complexity of the process model. It shows the number of places, transitions and edges in the petri net. However, size is not a conformance measure because it does not evaluate the conformance level between the model and the log (Bezerra and Wainer, 2011). Bezerra et al. believe that anomalies cause more complexity in the log because they add new paths to the model. The experiment result of the Bezerra and Wainer (2011) using size measure shows more accuracy.

In 2008, Bezerra, Wainer and Van Der Aalst present a combined work in [6]. It consists of five steps:

1. Determining the scope
2. Process discovery
3. Filtering the fitting models
4. Model selection
5. Log separation

Determining the scope is dependent on the domain and removes instances and activities that are out of scope. Two next steps select the appropriate models. In the fourth step, the most appropriate model is selected and in the final step, the instances are classified into normal and abnormal classes (Bezerra et al., 2009).

This approach has been implemented in the ProM framework and has been executed with a real log. ProM is a framework for process mining and provides the different algorithms for the analysis of three perspectives. The framework was presented in 2005 to unify the format of reading, storing and displaying process log files. It is an open-source framework and described in (Van Dongen, 2005) completely. Many researchers have added new capabilities in the form of a plug-in to the framework.

After the works presented in 2008, Ms. Jalali and Mr. Baraani (2010) have proposed a new approach based on the genetic algorithm. Their approach consists of three main steps: pre-processing, genetic process mining and log classification. Preprocessing is a scope-dependent step that removes irrelevant traces. The next step is finding the most appropriate model that discovers process models using the genetic algorithm. In genetic mining, fitness measure is assigned to each individual of population and aimed at discovering fittest individual. Fittest model describes the behavior of the log in the best way. Finally, in the third step, fittest model is used and the traces are classified in two classes: normal and abnormal. The authors state that anomalous traces are not covered by the model (Jalali and Baraani, 2010). They claimed that their approach has the following benefits:

- The classifier model is discovered during the anomaly detection. Thus, it does not need a normal log before the anomaly detection. In addition, a genetic algorithm-based system can train new changes to the system as it changes. In this way, the approach is flexible and responds quickly to new market strategy.
- Using genetic process mining, the most appropriate model is automatically discovered, and does not need to define precise metric for selecting appropriate model.
- Due to parallelism in the genetic algorithms, it is possible to evaluate several models in one time. Thus, it can be used for solving the large problem.
- Compared with process mining techniques, the genetic algorithms can support all possible structures such as sequence, parallel, selection, loops, free-choice tasks, hidden tasks and duplicated tasks. It is also resistant to noise.

There are many algorithms for genetic process mining. De Medeiros et al. have proposed an approach of genetic process mining in 2006 (De Medeiros et al., 2006). They stated the mined model covers additional behaviors that are never observed in the log (De Medeiros et al., 2006) due to fitness function. Therefore, they proposed an optimized genetic process mining in 2007 (De Medeiros et al., 2007). The new fitness function has focused on the completeness and precision. In this work, the individual that cover all traces and has less additional behaviors is fitter (Song and van der Aalst, 2008). They have verified the fitness function for completeness and precision using real logs. Jalali and Baraani have expressed this genetic algorithm that has implemented in ProM can be used in their approach but no experimental result was proposed.

The newer work of fraud detection in PAIS is the work of Bezerra and Wainer (2013) that have presented three algorithms. The new algorithms have been proposed based on the three algorithm in Bezerra and Wainer (2008).

First algorithm, sampling is shown in Figure 9.8. L as a log, T^A as a set of anomalous traces, s as the percentage of the sampling and *mine* as a process model miner are considered. L is the input and T^A is the output. In the algorithm, each trace has the less frequency than 2% (T^C) is selected among all traces (T). In order to determine whether a trace is anomaly or not, a small set of the traces (S) are considered according on the value of s and a process model is mined based on the set. Next, the desired trace is played in the model. If the model supports the trace, it is considered as a normal trace otherwise it is an anomaly.

There are two main parameters in sampling algorithm: process discovery algorithm and sampling factor. Bezerra et al. have expressed the best values for these parameters are heuristic miner and 70%.

The second algorithm is iteration algorithm that is shown in Figure 9.9. L as a log, T^A as a set of anomalous traces, x as a threshold value for conformance level, *mine* as a process model miner and *conformance* as an algorithm for measuring conformance variance are considered. L is the input and T^A is the output. In the algorithm, each trace has the less frequency than 2% (T^C) is selected among all traces (T). In order to determine whether a trace is anomaly or not, a process model is mined from all traces without the desired trace (T') and a trace that has lowest conformance level with the mined model (t_{min}) is selected. If the conformance level is less than the threshold, the desired trace is detected as an anomaly. The algorithm continues until there is no trace with a less conformance level than threshold. The authors have expressed the best algorithm for the process discovery is alpha algorithm, the best measure for conformance level is appropriateness and the best value for threshold is 0.9.

The third algorithm is threshold algorithm that is shown in Figure 9.10. L as a log, T^A as a set of anomalous traces, x as a threshold value for conformance level, *mine* as a process model miner and *conformance* as an algorithm for measuring conformance variance are considered. L is the input and T^A is the output. In the algorithm, each trace has the less frequency than 2% (T^C) is selected among all traces (T). In order to determine whether a trace is anomaly or not, a process model is mined from all traces without the desired trace (T') and the conformance level between the mined model and all traces (T) is measured. If the conformance level is less than the threshold, the desired trace is detected as an anomaly. It

```
Input: A log L

Output: A set of anomalous traces Tᴬ.

Parameter: A sampling size: s ∈(0, 1)

        Parameter: A process discovery algorithm: mine.

1 T ← the set of all classes of traces from the log L;

2 Tᶜ ← {} used to contain the anomalous candidate traces;

3 foreach t ∈ T do

4      If freqₗ(t) ≤ 2% then

5            Tᶜ= Tᶜ + {t};

6 foreach t ∈ Tᶜ do

7      S ← sample of s% of traces of L;

8      M ←mine(S);
```

FIGURE 9.8 Sampling Algorithm (Bezerra and Wainer, 2013).

is similar to iteration algorithm but it detects all anomalies in one iteration. For this reason, the execution time of threshold algorithm is less than iteration algorithm (Bezerra and Wainer, 2008). The authors have expressed the best algorithm for the process discovery is alpha algorithm, the best measure for conformance level is fitness and the best value for threshold is 0.9.

In general, the experimental result of these algorithms shows a high false positive rate. The authors believe false positives are more important than false negatives.

9.5.2 ORGANIZATIONAL VIEW

PAIS use role-based access control to separate duty and reduce the chance of committing fraud. Many researchers have proposed techniques for the role mining that can be used for fraud detection. Nevertheless, the study that investigates this perspective in fraud detection has not yet

```
Input: A log L

Output: A set of anomalous traces Tᴬ.

Parameter: A value x ∈ (0, 1) for conformance threshold.

Parameter: A process discovery algorithm: mine.

Parameter: A conformance assessment algorithm: conformance.

1 T ← the set of all classes of traces from the log L;

2 Tᶜ ← {};

3 foreach t ∈ T do

4     if freqₗ(t) ≤ 2% then

5             Tᶜ= Tᶜ + {t};

6 repeat

7     Cₘᵢₙ← 1;

8     foreach t ∈ Tᶜ do

9             T'= T - {t};

10            M ← mine(T');

11            cost ← conformance(M,t);

12            if cost < Cₘᵢₙ then

13                    Cₘᵢₙ ← cost;

14                    Cₘᵢₙ ← t;

15     if Cₘᵢₙ < x then

16            Tᴬ = Tᴬ + {tₘᵢₙ};

17            Tᶜ = Tᶜ - {tₘᵢₙ};

18 until Cₘᵢₙ ≥ x;
```

FIGURE 9.9   Iteration Algorithm (Bezerra and Wainer, 2013).

```
Input: A log L

Output: A set of anomalous traces Tᴬ.

Parameter: A threshold conformance value: x ∈ (0, 1).

Parameter: A process discovery algorithm: mine.

Parameter: A conformance assessment algorithm: conformance.

1 T = the set of all classes of traces from the log L;

2 Tᶜ = {} used to contain the anomalous candidate traces;

3 foreach t ∈ T do

4       If freqₗ ≤ 2% then

5               Tᶜ = Tᶜ +{t};

6 foreach t ∈ Tᶜ do

7       T' = T - {t}; M=mine (T');

8       if conformance(M,T) < x then

9               Tᴬ = Tᴬ + {t};
```

FIGURE 9.10   Threshold Algorithm (Bezerra and Wainer, 2013).

been presented. Thus, in this section, we discuss only the methods of organizational mining.

One of the useful works in this perspective is the study of Song and van der Aalst (2008). In this work, the authors propose three methods for organizational mining:

- organizational model mining;
- social analysis;
- information flow between organizational entities.

Organizational model mining discovers an organizational model from process log. Since the log includes only information of executed activities, it is impossible to achieve the real organizational model. However, this approach can discover the group of users who are similar in the executed activities and specify the relations between the activities and these groups.

In this approach, two organizational entities can be extracted from the log: task-based group and case-based group. Task-based group includes the users that work on the similar tasks and case-based group includes the users that work together on similar cases (Song and van der Aalst, 2008). Organizational model mining is done in four ways:

- Default mining: Here, users are grouped based on the tasks that they performed. Therefore, the number of organizational entities is dependent on the number of tasks.
- Using the shared tasks-based metrics: Here, it is assumed that the users that perform similar tasks are related together. Therefore, a user-task matrix is built that shows the frequency of tasks for each user. This matrix is known as a profile for the user. In the matrix, the tasks are placed as the columns and users are placed as rows. In this method, the distance between the rows is measured at first. The distance can be calculated based on the Minkowski, Hamming or Pearson Correlation Coefficient. Then a relational network of users is created based on this distance. Next, the less important edges of the network are removed according a threshold. Each remaining sub-network is an organizational entity.
- Hierarchical organizational mining: in two prior methods, the organizational model is flat while the organizational model should be hierarchical. In this method, a hierarchical clustering is used on the profile matrix. The closer rows are combined together and result in a dendrogram called hierarchical organizational model. It can be cut and change to a flat model.
- Using the case-based metrics: Here, a graph of users that work together is created. Next less important edges or more central nodes are removed.

The main idea of the social analysis is case routing between users. The social analysis uses the metrics that proposed in Van Der Aalst (2005). Different metrics can be used for this work. One of the most common metrics is handover of work. A handover of work from i to j is happen when a case covers two sequential tasks in such a way that first task is performed by user i and second task is performed by user j. In this way an edge is made from node i to node j (Song and van der Aalst, 2008). Other metrics is expressed as follow:

- Density: the number of elements in graph divided by the maximal number of elements
- Maximum geodesic distance in the graph: the distance of the shortest path between two nodes in the graph.
- Closeness: The reverse total of all geodesy paths that visit the desired node.
- Betweenness: a rate base on the number of geodesy paths that visit the desired node.

Discovering the information flow between organizational entities is done based on the mined social network in prior method. The difference is the network that shows the connections between organizational entities.

All these methods have implemented in ProM and can be used to compare the defined organizational model with the mined organizational model. It can be useful for fraud detection.

9.6 DISCUSSION

In this chapter, we investigate the need of Process Aware Information Systems (PAISs) in the organizations and the importance of the fraud detection in them. Today, PAIS is a necessity for organizations and its flexibility arise the need of fraud detection. Until now, many methods have been proposed for fraud detection in PAIS. We also investigate some of these works in this chapter. In this section, we present a compression of them and express some open problems.

All fraud detection method that is presented in this chapter is shown in Table 9.3. It can be seen that all methods focus on the process perspective. Most of them mine a process model and use it as a classifier for detecting fraud. The flexibility has more focus on the activities' order. Thus, in fraud detection, the process perspective is more important than other perspectives. In this way, the methods are performed in an unsupervised approach. It means there are no labeled instances and normal/abnormal behavior is not known. The nature of the PAIS forces to use unsupervised methods.

P: Process, S: Supervised, U: Unsupervised, Y: Yes, N: No, FP: False positive, FN: False Negative, A: Artificial, R: Real.

TABLE 9.3 Fraud Detection Methods Using Process Mining

Method	Year	perspective	Supervision	Implementation	Evaluation Measure	Evaluation Method	Log	challenges
Conformance Checking	2005	P	S	Y	-	-	-	It uses a normal model for fraud detection that is not always available.
Conformance Checking	2007	P	S	Y	FP and FN	ROC Curve	A	It is time consuming. Two process models are mined for each infrequent trace in the log.
Sampling	2007–2008	P	U	Y	FP and FN	ROC Curve	A	It is time consuming. A process model is mined for each infrequent trace in the log.
Iteration	2007–2008	P	U	Y	FP and FN	ROC Curve	A	It has high false positive rate due to process model.
Threshold	2007–2008	P	U	Y	FP and FN	ROC Curve	A	It has high false positive rate due to process model.
Combined	2009	P	U	Y	-	-	R	It is dependent on the domain.

TABLE 9.3 Continued

Method	Year	perspective	Supervision	Implementation	Evaluation Measure	Evaluation Method	Log	challenges
Genetic-based	2010	P	U	N	-	-	-	the genetic algorithm does not cause a structured output and in each run has different result.
New Sampling	2013	P	U	Y	FP and FN	F4-measure	A-R	It has high false positive rate due to process model. Detecting is directly related to the sample that is selected.
New Iteration	2013	P	U	Y	FP and FN	F4-measure	A-R	It has high false positive rate due to process model.
New Threshold	2013	P	U	Y	FP and FN	F4-measure	A-R	It has high false positive rate due to process model.

Despite the efforts in this domain, there are still some open problems. We express some cases as follow:

- Despite the importance of the process perspective, we should not ignore other perspective. An appropriate fraud detection method must be able to detect the frauds of all perspectives.
- For each log, there are several models that support all traces in the log. The mined model that is used for classifying and detecting frauds is select in more precision. If the model support the traces more than all traces in the log, it cause high false negative rate. If the model support the traces less than all traces in the log, it cause high false positive rate. Thus, selecting the best model for classifying and detecting frauds is a challenge. Some works mine more models to overcome the challenge. It might be time-consuming.
- The evaluation of the methods is often performed on an artificial log. Use of these logs provides researcher with more control on the log but does not provide enough confidence about the performance of the method.

- All traces are examined in the methods. Is all traces are worth to investigate?

The research on the fraud detection in PAISs has recently been initiated. There are many subjects that has not yet considered. Thus, researchers continue to study.

KEYWORDS

- **anomaly detection**
- **fraud detection**
- **process aware information system**
- **process mining**

REFERENCES

1. Bezerra, F., Wainer, J., "Algorithms for anomaly detection of traces in logs of process aware information systems," *Information Systems 38*(1), pp. 33–44, 2013.

2. Bezerra, F., Wainer, J., "Anomaly Detection Algorithms in Business Process Logs," *10th International Conference on Enterprise Information Systems*, pp. 11–18, 2008.
3. Bezerra, F., Wainer, J., "Anomaly Detection Algorithms in Logs of Process Aware Systems," *Proc. the ACM Symposium on Applied Computing*, pp. 951–952, 2008.
4. Bezerra, F., Wainer, J., "Fraud Detection in Process Aware Systems," *International Journal of Business Process Integration and Management* 5 (2), pp. 121–129, 2011.
5. Bezerra, F., Wainer, J., "Towards detecting fraudulent executions in business process aware Systems," *WfPM Workshop on Workflows and Process Management*, 2007.
6. Bezerra, F., Wainer, J., van der Aalst, W. M., "Anomaly Detection using Process Mining," *Enterprise, Business-Process and Information Systems Modeling*, pp. 149–161, 2009.
7. De Medeiros, A. A., Weijters, A. J. M. M., van der Aalst, W. M., "Genetic Process Mining: A Basic Approach and its Challenges," In *Business Process Management Workshops*, pp. 203–215, 2006.
8. De Medeiros, A. K. A., Weijters, A. J., van der Aalst, W. M., "Genetic process mining: an experimental evaluation," *Data Mining and Knowledge Discovery* 14 (2), pp. 245–304, 2007.
9. Dumas, M., Van der Aalst, W. M., Ter Hofstede, A. H., "Process Aware Information Systems: Bridging People and Software through Process Technology," Hoboken, New Jersey, *John Wiley & Son*, 2005.
10. Flegel, U., Vayssière, J., Bitz, G., "A State of the Art Survey of Fraud Detection Technology," *Insider Threats in Cyber Security*, pp. 73–84, 2010.
11. Jalali, H., Baraani, A., "Genetic-based Anomaly Detection in Logs of Process Aware Systems," *World Academy of Science, Engineering and Technology* 64, pp. 304–309, 2010.
12. Jans, M., van der Werf, J. M., Lybaert, N., Vanhoof, K., "A Business Process Mining Application for Internal Fraud Mitigation," *Expert Systems with Applications* 38 (10), pp. 13351–13359, 2011.
13. Khan, R. Q., Corney, M. W., Clark, A. J., Mohay, G. M., "Transaction Mining for Fraud Detection in ERP Systems," *Industrial Engineering and Management Systems*, 9(2), 2010.
14. Leitner, M., Rinderle-Ma, S., Mangler, J., "Responsibility-driven Design and Development of Process-aware Security Policies," *Sixth International Conference on Availability Reliability and Security*, pp. 334–341, 2011.
15. Reichert, M., Rinderle, S., Kreher, U., Dadam, P., "Adaptive Process Management with ADEPT2," *International Conference on Data Engineering – ICE*, pp. 1113–1114, 2005.
16. Reichert, M., Rinderle-Ma, S., Dadam, P.," Flexibility in Process-aware Information Systems," *Transactions on Petri Nets and Other Models of Concurrency II*, pp. 115–135, 2009.
17. Rinderle-ma, S., Reichert, M., Weber, B., "On the Formal Semantics of Change Patterns in Process-Aware Information Systems," *Object-Oriented and Entity-Z Modeling/International Conference on Conceptual Modeling/the Entity Relationship Approach - ER(OOER)*, pp. 279–293, 2008.
18. Rozinat, A., van der Aalst, W. M., "Conformance checking of processes based on monitoring real behavior," *Information Systems* 33 (1), pp. 64–95, 2008.
19. Rozinat, A., van der Aalst, W. M., "Conformance checking of processes based on monitoring real behavior," *Information Systems* 33 (1), pp. 64–95, 2008.

20. Rozinat, A., van der Aalst, W. M., "Conformance testing: Measuring the fit and appropriateness of event logs and process models," In *Business Process Management Workshops*, pp. 163–176, 2005.

21. Song, M., van der Aalst, W. M., "Towards comprehensive support for organizational mining," *Decision Support Systems* 46 (1), pp. 300–317, 2008.

22. The Association of Certified Fraud Examiners (ACFE), "Report to the Nation on Occupational Fraud and Abuse," Austin, TX, *ACFE*, 2012.

23. Van der Aalst, W. M., Weijters, A. J. M. M., "Process Mining: A Research Agenda," *Computers in Industry 53 (3)*, pp. 231–244, 2004.

24. Van Der Aalst, W. M., Reijers, H. A., Song, M., "Discovering Social Networks from Event Logs." *Computer Supported Cooperative Work (CSCW)*, pp. 549–593, 2005.

25. van der Aalst, W. M., van Dongen, B. F., Herbst, J., Maruster, L., Schimm, G., Weijters, A. J. M. M., "Workflow mining: A survey of issues and approaches," *Data and Knowledge Engineering 47*(2), pp. 237–267, 2003.

26. Van der Aalst, W. M., De Medeiros, A. K.A, "Process Mining and Security Detecting Anomalous Process Executions and Checking Process Conformance," *Electronic Notes in Theoretical Computer Science* 121, pp. 3–21, 2005.

27. Van Dongen, B. F., de Medeiros, A. K. A., Verbeek, H. M. W., Weijters, A. J. M. M., van der Aalst, W. M., "The ProM framework: A new era in process mining tool support," *Applications and Theory of Petri Nets*, pp. 444–454, 2005.

28. Weber, B., Reichert, M., Rinderle-Ma, S., "change patterns and change support features - enhancing flexibility in process-aware information systems," *Data and Knowledge Engineering*, vol. 66, no. 3, pp. 438–466, 2008.

29. Weber, B., Reichert, M., Wild, W., Rinderle, S., "Balancing Flexibility and Security in Adaptive Process Management Systems," In *On the Move to Meaningful Internet Systems*, pp. 59–76, 2005.

30. Wells, J. T., "Corporate Fraud Handbook: Prevention and Detection," Hoboken, New Jersey, *John Wiley & Sons*, 2011.

CHAPTER 10

E-ROBOTICS COMBINING ELECTRONIC MEDIA AND SIMULATION TECHNOLOGY TO DEVELOP (NOT ONLY) ROBOTICS APPLICATIONS

JÜRGEN ROSSMANN, MICHAEL SCHLUSE, MALTE RAST, and LINUS ATORF

Institute for Man-Machine Interaction, RWTH Aachen University, Germany

CONTENTS

Robotic applications are complex, require hardware prototypes, and thus are usually a costly endeavor. The research field of eRobotics has been established recently in order to cope with the inherent complexity, facilitate the development, and significantly cut down the cost of advanced robotics and mechatronics development. The objective is to effectively use electronic media—hence the "e" at the beginning of the term—to achieve the best possible advancements in the development of the robots respective fields of use. The aim of developments in eRobotics is to provide a comprehensive software environment to address the robotics-related issues for next generation robotic applications.

Advanced robotic applications in the field of space robotics, that is, planetary explorers, rovers, servicing satellites, and also for the International Space Station have made clear that combining robotic know-how with advanced simulation technology holds great promises not only for the development and optimization of the robotic system itself but also for the development of the man machine interface, operator training, and the dissemination of results.

The key idea of eRobotics has always been to not only to use electronic media to cope with the complexity of robotic applications but also to be able to efficiently represent and to provide the inherent capabilities of a robotic system to an application developer and to the user. Thus, the idea of "Virtual Testbeds" is closely connected to eRobotics. Actually, the notion of a "testbed" originates from the space community as engineers working in this field usually have to deal with environments like free space or planetary surfaces that are not easily accessible for system tests. In order to be able to test space hardware before the actual flight, engineers usually set up comprehensive—and mostly very expensive—testbeds. These testbeds are physical setups that try to replicate the geometric and most of the physical conditions, for example, a space robot has to face during its mission. Advanced testbed environments comprise pressure and temperature testing in a thermal-vacuum chamber, shock- and vibration testing devices, etc., all to make sure that a space robot does its job in the environmental conditions in space.

The problem today is that while the physical robot is being tested, it is not available for the development of, for example, the supporting software. And as those tests take months, if everything goes well—and years if it does not—this may cause significant delays especially for the software development and testing. Since recently, the only way to deal with the non-availability of the space device was to build multiple so-called engineering models: hardware replica of the flight hardware with a little less expensive components, but still quite expensive. These engineering models operate in a model environment (e.g., a sandbox for planetary landing missions) and make up the "testbed" for the developed software.

Nowadays, eRobotics applications step in to provide a substitute for physical engineering models by providing "Virtual Testbeds," advanced 3D simulation environments which realistically model all important geometrical and physical properties of a space application. Figures 10.1 and 10.2 show examples of two of the most advanced Virtual Testbeds available today, the Virtual International Space Station, the Virtual Crater for planetary exploration and Virtual On-Orbit Satellite Servicing.

After the first experiences were made with the use of Virtual Testbeds in Space, it became clear that these developments are not only useful to develop space applications but to communicate the use of robots in general. Figure 10.3 depicts a major problem robot developers face when

FIGURE 10.1 The Virtual Testbed of the ISS, a current eRobotics application for astronaut training and for the development and test of robotic concepts [62].

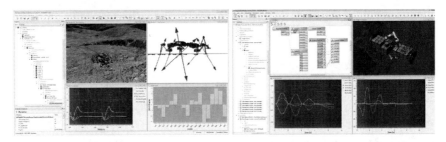

FIGURE 10.2 Two typical Virtual Testbed environments, here used for the development of walking robots [5] (left, robot model © DFKI Bremen) and satellite servicing [57] (right, satellite model © TU Berlin).

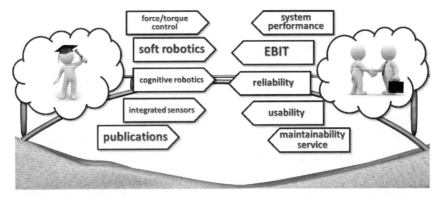

FIGURE 10.3 eRobotics supports communication between developers and potential clients.

they try to convince industrial clients of the capabilities of modern robotic developments. The problem is that researchers and businessmen speak "different languages," that is, they use different vocabulary *to describe and measure the success of a robotic application.*

Whereas researchers are proud of a robot's new capabilities, for example, in the field of soft robotics or advanced programming and control architectures, businessmen are usually more interested in the contribution of an envisaged robotic application to the company's profit as is measurable by its EBIT (earnings before interests and taxes). In addition, usability, maintainability, and service are usually way more important to the business side than to the research side (Figure 10.4).

Figure 10.5 depicts the main arguments for many businesses today to develop an eRobotics application before turning it into a real robotic application. One may ask what the difference is between to two pictures of Figure 10.5. To a businessman the difference is simple: It is approximately €130.000! This is the difference in the investment required to find out if the robot can "earn his money" in the envisaged application. In order to find out, if the depicted KUKA lightweight robot, one of the most advanced industrial robots in the world, provides a benefit for their

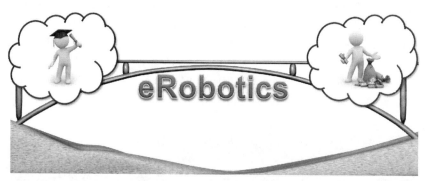

FIGURE 10.4   eRobotics builds a bridge between research and industry.

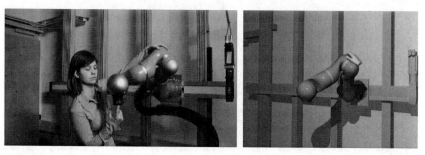

FIGURE 10.5   What is the difference between the two pictures above? Approximately €130.000!

business, the entrepreneurs would have to invest approx. €100.000 into the robot hardware and at least a few €10.000 into the human resources to set up an example application. On the right of Figure 10.5, an equivalent eRobotics based Virtual Testbed has been set up which required an investment of only a few thousand Euros.

The Virtual Testbed furthermore has the advantage that the robot can efficiently be tested under varying load conditions and in different setups (e.g., in multi-robot configurations) with almost no additional effort and cost. This makes the success of eRobotics in this field! The eRobotics application is easy to set up, easy to understand, and its results can serve as reliable basis for amortization calculations.

Space Robotics and industrial robotics (see Figure 10.6) are only two facets of current eRobotics developments. This paper will first introduce the term "eRobotics" and the basic eRobotics concepts (Chapter 1) like the Virtual Testbed approach (Chapter 2). This is the basis for new Simulation-based Optimization, Reasoning and Control concepts (Chapter 3). eRobotics makes great demands on the underlying simulation technology which leads to a new 3D simulation infrastructure (Chapter 4). Chapter 5 outlines some of the key aspects in 3D simulation from an eRobotics point of view before Chapter 6 then turns to current advanced eRobotics applications.

FIGURE 10.6 Trying to predict the week in which a robotic manufacturing application will amortize. That is one ultimate challenge to an eRobotics application.

10.1 WHAT IS E-ROBOTICS?

Formally, the term "eRobotics" can be defined as follows: "eRobotics is a branch of eSystem Engineering. eRobotics targets the development of concepts as well as the continuous and systematic provision of computer-based methods and processes addressing the entire lifecycle of complex systems. To this end, eRobotics combines the use of electronic media, state of the art simulation technology and robotics concepts." Initially, eRobotics targeted the development of robotic applications. However, in the past few years, the range of applications widened. Today, the eRobotics technology is used in the field of robotics—or more general mechatronics—as well as in various other fields like environment (forest, city) modeling and simulation, industrial automation, etc. (see Chapter 6).

Figure 10.7 illustrates the definition above. At first, eRobotics targets the development of new concepts for the development of complex systems as introduced above. To be able to use these concepts in everyday engineering tasks, eRobotics refines well-known engineering processes to provide new simulation-based agile development processes (see Chapter 1.1). To provide the necessary degree of computer support, eRobotics makes extensive use of 3D simulation technology (see also Chapter 1.2) such as semantic world modeling techniques to provide the necessary models for 3D simulation. These two form the basis for the realization

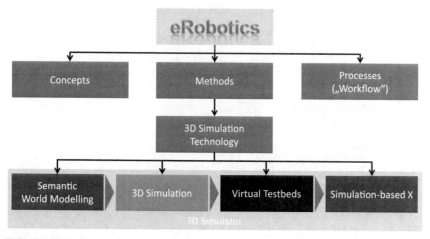

FIGURE 10.7 The definition of eRobotics.

of Virtual Testbeds which themselves provide the necessary basis for new "Simulation-based X"-concepts. All these simulation-based methods must be provided by a versatile 3D simulator (see Chapters 4 and 5).

10.1.1 SIMULATION-BASED ENGINEERING PROCESSES

The aim of eRobotics is to provide a comprehensive software environment for the development of complex technical systems. Starting with user requirements analysis and system design, support for the development and selection of appropriate hardware, programming, system and process simulation, control design and implementation, and encompassing the validation of developed overall systems, eRobotics provides a continuous and systematic computer support during the entire life cycle of complex systems (see Figure 10.8). 3D simulations are used right from the beginning of the development process to test first system design studies in the concept phase. During system development, fully functioning interactive virtual prototypes allow for an efficient and goal-oriented development, test, and

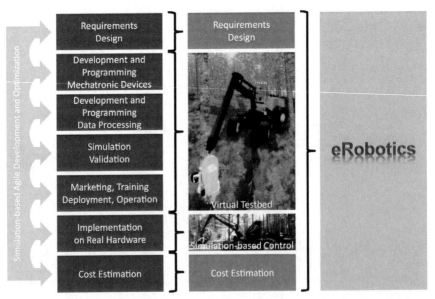

FIGURE 10.8 eRobotics supports the entire lifecycle of complex systems by an extensive use of 3D simulation technology.

verification both on component and on system level—at any point of time. Besides, 3D simulation technology not only allows to visualize, simulate, test, and experience the virtual prototype by providing so-called "virtual testbed" facilities [29], but can also be used as a development framework to implement both control and supervisor algorithms (such as motor controllers, robot programs, image processing algorithms) using concepts of "simulation-based control" [30].

The use of eRobotics is not restricted to certain development processes. eRobotics supports the waterfall model in software development, the technology readiness level oriented approach in space engineering, as well as the V model or modern agile development approaches.

Combining all these aspects, eRobotics enables new simulation-based agile development and optimization processes spanning the whole life-cycle of complex systems. In this way, the ever-increasing complexity of current computer-aided technical solutions will be kept manageable and know-how from completed work is electronically preserved and made available for further applications.

10.1.2   SIMULATION-BASED ENGINEERING METHODS

eRobotics provides various engineering methods (see Figure 9), which can be used in the different stages of a systems lifecycle. The methods all rely on 3D simulation technology.

Basis for this is a "3D model" of the system to be developed. According to Ref. [31], a model in general "is a simplified reproduction of a planned or existing system with its processes in a different conceptual or concrete system. Its differences from the real system in terms of its characteristics that are relevant to the investigation are within a given range of tolerance." One important aspect of eRobotics is the description and analysis of the spatial behavior of systems. That is why eRobotics follows the notion of a "3D model" which illustrates that one important aspect of the conceptual or concrete target system is the modeling of spatial system properties and behaviors (geometry, material, movement, etc.). New concepts of "Semantic World Modeling" [32] provide the methods necessary to set up these models considering all relevant components, systems, environments, controllers, supervisors, etc. The models can be built fully automatically,

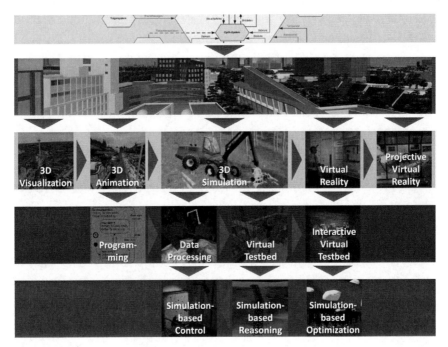

FIGURE 10.9   Classical and new 3D simulation based engineering methods provided by eRobotics.

manually or semi-automatically, using interactive modeling environments. Methods for system analysis and design support the developer even in the first development phases.

This 3D model is the basis for "3D simulation." In the context of eRobotics, "simulation is the representation of a system with its dynamic processes in an experimentable model to reach findings which are transferable to reality. In particular, the processes are developed over time. In the broader sense, simulation refers to the preparation, execution, and evaluation of targeted experiments with a simulation model" [31]. That is why eRobotics provides all the methods necessary for 3D visualization, 3D animation, 3D simulation, and Virtual Reality (VR). "Projective Virtual Reality" methods link the VR system to the real world to intuitively monitor and interactively command physical systems [40].

The integration of data processing algorithms like image processing or generic controllers or supervisors leads to a comprehensive virtual testing facility, the "Virtual Testbed," allowing for detailed (interactive or preprogrammed) tests at system level (see Chapter 2).

In addition to this, Virtual Testbeds are the basis of various simulation-based engineering concepts (see Chapter 3). "Simulation-based Optimization" enables an engineer to optimize system properties or controller parameters. "Simulation-based Reasoning" concepts allow the controllers of technical systems to "think" using the Virtual Testbed as a detailed model of their behavior. "Simulation-based Control" concepts bridge the gap between simulation and reality by attaching the 3D simulation system to physical systems, allowing the virtual data processing algorithms to not only work on virtual devices, but also in real world scenarios in the context of "Rapid Control Prototyping" [33].

10.2   THE VIRTUAL TESTBED CONCEPT

Generally speaking, Virtual Testbeds are advanced 3D simulation environments, which realistically model all important aspects of an application. To be more precise, a "Virtual Testbed" can be defined as follows: "A Virtual Testbed is a 3D simulation software environment for the integrated cross-system, -discipline, and -application development of complex systems based on experimentable 3D models. It provides (interactive) methods for the reproducible test and for the verification and validation of such systems under arbitrary boundary conditions. For this, it integrates 3D models, simulation algorithms, and real world devices for a holistic view of the dynamic overall system including internal interdependencies in a user-defined granularity."

10.2.1   FROM SIMULATION TO VIRTUAL TESTBEDS

Virtual Testbeds greatly enhance the scope of the "classical simulation approach" (see Figure 10.10). While typical simulation applications examine only specific aspects of an application (e.g., a laser scanner scanning some obstacles), a Virtual Testbed enables the development engineer to examine the entire system in its environment (the laser scanner is mounted on a harvester cutting trees in a virtual forest environment). To be

FIGURE 10.10   Distinguishing simulation from Virtual Testbeds from Simulation-based Control.

able to replicate the entire harvester, the Virtual Testbed also contains the necessary control algorithms (for example for harvester localization, see Chapter 6.2). These algorithms cannot only be used for simulation but also in the real word (connected to real laser scanners) using Simulation-based Control approaches or to set up versatile user interfaces.

10.2.2   A MATHEMATICAL VIEW ON VIRTUAL TESTBEDS

Figure 10.11 outlines the basic structure of Virtual Testbeds, the components involved, and their state and parameter vectors. Each system vector $\underline{s}(t) = [\underline{x}(t), \underline{a}]$ consists of a dynamic part $\underline{x}(t)$ (any time-dependent state variables such as positions of moving parts, joint values, or motor currents) and a static part $\underline{a}$ (e.g., controller parameters, fixed structural dimensions, or any other constants).

The core of the figure is the data processing system (DPS) processing sensor data or commanding the robots' movements. The internal system vector of this DPS is called $\underline{s}_{impl}^{dps}(t)$. All sensor data input to the DPS is contained within $\underline{s}_{sense}^{dps}(t)$, whereas $\underline{s}_{act}^{dps}(t)$ contains output to actuators. Sensory input to a DPS is usually evaluated and interpreted in a certain way, which leads to an estimate about some physical quantities of the real environment, the "perceived" environment $\underline{s}_{env}^{dps}(t)$ of the DPS. For example, when a mobile robot system uses a SLAM algorithm to maintain its current location and an estimated map of its environment, then this internal map representation would be part of $\underline{s}_{env}^{dps}(t)$. In conclusion, the full system vector of the DPS is:

$$\underline{s}^{dps}(t) = [\underline{s}_{sense}^{dps}(t), \underline{s}_{impl}^{dps}(t), \underline{s}_{act}^{dps}(t), \underline{s}_{env}^{dps}(t)] \tag{1}$$

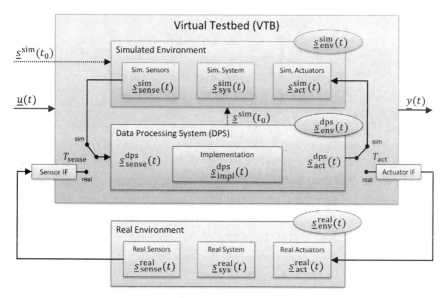

FIGURE 10.11   The structure of a Virtual Testbed.

During regular operation of the DPS in production or for evaluation, both switches T_{senses} and T_{act} are set to position "real," that is, is $\underline{s}_{sense}^{dps}(t)$ fed from the physical sensors with $\underline{s}_{sense}^{real}(t)$ and the DPS's actuator commands $\underline{s}_{act}^{dps}(t)$ are forwarded to the physical actuators $\underline{s}_{act}^{real}(t)$.

To model the physical system (e.g., vehicle, robot, or assembly line), all its dynamic and static properties are aggregated in $\underline{s}_{sys}^{real}(t)$, which covers mechanical structure, joints, and software modules not included in the DPS—any relevant state of the real system. Finally, we have the physical environment the system is operating in. All relevant physical properties, for example, geometric shapes of the ground and surroundings, lighting conditions, gravity, and so on, are represented by $\underline{s}_{env}^{real}(t)$. Hence, similar to Eq. (1), we have the full system vector of the physical system and its environment in

$$\underline{s}^{real}(t) = [\underline{s}_{sense}^{real}(t), \underline{s}_{sys}^{real}(t), \underline{s}_{act}^{real}(t), \underline{s}_{env}^{real}(t)] \qquad (2)$$

In order to operate the DPS in a computer simulation with both T_{senses} and T_{act} switches and set to position "sim," we need to provide realistic sensor data $\underline{s}_{sense}^{dps}(t)$ and have the simulation react to the output $\underline{s}_{act}^{dps}(t)$. This is only possible in a simulated environment, which mimics all relevant aspects of the real world $\underline{s}_\alpha^{real}(t)$. This is why we introduce a corresponding $\underline{s}_\alpha^{sim}(t)$ for each $\underline{s}_\alpha^{real}(t)$. Together, they make up the full system vector $\underline{s}^{sim}(t)$ (similar to Eq. (2)). Thus, our full VTB consists of the DPS's system vector $\underline{s}^{dps}(t)$ and the simulation's system vector $\underline{s}^{sim}(t)$, so

$$\underline{s}^{vtb}(t) = [\underline{s}^{sim}(t), \underline{s}^{dps}(t)] \tag{3}$$

Key idea of the VTB concept is that the DPS is a component within the whole VTB. This enables development, evaluation, optimization, and productive operation in the very same hard- and software environment. By turning only a single switch—i.e., T_{sense} and T_{act} combined—one can alternate between virtual and real operation. By design, the DPS is not aware whether it is being operated in a real or simulated environment.

The input signal $u(t)$ consists of commands for robot operation, controller set points, trajectories, or other input. It may also be required to initialize the simulated environment and its components $\underline{s}^{sim}(t)$ with $\underline{s}^{sim}(t_0)$, which can usually be loaded from file or sometimes even be generated from $\underline{s}^{dps}(t)$, depending on the scenario.

10.3   SIMULATION-BASED X

Virtual Testbeds are able to efficiently represent and make available the inherent capabilities of a robotic system to an application developer, to the user—and to the robot itself. Such Virtual Testbeds provide the methodic foundation for newly developed Simulation-based engineering processes and methods as well as for the optimization of such systems or the realization of flexible and intelligent systems using the simulation-based X concepts provided by the eRobotics approach.

10.3.1   SIMULATION-BASED OPTIMIZATION

During design and implementation phases, engineers may need to optimize certain aspects $\underline{a}_{impl}^{dps}(t)$ of the DPS or of the system the DPS is part of, that is, $\underline{a}_\alpha^{sim}(t)$ with respect to different environmental properties $\underline{a}_{env}^{sim}(t)$, such as, for example, gravity. By using simulation-based optimization, an optimization controller varies relevant parameters and runs its optimization loop inside the VTB while keeping $\underline{u}(t)$ and $\underline{x}_0^{vtb}$ unmodified. The example below shows how the maximum mass $\underline{a}_{sys,mass}^{sim}$ of a component for the "Space Climber" can be determined before the robot flips over.

10.3.2 SIMULATION-BASED REASONING

The idea of Simulation-based Reasoning is similar to simulation-based optimization. But as the system is already in production, the system properties $\underline{a}^{vtb}$ cannot be changed anymore. However, the input $\underline{u}(t)$ (e.g., trajectories or commands) is still open for optimization. When the optimization controller is part of the VTB, the system can find the best input (i.e., decision) for the current state $\underline{x}_{env}^{sim}(t_0) = \underline{x}_{env}^{dps}(t_0)$ itself, that is, the VTB serves as "mental model" [60]. The example below shows alternative trajectories for the robot "Space Climber" which is trying to reach a crater's rim (top picture: success).

10.3.3 SIMULATION-BASED CONTROL

Virtual Testbeds enable the development of controllers when the real system is not available or when the risks of testing on the real system are too high.

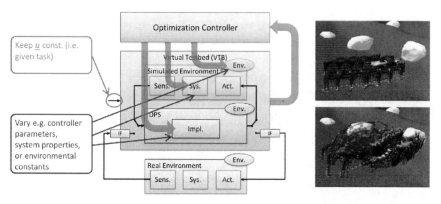

FIGURE 10.12 Basic idea of simulation-based optimization using Virtual Testbeds (left), application to mobile robotics (right, robot model © DFKI, Bremen). Shown is $\underline{s}(t)\forall t \in [0s,15s]$ and $\underline{a}_{sys,mass}^{sim} = [0kg, 45kg]$.

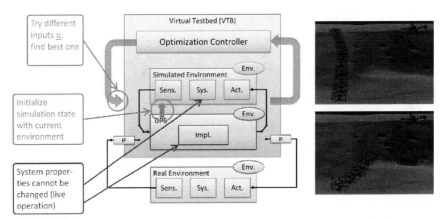

FIGURE 10.13 Basic idea of simulation-based reasoning using Virtual Testbeds (left), application to walking robot trying to reach the rim (right, robot model © DFKI, Bremen). Shown is and $\underline{s}(t) \forall t \in [0\,\mathrm{s}, 300\,\mathrm{s}]$ and $u(t) = const = \{0°, 30°\}$.

A DPS which has been successfully tested in simulation can seamlessly be run in production. No code generation is necessary as the DPS used for the simulated as well as for the physical system runs within the same single framework.

10.4.3D SIMULATION TECHNOLOGY FOR EROBOTICS

To be able to use the concepts introduced so far, the availability of a "3D simulator" offering the performance characteristics necessary to set up 3D models, to simulate these models in Virtual Testbeds and to use the result in simulation-based X approaches is crucial. Formally, a simulator "is a software program which can be used to build a model reproducing a system's dynamic behavior and processes and to make this model executable" [31]. Due to the complexity of the underlying model (with parameters $\underline{a}^{vtb}$ and state variables $\underline{x}^{vtb}(t)$) being executed by different simulation and data processing algorithms simultaneously—sometimes in a direct connection to real world devices—the use of eRobotics concepts leads to new requirements for 3D simulators. For eRobotics applications, we need one single but comprehensive and integrated 3D simulation framework, which is able to implement all the methods and support all the processes outlined above.

10.4.1   REQUIREMENTS

This leads to various requirements regarding the underlying 3D simulation framework:

Overall Flexibility: The simulation system must support a broad range of applications and usage scenarios (see Figure 10.27). It must be usable as an engineering tool on the desktop, as an interactive and immersive "Virtual Reality" system, and as a tool for realizing control algorithms on real-time capable systems. Hence, it must separate simulation algorithms from user interface implementation.

Performance: To allow hands-on interaction, the simulation must perform in real-time. The simulation must also be capable to perform in a hard real-time mode to enable hardware-in-the-loop (HIL) scenarios.

Freely Configurable Database: New usage scenarios require new data structures. Therefore, the data model must be adaptable to new simulation models, even at run time. Thus, a metadata system and a reflection API are necessary. Such a flexible database can then be used in all kinds of data storage and data manipulation scenarios, not only 3D simulation, but any other kind of simulation. This way, all methods can use the same model that contains (on an equal level) geometric information, as well as, for example, sensor configurations or controller programs.

Distributed and Parallel Simulation: In order to separate user interaction from (hard) real-time simulation, and to allow the parallel and distributed execution of the simulation, it is necessary to distribute the simulation state across computer nodes. The mechanisms for an efficient, recursion-free simulation state transfer have to be an integral part of the simulation database.

Modular Simulation Model and Adjustable Level of Detail: Since the level of detail of the model will increase during the development process, the simulation models must be adaptable and interchangeable. Furthermore, some tests may require a more detailed simulation model, but no real-time performance. So it must be possible to record the simulation state changes and play them back later in real-time.

Realism: For many applications (such as games), a computationally fast and visually plausible model is sufficient. However, for the use in engineering processes the simulation error must be quantifiable. The necessary degree of realism of different aspects (visualization, physics, sensors, etc.) depends on the actual use case and must be adjustable.

Calibrated Simulation Algorithms: In order to obtain valid and reliable results, the simulation algorithms have to be calibrated against real systems – the simulation error must be quantified. Depending on the requirements of the application, the simulation parameters have to be adjusted such that the simulation error lies within a specified limit.

Flexible and Standardized Interfaces: To couple and interchange various simulation components, internal interfaces of the simulation components need to be standardized. To use external components, such as special simulation models (co-simulation), control software (software-in-the-loop) or hardware (hardware-in-the-loop), external interface specifications must be adhered to. By mapping the external interfaces to the internal interfaces, it is then possible to easily interchange data between the 3D simulation and external co- and subsystems.

Integrating Data Processing Algorithms: Besides the simulation of its physical behavior, the complete reproduction of a system needs to incorporate control algorithms. Only then it is possible to simulate complex interactions, for example, the influence of actuator control commands on sensor data.

Seamless Transition from Simulation to Reality: The use of block oriented data models and subsequent code generation for controller implementation is standard in "Rapid Control Prototyping." This well-established workflow should also be available in the 3D simulation system, such that data processing algorithms developed with and integrated into the system can be used on the real hardware.

Cross Platform Support: The core of the simulation system should be platform independent, in order to be executable on a variety of desktop OS (Windows, Linux) and mobile devices. It should also be possible to run the simulation without any graphical user interface on embedded devices, using, for example, QNX.

10.4.2 STATE OF THE ART

Taking a look at the state of the art of simulation technology reveals various approaches to simulation technology. Discrete event simulation systems [41], block oriented simulation approaches like the MATLAB/ Simulink framework [43] or the declarative Modelica modeling language [42] as well as various FEM-based simulation tools (e.g., [44]) are probably the most well-known ones. But even when focusing on (quasi continuous) 3D simulation technology, various approaches can be found, starting with game engines (e.g., [45]) or generic mechatronic systems (e.g., [46]).

In the field of robotics, the simulation of robotic systems is a well-known and powerful source of information for robotic applications. A large variety of software tools is available to plan, predict, scale, and safely test different scenarios in various disciplines in a time- and cost-efficient manner. The methodology of robot simulation is usually based on kinematics and/or rigid body dynamics. Whereas kinematic approaches just focus on the path (a sequence of frames or joint angles) of a serial-link manipulator, a rigid body dynamic simulation goes one step further and adds physical interaction and realistic physical properties. Robot dynamics model the robots' reactions under the influence of applied torques and forces by integrating the equations of motion. Today, the most commonly used motion-, contact- and friction models are simplified descriptions of the physical processes [8].

Examples of the use of rigid body models for robotics are the design, analysis, and optimization of conveyor systems [12] and the virtual robot program development for assembly processes [13]. Two essential aspects of rigid body simulations are the modeling of the relevant physical effects on the one hand and the parameterization of models on the other hand. The identification and validation of parameters is achieved in different experiment setups [9–11].

In the field of robotics there is a variety of efficient simulation software available. Some known and currently used simulation software tools are ROBCAD [14], ROBOTEXPERT [15] and PROCESS SIMULATE [16] from Siemens PLM software, DELMIA from Dassault Systems [17] and CIROS [18] (all used for simulation and decision making of robot-based automation), EASY-ROB [19] (robotic simulation down to joint level), V-REP [20] and GAZEBO [21] (both specialize on sensors and image-based assembly systems), and ROBOTRAN [22] (physics-based symbolic modeling of rigid body systems, used for symbolic regression models in the identification of target-robotic-systems, like in [23]). The application of such simulation tools led to efficient and flexible possibilities for planning and optimizing robotic factories, sensors, assembly systems and much more.

10.4.3   A NEW 3D SIMULATION INFRASTRUCTURE FOR EROBOTICS

Still missing is a holistic and encompassing approach that enables and encourages the synergetic use of simulation methods on a single database throughout the entire lifecycle of a technical system. This is crucial to eRobotics, but most approaches focus on dedicated application areas, dedicated disciplines (electronics, mechanics, electronics, thermodynamics, etc.), or are restricted to the development of single components.

To overcome these limitations and to fulfill the requirements listed above, we developed a new architecture for simulation systems (which is not restricted to 3D). This framework itself is purely abstract, so that it can act not only as the basis for 3D simulation, but also for virtually any other type of simulation, such as block oriented simulations, discrete event simulations, or FEM analysis methods. In addition, it can be linked to other simulation systems for co-simulation or hard-/software in the loop simulations.

Micro-Kernel Architecture: The key idea is the introduction of a micro kernel, the "Versatile Simulation Database" (VSD, Figure 10.15). Basically, the VSD is an object-oriented real-time database holding a description of the underlying simulation model. Fully implemented in C++, it provides the central building blocks for data management, meta information, communication, persistence, and user interaction. The VSD is called "active" because it is not simply a static data container, but also contains the algorithms and interfaces to manipulate the data. Furthermore, the VSD incorporates an intelligent messaging system that notifies subscribed listeners of data creation, change, and deletion events. In addition, the VSD provides essential functionalities for parallel and distributed simulation. Depending on the application, the performance of the simulation or the controller can thus be enhanced by being parallelized on multi-core processors or the distribution in a network of computers. The mechanisms for distribution allow for separation of real-time processes from graphical user interfaces and monitoring tools (see Figure 10.14, right). All simulation functionality of the framework is achieved by creating specialized plugins, which build upon and interact with the VSD core.

Model Representation: The tiered data model consists not only of the simulation model itself, but also incorporates a meta model layer.

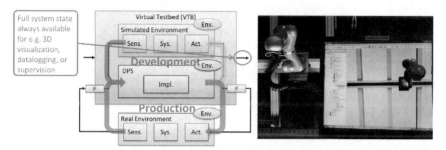

FIGURE 10.14 Basic idea of simulation-based control using Virtual Testbeds (left), application to robot control (right).

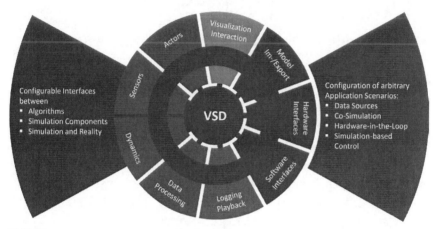

FIGURE 10.15 The micro kernel architecture.

The meta model is essential for the flexibility, as well as the developer and end user friendliness of the database and the simulation system. The design shown in Figure 10.16 is inspired by the Object Management Group (OMG) meta model hierarchy [34]. The uppermost layer (labeled M2) is the meta information system, the basis for persistence, user interface, parallel and distributed simulation, scripting, and communication. It mainly consists of meta types, meta instances, meta properties, and meta methods. In addition to "build-in" classes, it is also possible to generate meta instances with the corresponding meta properties and meta methods during runtime (e.g., for object oriented scripting [47] or new data models [48]). Such "run time meta instances" are treated in exactly the same way as the build in meta instances without any performance overheads in the data management.

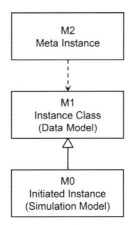

FIGURE 10.16   The meta model hierarchy of the VSD.

The middle layer (labeled M1) describes the data model of the simulation. In order to be able to retain semantic information and integrate data and algorithms into one single database, the VSD data model is an object oriented graph database [35], consisting of nodes and node extensions. A simplified class hierarchy of the VSD core is shown in Figure 10.17. All nodes in the graph database, the database itself and even the simulation environment are derived from a single base class called "Instance." This base class provides mechanisms for inter-instance communication, as well as access to the meta information system which allows introspection of class hierarchy, properties, and methods.

Following this approach, the database is able to integrate standard geometric models as well as block-oriented simulation models using input/output connections (see Figure 10.25), the intermediate representation of scripting languages [36], or entirely different types of information like forest inventory data [37].

System Integration: The functionality of the micro kernel is extended by various plugins implementing simulation or data processing algorithms, interfaces to hard- or software systems, user interfaces, etc. (see Figure 10.15). Using the VSD, the plugins can communicate with the database, as well as establish directed communications between themselves. One crucial point is the combination of different simulation algorithms and the communication between them. This can be challenging in complex

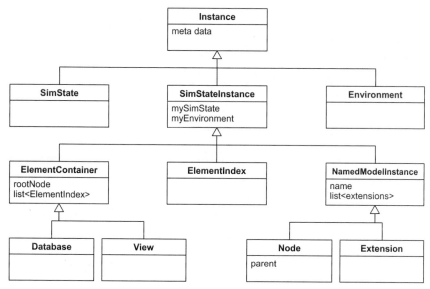

FIGURE 10.17   The core database class hierarchy.

scenarios that incorporate different application domains and require a mutual interaction between the different domains for realistic simulation results (see Chapter 5.2).

10.5   KEY ASPECTS FOR 3D SIMULATION

The 3D simulation infrastructure introduced above acts as a framework which has to be filled with various software components for data storage, visualization, interaction, scheduling, kinematics, rigid body dynamics, process simulation, sensor/actuator simulation, data processing, parallel and distributed simulation, geographic information systems, graphical user interface, etc., just to name the most important ones. In the following we give some examples ranging from rigid body dynamics to visualization techniques to advanced sensor simulation.

10.5.1   ADVANCED MODELING TECHNIQUES

In order to provide a "natural" behavior of simulated objects in the automation and robotics domain, a minimum requirement an eRobotics system has to meet, is to be able to deal with rigid multi-body systems correctly. The following is a short overview of the basic mathematics in this field [1, 2].

All constraints in the multi-body system can be formulated in the following form:

$$\underline{J}\,\underline{v}(t) = \underline{b},\ \underline{b} \geq \underline{0} \tag{4}$$

Herein $\underline{J} \in \mathbb{R}^{m \times 66\prime}$ is the Jacobean matrix of the constraints encoding the m velocity based constraint (in-) equations and $\underline{v}(t) \in \mathbb{R}^6$ is the multi-body velocity vector containing all n bodies' linear and angular velocities. The velocity based constraints result from derivation of classical holonomic constraints $\dfrac{d}{dt}(C(\underline{x}) = 0)$ or can be directly formulated. Thus all holonomic as well as a wide range of nonholonomic constraints match this formulation.

The equations of motion of a multi-body system discretized with an explicit Euler integration step are:

$$\underline{M}\,\frac{\underline{v}(t + \Delta t) - \underline{v}(t)}{\Delta t} = \underline{J}^T\,\underline{\lambda} + \underline{f} \tag{5}$$

$$\underline{v}(t + \Delta t) = \underline{v}(t) + \underline{M}^{-1}\underline{J}^T\,\underline{\lambda}\Delta t + \underline{M}^{-1}\underline{f}\Delta t \tag{6}$$

Herein the vector $\underline{J}^T\underline{\lambda}\Delta t$ describes the "constraint impulses." Due to d'Alembert's principle of virtual work their directions are given by the constraints themselves, $\underline{J}^T$; only their absolute values $\underline{\lambda}$ are unknown. Note that the vector $\underline{f}$ of the torques and forces not only contains explicit external forces like gravity, but also the gyroscopic term of Euler's equation of angular motion $\underline{M} \in \mathbb{R}^{6n \times 6}$ is the multi-body system's combined mass matrix, simply containing all the body's masses and inertia tensors along its main diagonal. Because of its block diagonal structure, it can be inverted block wise. Mass scalars can always be inverted, as they may not be zero. Inertia tensors can always be inverted, as their eigenvalues correspond to the *principal moments of inertia* and thus are positive. That is why each inertia tensor is positive definite.

In this kind of formulation the vector $\underline{\lambda}$ is subject to some additional complementarity conditions: $\underline{\lambda}$ is asked to lie within any specified limits:

$\underline{\lambda}_{low} \leq \underline{\lambda} \leq \underline{\lambda}_{high}$. The most popular example for the usage of these limits is a contact's normal impulse: a contact prevents the bodies from interpenetrating each other, but it may not *pull* the bodies in contact towards each other. Therefore, in this example the Lagrange multiplier of the *i*-th contact's interpenetration impulse is limited by $0 \leq \lambda_i \leq \infty$. Note that λ_{low}, λ_{high} may take any rational number allowing also "technical" limits such as maximum motor torques, etc.

Bringing together (4) and (6) leads to:

$$\underline{J}\,\underline{v}(t + \Delta t) = \underline{b}, \; \underline{b} \geq \underline{0} \tag{7}$$

$$\underline{J}\,\underline{v}(t) + \underline{J}\underline{M}^{-1}\underline{J}^T \lambda t + \underline{J}\underline{M}^{-1}\underline{f}\Delta t = \underline{b}, \; \underline{b} \geq \underline{0} \tag{8}$$

Complementary to:

$$\underline{\lambda}_{low} \leq \underline{\lambda} \leq \underline{\lambda}_{high} \tag{9}$$

Equations (8) and (9) do not exactly form *a mixed linear complementarity problem*, but it is very similar to this class of mathematical optimization problems and can usually be solved for the unknown absolute values of the constraint impulses $\underline{\lambda}\Delta t$. Details of this approach are given also in Refs. [3, 4].

10.5.2   MULTIDISCIPLINARY SIMULATION

The rigid body dynamics described in Eqs. (4)–(9) form an effective base for more advanced physical simulation algorithms. But when setting up Virtual Testbeds containing the systems to be developed, their environment as well as their data processing systems, it is necessary that algorithms (simulation, data processing, etc.) from multiple domains work together to mimic all relevant aspects of the real world. The goal is that the system as well as its environment evolve almost identically ($\underline{s}^{real}(t) \approx \underline{s}^{sim}(t)$) with respect to the controller implementation $\underline{s}_{impl}^{dps}$.

Especially when it comes to advanced eRobotics systems which involve not only rigid body dynamics but also soil mechanics (e.g., for work machine simulations or planetary exploration [54]) or fluid dynamics (e.g., for underwater robotics [55]) an aspect named "glueing," that is, combining the results of different process simulations, becomes the major issue. In order to "glue" different process simulations for, for example, soil dynamics, fluid dynamics, fire propagation, hydraulics, pneumatics together in a way that the different processes can interact in a physically correct manner, further research is required. So far, it is not even clear which physical quantity such as, for example, momentum, force, work, power, etc. is to be chosen to describe best the interaction between different processes. Besides solving the principal problems, research in this field has to make sure that the process interaction can be dealt with in real-time, because the real-time capability and the responsiveness of an eRobotics application has proved to be a key factor for its success.

Methods for the multidisciplinary simulation of physical processes can be categorized into single-domain and multi-domain methods (see Figure 10.18). If the system is modeled with a multi-domain language, an interpreter converts the model into a system of differential-algebraic equations (DAEs), which are then numerically solved. Another class of

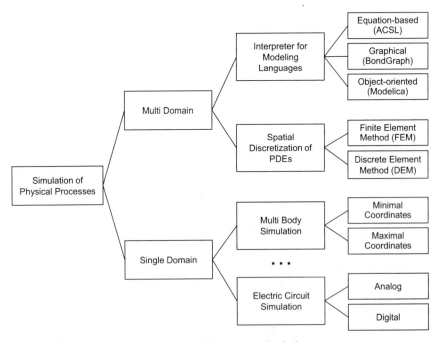

FIGURE 10.18   Categorization of physical process simulation.

multi-domain simulation methods is based on the spatial discretization of the considered system. In multi-domain modeling, it is convenient for the user that he can model the overall system with a single user interface. However, the automatically generated DAE-System can be complex and unstructured and may have to be transformed to allow for a stable numerical integration.

Single-domain simulation methods that are implemented to model one specific physical process are typically more powerful in this specific domain than multi-domain tools, that is, they cover more effects closer to reality or are computationally more efficient. On the other hand the coupling of domain-specific simulation tools requires more effort, the user has to deal with various systems, user interfaces, programming languages, and software licenses.

Bond graphs are a graphical description language for physical systems [25, 26]. They can be used to model significant parts of the simulation model $\underline{s}^{sim}$ and deliver basic concepts for the integration of different algorithms. The nodes of bond graphs are physical subsystems and the edges (called bonds) describe their interaction [25, 26]. A bond exchanges two physical quantities (called effort and flow) whose product is power (hence also called power-coupling). A bond is an ideal connection that guarantees energy conservation between the subsystems. The basic elements of a bond graph can be generalized, since the physical concepts of the various domains (electric, mechanic, hydraulic, acoustic, thermodynamic) are the same. Bond graph models can be reused and hierarchically assembled to larger models. Not until the system should be simulated, the calculation direction in the bonds is determined in the so-called causal analysis. This corresponds to the transformation of the subsystems' equations to a DAE-system that can be numerically integrated.

For illustration, Figure 10.19 shows a common DC motor model in classical block diagram notation and as bond graph.

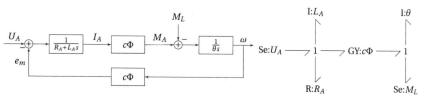

FIGURE 10.19   Block diagram of a DC motor (left), Bond graph of this system (right).

In the modeling of a multi-domain system using several domain-specific simulation tools, the coupling points, the exchanged physical quantities and the numerical coupling scheme have to be chosen. Let the overall system be described with a DAE of the form:

$$\underline{A}\,\underline{\dot{x}}(t) = \underline{f}(\underline{x}, t) \tag{10}$$

To describe one part of the system with a more efficient domain specific simulation system, we could break it apart:

$$\underline{A}_1\,\underline{\dot{x}}_1(t) = \underline{f}_1(\underline{x}_1, \underline{u}_1, t) \qquad \underline{A}_2\,\underline{\dot{x}}_2(t) = \underline{f}_2(\underline{x}_1, \underline{u}_2, t)$$
$$\underline{y}_1 = \underline{g}_1(\underline{x}_1, \underline{u}_1) \qquad\qquad \underline{y}_2 = \underline{g}_2(\underline{x}_2, \underline{u}_2) \tag{11}$$

To arrive at an equivalent overall system, we have to couple input and output quantities:

$$\begin{pmatrix} \underline{u}_1 \\ \underline{u}_2 \end{pmatrix} = \begin{pmatrix} \underline{L}_{11} & \underline{L}_{12} \\ \underline{L}_{21} & \underline{L}_{22} \end{pmatrix} \cdot \begin{pmatrix} \underline{y}_1 \\ \underline{y}_2 \end{pmatrix} \tag{12}$$

For the coupling of mechanical simulation models, several approaches for the exchange of physical quantities are used. That can be purely kinematic properties like positions and velocities or only forces or a combination of both. In the displacement/displacement coupling, that is, the bidirectional exchange of positions or displacements, the system does not have to be separated into disjoint parts at one coupling point, but both subsystems have to contain an overlapping part [27]. The coupling has to be modeled carefully to avoid a distortion of the system properties. Furthermore, the overlapping part can become inconsistent due to integration errors and has to be synchronized. If a bidirectional exchange of velocity and force (i.e., power-coupling) is used, a single coupling point can be chosen. This can be interpreted as a bond that is not coupling two subsystems in one DAE, but between two DAEs. This power-coupling is again energy-conserving by definition and is generic, such that it can be applied to all domains.

Besides the question of the coupling quantities, there are several approaches to the numerical integration of the coupled system. Let the system be modeled in two different simulation systems in the form of DAEs. In the strong coupling, the DAE is exported from one system to the other and the complete system is solved with one integrator. This is the most stable solution. Sometimes it can be sensible to solve the systems separately (weak coupling), for example if the system be separated into a stiff and a non-stiff part. The non-stiff part can be solved with larger time steps and the overall simulation is more efficient. The weak coupling can be executed by embedding the first DAE with dedicated integrator into the second simulation system or with classical co-simulation. The Functional Mockup Interface [28] provides an implementation for strong and weak coupling.

There are several schemes for the coordination of the numerical integration in the weak coupling. These can be explicit (e.g., Jacobi scheme), semi-implicit (e.g., Gauss-Seidel scheme) or implicit, analogous to the integration schemes. Since the coordination scheme has no information on the DAEs, implicit coordination schemes can only work iteratively (e.g., waveform relaxation [58]).

10.5.3 ADVANCED RENDERING TECHNIQUES

Another key factor for the success of an eRobotics application is a rendering and visualization component, which delivers close-to-reality pictures at a high frame rate. Modern eRobotics applications, especially those incorporating data from geographic information systems (GIS, for example, for the simulation of vegetation and/or cities) or from product data management (PDM) systems, need to be able to work on unrefined data without extensive offline conversion. The key idea to deal with large scale eRobotics applications from a visualization standpoint is to base the rendering on the extendible object oriented graph database VSD introduced in Chapter 4.3) (Figure 10.20).

Whole scene descriptions including all functionalities can be described within this single database. Optimization techniques are introduced, which are automatically applied to the simulation data in order to extract a render-friendly structure: Specific semantic objects can be interpreted by the render framework to enhance the simulation, in both, function and visual

FIGURE 10.20   The rendering of vegetation requires to deal with an enormous amount of data [61, 62].

representation [6]. This describes a significant shift from a rather inflexible scene graph centric representation to a database oriented approach where the scene graph is "just another view" onto the database. This does not only make the architecture more flexible, it also supports the distribution of the virtual worlds over arbitrary many PCs by using state-of-the-art distributed database approaches.

Last but not least, the semantic data in the database can be used to further enhance the visual representation. As depicted in Figure 10.21, even the visualization of a city can get fascinating—if the current time of the day, the current weather, and advanced lighting effects are incorporated. This information, together with hints how to visualize the various physical effects, are all stored in the database and are used to improve the user experience with the application.

FIGURE 10.21   A simple city visualization (left) compared to advanced visualization using semantic data interpretation (right) [61, 62].

10.5.4   ADVANCED SENSOR SIMULATION

Besides being used for the visualization, the novel renderer developed for eRobotics applications supports the efficient simulation of optical sensors through the hardware of the GPU. It is based on the data provided for rendering purposes and capable of delivering more precise results than common CPU-based sensor simulations [7]. This is achieved by employing render-specific calculations like normal, displacement or gloss-mapping—originally only used to simulate more complex scene geometry and to improve the visual appearance. In contrast to common laser scanner simulations, which deliver results on a per-vertex accuracy, the developed render-supported approach allows for per-pixel accuracy. Thus, good simulation results can also be achieved for low-polygon models. Moreover, additional data-like textures containing optional physical properties that allow for even more precise calculations can easily be added. A major advantage of the chosen approach is the fact that it is possible to use the render component as well as known rendering techniques to significantly improve the sensor simulation components. In addition, the realistic simulation of different camera models including effects such as lens refraction, distortion, and CCD-dependent properties are incorporated. With these new features, the rendered scene can also be used as a basis for image-based simulation modules, which handle single images or videos of real-world cameras. This capability becomes very important in the fields of space robotics and mission planning because it allows to test the algorithms against real and artificially degraded images in order to evaluate the robustness of the developed image recognition algorithms.

FIGURE 10.22   The GPU supported simulation of laser- und camera based sensors also enables sensor-heavy eRobotics applications in real-time [61].

10.6 NEXT GENERATION EROBOTICS APPLICATIONS

It has been really amazing to see how quickly the word has spread that eRobotics has the potential to cut the cost and to improve the development speed of current advanced robotic applications. In this chapter some of the major applications will be listed. Related videos can be found in Ref. [56].

Our 3D simulation framework introduced above has been used to realize a large variety of different eRobotics applications so far (see also Figure 10.27). Currently, eRobotics applications mainly focus on three application areas (see Figure 10.23), environment (e.g., forest inventory or forest machines), industry (e.g., industrial automation), and space (e.g., space robots), for which it provides a common development approach. In this chapter we shortly outline the results of three applications, one for each application area.

10.6.1 DEVELOPMENT OF SPACE ROBOTS

eRobotics has its roots in space robotics, be it mobile robots, landers for planetary exploration, robot manipulators on satellites, or the International Space Station (see Figure 10.24). Space is still a major driving force for

FIGURE 10.23   Major application areas for eRobotics.

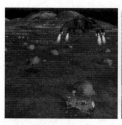

FIGURE 10.24   Developing space robots: image processing for landers [63], sensors for exploration rovers [53], and manipulators for debris removal satellites.

the development of eRobotics, especially as the agencies have begun to appreciate the potential to cut down development costs by up to 50% and also by having the chance to disseminate the results more quickly: Give the users the simulation model so they can experiment and see if they can use space technology in their own applications.

Here, the eRobotics framework is used to its full extent, starting with design studies over virtual testing environments for prototype testing and validation at systems level and ending with the development of intuitive and interactive user interfaces. One impressive example is the simulation of robot manipulators on satellites in space (see Figure 10.24, right). Here, the Virtual Testbed comprises the dynamic simulation of the robot [38] and the satellite, the simulation of the corresponding control algorithms (like robot programs or satellite attitude control algorithms) commanding virtual actuators, as well as the simulation of various sensors.

10.6.2   LOCALIZATION OF FOREST MACHINES

In most application areas it is crucial to know the current position, orientation, and movement direction of mobile systems as a prerequisite to apply various assistance and optimization algorithms like navigation or autonomy concepts. Most of the time GPS is used for this. However, in forest environments the GPS accuracy is often off by 10 to 50 m. We therefore developed the Visual GPS approach which compares a global tree map (pre-calculated using remote sensing data) to a local tree map using local sensors like laser scanners [53]. Both maps are processed using Semantic World Modeling techniques. The position estimation is the result of a pattern matching process using particle filters. Here, 3D simulation technology has been used to test and verify the newly developed localization algorithms in an interactive forest machine simulator consisting of a fully operable virtual forest machine equipped with simulated sensors like GPS, compass, laser scanners, and stereo cameras operating in a close-to-reality forest environment (see Figure 10, middle).

In addition to this, the same simulation has been used to design and implement the localization algorithm itself. The simulation database, able to store and update the world model, and simulation components like collision detection or input/output networks, were used to implement the algorithms and to interface them to virtual or real sensors or onboard computer hardware (see Figure 10.25). In addition, the 3D simulation was the basis for the implementation of the user interface (see Figure 10.10, right). Following this approach, after approx. two years of development, the source selection switch in Figure 10.25 has been switched from virtual to real sensors. It then took about two days before the real system was fully operational.

10.6.3 PROGRAM AND CONTROL ROBOT MANIPULATORS

The third example focuses not only on the programming of robot manipulators but also on their control. The programming task has been

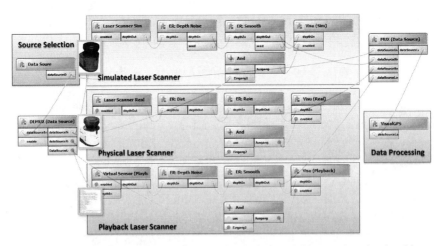

FIGURE 10.25 Block oriented view on a simulation model integrating simulated laser scanners as well as interfaces to physical laser scanners, error model, source selection, and data processing blocks [39].

well investigated during the last years. Here, 3D simulation technology is used to derive movement paths and to program and test the application logic. This has been carried out based on our 3D simulation framework, following the Native Language Programming approach, enabling the user to program a robot using the manufacturer-specific robot programming language [51, 52]. The robot programs are compiled into the State Oriented Modeling Language (SOML++ [59]), which then provides the basis for this simulation. For a close-to-reality simulation of the robots movements, the kinematics framework is used [49, 50]. Normally, at the end of the programming work, the robot program is downloaded to the physical robot controller. Here, the eRobotics approach offers a new alternative—to apply the 3D simulation technology to the controller itself. This allows for the realization of intelligent control structures for complex multi-robot environments, offering 3D user interfaces, automatic action planning components, a multi-robot coordination layer or online collision avoidance (see Figure 10.26). To implement the controller, the simulated robots are simply replaced

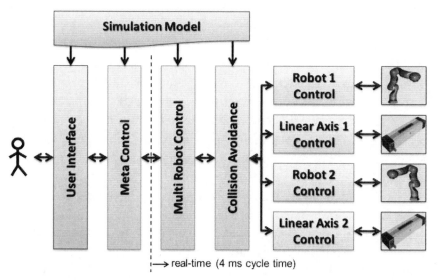

FIGURE 10.26 Simplified structure of an intelligent multi-robot control system implemented using 3D simulation technology.

by interfaces to the real ones. The resulting simulation model is then executed using a stripped down version of the same 3D simulation system running under a real-time operating system like QNX Neutrino for a coordinated control of four separate devices [30].

10.7   CONCLUSION

The aim of the developments in the field of eRobotics is to provide a comprehensive software environment to address robotics-related issues. Starting with user requirements analysis over to the system design, support for the development and the selection of appropriate robot hardware, robot and mechanisms programming, system and process simulation, control design, and encompassing the validation of developed models and programs, eRobotics relies on a continuous and systematic computer support. In this way, the ever-increasing complexity of current computer-aided robotic solutions will be kept manageable and know-how from completed work is electronically preserved and made available for further applications. The connection to the fast-paced field of "eSystems engineering" is established via the development approach: eRobotics applications are characterized by their use of extensive computing resources in order to solve the robotics related scientific and technical problems, their environment enabling an interdisciplinary and collaborative work to address different issues, as well as by making sure that each phase of the work can be carried out by interdisciplinary teams over distributed and efficiently interconnected workspaces. Various applications have been designed and developed for space activities and the gained experiences have been gradually transferred to the areas of forestry and environment management, into the factory, as well as to construction sites in more than 30 applications as depicted in Figure 10.27. Last but not least, eRobotics has also proved to be a key for technology transfer, for example, from the space (robotics) industry. The traditionally high demands on the accuracy of the simulation of mechatronic components and their interaction with the environment, the integration between different process simulations and environmental conditions (rigid body simulation, terra mechanics, hydraulics, pneumatics,

FIGURE 10.27   More than 30 applications have been realized so far based on the ideas of eRobotics. Videos of some of the applications can be found in Ref. [56].

interaction with water and fire, etc.), as well as the realistic simulation of sensors result in an excellent basis for Virtual Testbeds for terrestrial applications. Most applications were realized based on a software framework named VEROSIM®.

ACKNOWLEDGEMENTS

This paper contains contributions of other MMI staff members namely M.Sc. Thorben Cichon (parts of Chapter 4.2), M.Sc. Ralf Waspe (parts of Chapter 4.3), Dipl.-Inform. Nico Hempe (Figures 22 middle/right), Dipl.-Inform. Markus Emde (Figures 1 left), Dipl.-Inform. Thomas Steil (Figures 1 right, 22 left, 24 right) and Dipl.-Ing. Eric Kaigom (Figure 2 right). Parts of the presented work was funded by the German Aerospace Center (DLR) with funds provided by the Federal Ministry of Economics and Technology (BMWi) und grant numbers 50RA0913 (Virtual Crater), 50RA0911 (SELOK), 50RA1306 (INVIRTES) and 50RA1304 (ViTOS).

KEYWORDS

- **3D Simulation**
- **eRobotics**
- **Industrial Robotics**
- **Physics Simulation**
- **Space Robotics**

REFERENCES

1. Abel, D., Bollig, A. *Rapid Control Prototyping*, Springer: 2006.
2. Atorf, L., Krehel, M., Roßmann, J., Sukhatme, G. A Virtual Testbed for Underwater Robotics—Application to Control Design for AUVs, ISC, 2014, Skövde, Sweden.
3. Banks, J. *Discrete-Event System Simulation*, Prentice-Hall: 2010.
4. Blochwitz T., Otter M., Arnold M., Bausch C., Clauß C., et al. The functional mockup interface for tool independent exchange of simulation models. 8th International Modelica Conference, 2011, pp. 20–22.
5. Borutzky W. Bond graph modeling and simulation of mechatronic systems: An introduction into the methodology, 20th European Conference on Modeling and Simulation (ECMS), 2006.
6. Broenink J. F. Introduction to physical systems modeling with bond graphs [Online], 1999, http://www.ce.utwente.nl/bnk/papers/BondGraphsV2.pdf (accessed Oct 27, 2014).
7. Busch M. *Zur effizienten Kopplung von Simulationsprogrammen*, Kassel University Press GmbH: 2012.
8. COMSOL. www.comsol.com (accessed Oct 27, 2014).
9. Coppelia Robotics Software, V-rep. http://www.coppeliarobotics.com (accessed Oct 27, 2014).
10. Dassault Systems, Delmia. http://www.3ds.com/de/produkte-und-services/delmia/oesungen/alle-delmia-loesungen/ (accessed Oct 27, 2014).
11. Dolinsky J. U. The development of a genetic programming method for kinematic robot calibration, PhD Dissertation, Liverpool John Moores University, 2001.
12. Emde, M., Rossmann, J., Sondermann, B., Hempe, N. Advanced Sensor Simulation In Virtual Testbeds: A Cost-Efficient Way To Develop And Verify Space Applications, AIAA SPACE, 2011, Long Beach, California.
13. Featherstone, R. *Rigid Body Dynamics Algorithms*, Springer New York: 2008.
14. Fisette P., Samin J. *Robotran. Symbolic generation of multi-body system dynamic equations*, In: *Advanced Multibody System Dynamics*, Springer, 1993, pp. 373–378.
15. Freund, E., Rossmann, J. *Projective Virtual Reality: Bridging the Gap between Virtual Reality and Robotics*, In: *IEEE Transactions on Robotics and Automation*, Vol. 15, No. 3, 1999.

16. Freund, E., Schluse, M., Rossmann, J. State Oriented Modeling as Enabling Technology for Projective Virtual Reality, IROS, 2003, Hawaii.
17. Fritzson, P. *Principles of object-oriented modeling and simulation with modelica 2.1*, Wiley: 2003.
18. Gyssens, M., Paredaens, J., van den Bussche, J., van Gucht, D. *A graph-oriented object database model.* In *IEEE Transactions on Knowledge and Data Engineering* Vol. 6, 1994.
19. Hempe, N., Roßmann, J. A Semantics-Based, Active Render Framework to Realize Complex eRobotics Applications with Realistic Virtual Testing Environments. EMS, 2013, Manchester, UK.
20. Hempe, N., Roßmann, J. Taking the Step from Edutainment to eRobotics - A Novel Approach for an Active Render-Framework to Face the Challenges of Modern, Multi-Domain VR Simulation Systems. IEEE IC3e, 2013, Kuching, Malaysia.
21. Hempe, N., Rossmann, J., Waspe, R. Geometric interpretation and optimization of large semantic data sets in real-time VR applications, Proceedings of the ASME 2012 International Design Engineering Technical Conferences & Computers and Information in Engineering Conference (IDETC/CIE), August 12–15, 2012, Chicago, Illinois, USA.
22. Hoppen, M., Schluse, M., Rossmann, J., Weitzig, B. Database-Driven Distributed 3D Simulation, WSC, 2012, Berlin.
23. Jochmann, G., Blümel, F., Stern, O., Roßmann, J. *The Virtual Space Robotics Testbed: Comprehensive Means for the Development and Evaluation of Components for Robotic Exploration Missions Mental Models for Intelligent Systems.* In: *KI-Künstliche Intelligenz*, Vol. 28, No. 2, 2014, pp. 85–92.
24. Jung, T. Methoden der Mehrkörperdynamiksimulation also Grundlage realitätsnaher Virtueller Welten, PhD Dissertation, RWTH Aachen University, 2011.
25. Jung, T., Rast, M., Guiffo Kaigom, E., Roßmann, J. Fast VR Application Development Based on Versatile Rigid Multi-Body Dynamics Simulation, Proceedings of the ASME 2011 International Design Engineering Technical Conferences and Computers and Information in Engineering Conference (IDETC/CIE), Washington, DC, pp. 1–10, August 28–31, 2011.
26. Kaigom, E., Rossmann, J. Simulation of electrically-driven robot manipulators. ICMA, 2012, Chengdu, China.
27. Kurtev, I., van den Berg, K., Mistral: A language for model transformations in the MOF meta-modeling architecture. MDAFA, 2004, Linköping, Sweden.
28. Lelarasmee, E., Ruehli, A. E., Sangiovanni-Vincentelli, A. L. *The waveform relaxation method for time-domain analysis of large-scale integrated circuits.* In: *IEEE Transactions on Computer-Aided Design of Integrated Circuits and Systems*, Vol. 1, 1982.
29. MATLAB/Simulink. www.mathworks.de/products/simulink/ (accessed Oct 27, 2014).
30. Müller R., Esser M., Janßen C., Vette M. Systemidentifikation für Montagezellen-Erhöhte Genauigkeit und bedarfsgerechte Rekonfiguration. wt Werkstattstechnik online 2010, 100(9), pp. 687–691.
31. Open Source Robotics Foundation, Gazebosim. http://gazebosim.org/ (accessed Oct 27, 2014).

32. Roos E., Behrens A., Anton S., RDS - realistic dynamic simulation of robots., Int. Symp. on Industrial Robots, Int. Fed. of Robotics & Robotic Industries, 1997, Vol. 28, pp. 17–27.

33. Rossdeutscher M., Zuern M., Berger U. Virtual robot program development for assembly processes using rigid-body simulation, Int. Conf. on Computer Supported Cooperative Work in Design, IEEE, 2010, pp. 417–422.

34. Rossmann J., Schluse M., Schlette C., Waspe R. A new approach to 3D simulation technology as enabling technology for eRobotics, Int. Simulation Tools Conf. & EXPO (SIMEX), 2013, pp. 39–46.

35. Rossmann J., Steil T., Springer M. Validating the camera and light simulation of a virtual space robotics testbed by means of physical mockup data, Int. Symp. on Artificial Intelligence, Robotics and Automation in Space (i-SAIRAS), 2012.

36. Rossmann J., Wischnewski R., Stern O. A comprehensive 3-d simulation system for the virtual production, Int. Industrial Simulation Conference (ISC), 2010, pp. 109–116.

37. Rossmann, J., Eilers, K. Translating Robot Programming Language Flow Control into Petri Nets, ETFA, 2011, Toulouse.

38. Rossmann, J., Eilers, K., Schlette, C., Schluse, M. A Uniform Framework to Program, Animate and Control Objects, Kinematics and Articulated Mechanisms in a Comprehensive Simulation System, ISR/Robotics, 2010, Munich, Germany.

39. Roßmann, J., Hempe, N., Emde, M. New Methods of Render-Supported Sensor Simulation in Modern Real-Time VR-Simulation Systems, In: N. Mastorakis, V. Mladenov et al. (Eds.) Proceedings of the 15th WSEAS International Conference on Computers – Recent Researches in Computer Science (Part of the 15th WSEAS CSCC Multiconference), July 15–17, 2011, Corfu Island, Greece, pp. 358–364.

40. Rossmann, J., Jung, T., Rast, M. Developing virtual testbeds for tasks in research and engineering, WINVR, 2010, Ames, USA.

41. Roßmann, J., Kaigom, E., Atorf, L., Rast, M., Grinshpun, G., Schlette, C. *Mental Models for Intelligent Systems: eRobotics Enables New Approaches to Simulation-Based AI*. In: *KI-Künstliche Intelligenz*, Vol. 28, No. 2, 2014, pp. 101–110.

42. Rossmann, J., Schlette, C., Emde, M., Sondermann, B. *Advanced Self-Localization and Navigation for Mobile Robots in Extraterrestrial Environments*, In: *Computer Technology and Application*, Vol. 2, No. 5, 2011.

43. Rossmann, J., Schluse, M. Virtual Robotic Testbeds: A foundation for e-Robotics in Space, in Industry—and in the woods, 2011, DeSE, Dubai.

44. Rossmann, J., Schluse, M., Bücken, A., Krahwinkler, P., Hoppen, M. Cost-efficient semi-automatic forest inventory integrating large scale remote sensing technologies with goal-oriented manual quality assurance processes, IUFRO Division 4 Conference, 2009, Quebec City, Canada.

45. Rossmann, J., Schluse, M., Hoppen, M., Waspe, R. Integrating semantic world modeling, 3D-simulation, virtual reality and remote sensing techniques for a new class of interactive GIS-based simulation systems, International Conference on Geoinformatics, 2009, Fairfax, USA.

46. Rossmann, J., Schluse, M., Jung. T. Introducing intuitive and versatile multi modal graphical programming means to enhance virtual environments, IDETC/CIE, 2008, New York.

47. Rossmann, J., Schluse, M., Schlette, C., Waspe, R. Control by 3D Simulation—A New eRobotics Approach to Control Design in Automation. ICIRA, 2012, Montreal, Canada.

48. Rossmann, J., Schluse, M., Waspe, R. Integrating object oriented petri nets into the active graph database of a real time simulation system. WSC, 2012, Berlin.

49. Rudolph J., Woittennek F. An algebraic approach to parameter identification in linear infinite dimensional systems, In: Mediterranean Conf. on Control and Automation, IEEE, 2008, pp. 332–337.

50. Schlette, C. Multi-Agentensysteme zur Simulation, Analyze und Steuerung von anthropomorphen Kinematiken, PhD Dissertation, RWTH Aachen University, 2012.

51. Schluse, M. Zustandsorientierte Modellierung in Virtueller Realität und Kollisionsvermeidung, PhD Dissertation, TU Dortmund, VDI-Verlag: 2002.

52. Siemens, Processsimulate. http://www.plm.automation.siemens.com/de_de/products/tecnomatix/robotics_automation/robotexpert.shtml (accessed Oct 27, 2014).

53. Siemens, Robcad. http://www.plm.automation.siemens.com/de_de/products/tecnomatix/robotics_automation/robcad/ (accessed Oct 27, 2014).

54. Siemens. Robotexpert. http://www.plm.automation.siemens.com/de_de/products/tecnomatix/robotics_automation/robotexpert.shtml (accessed Oct 27, 2014).

55. Simmechanics. www.mathworks.de/products/simmechanics/ (accessed Oct 27, 2014).

56. Song P., Trinkle J. C., Kumar V., Pang J. S. Design of part feeding and assembly processes with dynamics, Int. Conf. in Robotics and Atomation (ICRA), IEEE, 2004, Vol. 1, pp. 39–44.

57. Stewart, D., Trinkle, J. An Implicit Time Stepping Scheme for Rigid Body Dynamics with Inelastic Collisions and Coulomb Friction. *International Journal for Numerical Methods in Engineering*, 1996, Vol. 39, pp. 2673–2691.

58. Stewart, D., Trinkle, J. An Implicit Time Stepping Scheme for Rigid Body Dynamics with Coulomb Friction, IEEE International Conference on Robots and Automation, 2000, pp. 162–169.

59. The Unreal game engine. www.unrealengine.com (accessed Oct 27, 2014).

60. VDI-Standard: VDI 3633 Blatt 1, Simulation of systems in materials handling, logistics and production – Fundamentals [online]. www.vdi.eu/3633 (accessed Oct 27, 2014).

61. VEROSIM YouTube channel. www.youtube.com/user/verosimsimulations (accessed Oct 27, 2014).

62. Weise, J., Briess, K., Adomeit, A., Reimerdes, H.-G., Göller, M., Dillmann, R. *An Intelligent Building Blocks Concept for On-Orbit-Satellite Servicing*, iSAIRAS, 2012, Turin, Italy.

63. Yoo, Y., Jung, T., Römmermann, M., Rast, M., Kirchner, F., Roßmann, J. Developing a Virtual Environment for Extraterrestrial Legged Robots with Focus on Lunar Crater Exploration, The 10th International Symposium on Artificial Intelligence, Robotics and Automation in Space (i-SAIRAS), August 29–September 1, 2010.

CHAPTER 11

TACKLING FRAUD: A COLLABORATIVE MODEL FOR INSURANCE COMPANIES

ASHIS PANI[1] and RAKESH TIWARI[2]

[1]*Chairman, Information Systems Area, XLRI Jamshedpur, Jharkhand, India – 831001, Tel: 91-0657-3983144, Fax: 91-657-2227814, E-mail: akpani@xlri.ac.in*

[2]*Senior Manager, SSP India Private Limited, Gurgaon, Haryana, India, 122002*

Research Fellow, Information Systems Area, XLRI Jamshedpur, Jharkhand, India, 831001, Tel: 91-9971145616, E-mail: r10014@astra.xlri.ac.in

CONTENTS

11.1 ABSTRACT

With rapid advances in information and communication technologies, coupled with business intelligence and data mining tools and techniques, there has been dramatic increase in the availability and importance of information in today's business environment. These advances not just have impact on the decisions made by consumers or firms, but in some cases, have impact on the structure of overall industry.

With difficult financial conditions leading to increased financial crimes, fraud risk is posing a huge challenge for the insurance industry. The individual insurance company's effort to build in checks, using latest technologies, tools and techniques has proven to be insufficient.

The extraordinary loss due to insurance fraud is not just limited to insurance companies but indirectly affects the national budgets and citizens. It thus requires coordinated effort by insurance companies, government and citizens. The alliance between all the stakeholders can work together in development of standards, which will enable seamless information sharing between the insurance companies and government departments so that the aggregated historical data can be used for more accurate assessments of risk and fraud.

This chapter examines various aspects of alliances with particular focus on the network externalities. It then analyzes standardization with special focus on financial institutions. The extent of insurance fraud prevailing in various countries is highlighted with current alliances working on different mechanism to tackle the fraud.

A mathematical model is proposed to analyze the decisions made by firms for the level of participation in the alliance. It will also demonstrate how tackling of fraud using alliances can have positive impact on the overall industry.

The chapter concludes with list of recommendations for increasing the participation of insurance companies based on the mathematical model.

11.2 INTRODUCTION

By middle of 19th century, US and Europe underwent a sea change in economy driven by economies of production and transportation (Malone, 1991). This period was known as industrial revolution. The industrial economy is undergoing another phase of radical transformation, which is driven by change in coordination. The information technology has dramatically reduced the cost of coordination and increased the speed and quality of coordination resulting in coordination intensive business structures, thus enabling people and companies shape coordination in the form of redefining coordination processes, improving the efficiency and effectiveness of coordination and evolving new coordination structures (Bodendorf and Reinheimer, 1997; Malone and Rockart, 1991; Malone and Crowston, 1994).

The second transformation can be understood by analyzing the first transformation. In the first transformation, when the technology emerged, it was initially used as substitute to old means of transport (carts and carriages, etc.). With its improvement, the transportation had second order effect. Firms started using the transportation technology more and more and started reaching new geographies. With people using more and more transportation, resulted in third order effect – emergence of transport intensive social and economic structure.

The information technology too is having similar cycle. It started with it being used as a substitute for human coordination. Thus back office data processing jobs were replaced by Information Technology, later eliminating the middle manager layer. The second order effect of Information Technology resulted in business transformation. New business models in airlines and banks are the example of this effect. The third order effect is the coordination intensive economic structure. For example, the U.S. textile industry implemented a series of electronic connections among companies as part of the Quick Response program which linked companies in the production chain, from suppliers of fibers (such as wool and cotton) to the mills that weave these fibers into fabric, to the factories that sew garments and to the stores that sell the garments to consumers thereby saving approximately half of $25 billion of inventory cost (Hammond, 1990). Another third order effect of wide use of technology by broad spectrum of companies is the network effect.

11.2.1   NETWORK ECONOMIES

The theory of network economics is a relatively new topic of research, which emerged in early 80's when a growing number of contributions on the field of standards were recognized in the literature.

The differences between the old and the new economy lie on economics of networks - the old industrial economy was driven by the economies of scale; the new information economy is driven by the economics of networks (Shapiro and Varian, 1998).

The basic concept of network economics is the positive feedback (Shapiro and Varian, 1998). Their overall value as well as the value for the individual participant depends on the number of other participants in the same network. Thus, greater the number of people in the network the more likely it is that other participants will join the network. These network effects are also called network externalities or demand-side economies of scale.

11.2.2   SCOPE OF NETWORK

The scope of the network that gives rise to the consumption externalities will depend upon markets (Katz and Shapiro, 1985). In cases like automobile, the sales of only one firm constitute the relevant network. In some cases like stereo phonographs, the relevant network will comprise the outputs of all firms producing the good. In still other cases like hardware manufacturers adopting common operating system, the network may be a coalition of firms. In cases like information network, which is the case studied in the chapter, the number of firms updating the information determine the network size.

11.2.3   STANDARDIZATION

Standards relating to information, quality and compatibility help to determine the ways in which the costs and benefits of innovation are distributed and the speed at which the innovation occurs. There are three types of processes through which standards may be developed and adopted (Besen and Johnson, 1986). Noncooperative behavior, commonly called as marketplace approach wherein firms adopt standards independently and the industry may or may not follow that standard. In cooperative behavior the representatives of private firms and other interested parties use formal procedures of committee participation, meet to develop, recommend, and adopt industry standard. The resulting standard may or may not be unanimously followed. The best example of this category is American National Standards Institute (ANSI) [http://www.ansi.org/], which has about a thousand corporate members and several hundred affiliated professional organizations. It collects, disseminates, and coordinates standards set by member organizations and by industry groups.

Third category is the Government agencies, which sometimes mandate the adoption of certain standards. It differs from the private standard in that it is mandatory.

It was found that although the committee system is slower, it outperforms the market mechanism (Farrell and Saloner, 1988).

11.2.4   STANDARDIZATION BENEFITS

There are several theoretical studies, which examine the sources of benefit from standardization (Katz and Shapiro, 1985; Farrell and Saloner, 1988; Berg, 1984). Effect of the number of users can have benefit on the product quality-consumption externalities generated through direct physical effect of number of purchasers. For example the utility of telephone, facsimile will depend on number of other customers in the network. Effect of number of users can have benefit on the availability and prices of complementary inputs. For example, the number of software available for particular hardware is dependent on number of hardware units sold, same with video games and video players. Effect of number of users can have benefit on the post purchase service. For example the automobiles sales of new and less popular brands are low on account of thinner server network. Some other benefits include easy access of information of popular brands, psychological, bandwagon effects (Katz and Shapiro, 1985). Similar benefits have been attributed to "market mediated" effect wherein the standardization permits a thick second hand market and may promote price competition among sellers (Farrell and Saloner, 1988). Economies of scale in production of components and complementary inputs can also result in reduced price (Berg, 1984).

11.2.5 EXAMPLES

Web services are one of the best examples of implementation of standards. Web services is defined as an application interface that conforms to specific standards in order to enable other applications to communicate with it through that interface regardless of programming language, hardware platform, or operating system (Hansen et al., 2000). Various standards complied by interfaces are:

(a) XML (Extensible Markup Language) is a simple text-based format for representing structured information: documents, data, configuration, books, etc. [World Wide Web Consortium (W3C)].
(b) SOAP (Simple Object Access Protocol) standard specifying how documents are exchanged.
(c) WSDL (Web Services Description Language) standard to provide description of the input and output parameters to use of the service
(d) UDDI (Universal Description, Discovery, and Integration) to register the web service.

11.2.6 STANDARDS IN FINANCIAL INSTITUTIONS

Financial Information Exchange: Financial Information Exchange protocol (FIX) is a messaging standard developed specifically for the real-time electronic exchange of securities transactions. It has been developed through the collaboration of banks, broker-dealers, exchanges, industry utilities and associations, institutional investors, and information technology providers from around the world (http://fixprotocol.org/).

Interactive Financial Exchange Forum: IFX is a financial messaging protocol for interoperability of systems seeking to exchange financial information internally and externally, as in Cash Management, Electronic Bill

Presentment and Payment and Business-to-Business Payments. The IFX Forum is made up of industry-leading financial institutions, service providers and independent software vendors, who all believe in the process of open, industry-driven standards creation (http://www.ifxforum.org).

International Swaps and Derivatives Association: Financial products Markup Language (FpML) is an information exchange standard for electronic dealing and processing of financial derivatives instruments, establishing protocol for sharing information on, and dealing in swaps, derivatives and structured products (http://www.fpml.org/).

FinXML Consortium: FinXML is an XML-based framework within which vocabularies for capital markets (including interest rate, foreign exchange and commodity derivatives, bonds, money markets, loans and deposits, and exchange traded futures and options) can be defined and within which applications using these vocabularies can be developed and deployed. The FinXML is supposed to be interoperable with other standards as FIX and SWIFT - Society for the Worldwide Interbank Financial Telecommunication (http://www.finxml.org/).

ACORD (Association for Cooperative Operations Research and Development) XML-based standards to be used in transactions related to Property and Casualty and Life insurance sectors (http://www.acord.org).

11.3 STRATEGIC ALLIANCE

As information technology laid the foundation for coordination intensive business structure, the state of "hyper-competition" (D'Aveni, 1994) between individual companies transformed into one of hyper-cooperation. The motivations to enter into alliance have been described by various studies, which are described in the following subsections.

11.3.1 REGULATORY REQUIREMENT

One of the motivations for alliance is regulatory requirement (Whetten, 1981). Sometimes organizations are forced to enter into alliance (Oliver, 1990), as many developing countries insist that the access to local market is only through local partners (Beamish, 1988).

11.3.1.1   Transaction Cost

Here the motivation for entering an alliance is cost savings from the alliance. Thus firms enter into alliances to economize on the production and transaction cost (Jarillo, 1988; Jarillo, 1990; Jarillo and Stevenson, 1991; Madhok, 1998).

11.3.1.2   Resource Based View

Firms enter into strategic alliances to generate value through potential synergies (Madhok, 1998). At times, firms need to enter into relationships because they cannot generate all the necessary resources internally (Child, 1974; Pfeffer and Salancik, 1978). Alliances give small firms access to complementary assets that are necessary to commercialize innovations (Hobday, 1994; Teece, 1986). In technology intensive industries such as biotechnology, this form of strategic technology alliances has been extensively found (Pisano, 1991; Pisano and Mang, 1993).

11.3.2   ORGANIZATIONAL LEARNING

Here learning is the motivation for entering into alliance. (Badaracco, 1991; Lei and Slocum, 1992; Mowery et al., 1996). Some authors argue that firms enter into alliances to acquire new skills or technologies from the partner (Hamel et al., 1989; Harrigan, 1985). It has been pointed out in the resource-based literature that building new resources and capabilities suffers from time compression diseconomies (Dierickx and Cool, 1989). Also, transferring the tacit knowledge will be easier in alliances that foster intense interaction and collaboration (Kogut and Zander, 1992). In some industries, firms enter into alliance to draw upon technologies in which they have no ore only very weak capabilities (Doz and Hamel, 1998).

11.3.3   STRATEGIC POSITIONING

Standard setting alliances are the examples wherein the firms, in the industry characterized by network externalities, enter into alliance, wherein the competition shifts from the firms to the alliance coalitions (Gomes-Casseres, 1996; Moore, 1996). However, firms should balance the benefits of standardization with the problems resulting from collaborating (Axelrod, 1984).

In our context, the alliance is being considered for strengthening the insurance industry by protecting it from ever increasing fraud.

11.4   INSURANCE FRAUD

Insurance fraud is any act committed with the intent to fraudulently obtain payment from an insurer.

The insurance fraud causes extraordinary losses to national budgets, citizens and insurance companies. Most common form of general insurance fraud is opportunistic retail fraud wherein the individual or firms exaggerate or inflate genuine claims to increase the value of payout. In organized fraud, the criminals stage the insurable event to make the fraud claim.

11.4.1   EXTENT OF FRAUD

The value of fraudulent insurance claims uncovered by insurers rose to a record £1.3 billion in 2013, up 18% on the previous year according to figures published by the Association of British Insurers (ABI, 2014). This figure is more than double the cost the UK's shoplifting bill (BRC Retail Crime Survey, 2013). Insurers detected a total of 118,500 bogus or exaggerated insurance claims, equivalent to 2,279 a week. The average fraud detected across all types of insurance products was £10,813. Fraudulent motor insurance claims were the most expensive and common, with the number of dishonest claims at 59,900 claims up 34% on 2012 and their value at £811 million up 32%. Since 2007 the value of dishonest general insurance claims detected has more than doubled, with the number detected up 30% over the same period.

Examples of insurance cheats include crash for cash staged accident fraud. In one case, a professional golfer who claimed £8,000 on his income protection policy for a knee injury, which left him unable to work was caught on camera giving golf lessons. A woman was jailed for twenty two months following a series of invented street robberies for items including laptops and designer clothes. A vet was jailed for two years for inventing veterinary claims nearly £200,000 for treating non-existent pets (ABI, 2014).

In Italy, according to estimates of Italian insurance association, (ANIA, 2012), a total of 54,502 fraudulent claims were detected, equal to 2.04% of all claims incurred and reported. In terms of value, the 2.42% of claim value was attributed to fraud. The fraud varies from region to region with North attributing to 0.94%, Center 1.34%, Islands 2.29% and south attributing to majority of the fraudulent claims of 6.53%. The General Insurance Association of Singapore estimated that the percentage of fraudulent motor claims paid in the country was 20%, a surprisingly high rate since Singapore is widely believed to be one of the least corrupt countries globally (EY, 2104).

The fraudulent claims extent is difficult to estimate, but a broad international estimate of general insurance claim fraud was compiled by Association of British Insurers (ABI, 2009) in the Table 11.1.

Some of the precise figures available from the European markets suggest widespread claims fraud. Figures from the Association of British Insurers (ABI) show that despite insurers detecting more fraud, it is estimated that around £1.9bn (€2.2bn) of fraud goes undetected each year. The value of detected fraud in 2011 rose 7% to £983m (€1148m)

TABLE 11.1 Fraud Estimates (Association of British Insurers)

Country	Product line	Estimate	Source
United Kingdom	Retail	7% of claims (by value)	ABI, 2007
Australia	General	10% of claims (by value)	ICA, 1994
United States (Arizona)	General	10% of claims (by volume)	ADI, 2009
United States	General	10% of claims (by volume)	Hoyt, 1988
United Kingdom	General	10% of claims (by value)	ABI, 2009
Canada	General	10–15% of premiums	IBoC, 2006
United States	Motor	11–15% of claims (by value)	IRC, 2008
Germany	Motor	11% of claims (by volume)	Clarke, 1990
United Kingdom	Retail	11% of claims (by volume)	ABI, 2007
United Kingdom	General	13% of claims (by volume)	ABI, 2009
United States	General	15% of claims (by value)	Hoyt, 1988
Spain	Motor	22% of claims (by volume)	Artis et al., 1999

from £919m in 2010. In 2011 insurers uncovered 138,814 fraudulent insurance claims equivalent to 2670 claims every week up 5% on 2010. The value of savings for honest customers from detected frauds represented 5.7% of all claims, compared to 5% in 2010 (Insurance Europe, 2013). In Germany, a study conducted by the insurance association (GDV) concluded that more than half of all claims arising from loss or damage to smartphones or tablet PCs could not have arisen and therefore must have been fraudulent to some extent. Figures from Insurance Sweden (Larmtjänst) suggests that Insurance fraud investigators, established by insurance companies, conducted 6200 investigations into suspected fraud in 2011 and detected a total of €40m of fraud. It was found that 10–20% of all fraudulent claims are claims for losses arising from events that never occurred (i.e., untruthful claims) and 80–90% of all fraudulent claims are exaggerated claims. Figures from the insurance association (FFSA) reveal that 35,042 fraudulent insurance claims were recorded in 2011, leading to €168m not being paid out to dishonest individuals.

11.4.2   IMPACT OF RECESSION

There have been several studies linking financial crime and economic indicators (Osborne, 1995; Scorcu and Cellini, 1996; Tsushima, 2002; Buonanno and Montolio, 2008). Compounding this, the difficult financial conditions reduce the firm's capacity to mitigate operational and financial crime risks. This warning has come from various consulting and advisory panels in 2009 (BDO Stoy Hayward, 2009; PricewaterhouseCoopers, 2009; KPMG, 2009, Financial Services Authority, 2009; Fraud Advisory Panel, 2009; CIFAS, 2009).

Some indicators are already showing this trend (ABI, 2009):

- CIFAS (2009) estimate that there has been a 44% increase in false insurance claims in Q1 2009 compared to Q1 2008.
- Research by Royal and SunAlliance (RSA, 2009) showing that the number of people in Britain who think insurance fraud is acceptable increased dramatically (from 3.6 to 4.6 million people) between March 2008 and January 2009.
- An increase in calls to the Insurance Fraud Bureau (IFB, 2009) 'Cheatline.'

The above suggests the recession (and the consequent rise in unemployment) will increase the risk of fraud for insurers, and hence becomes important for Insurance companies to get geared up to tackle the fraud.

11.4.3 TACKLING FRAUD

The insurance industry's responses to fraud vary between countries and the initiatives are wide-ranging.

- **Exchange of Information:** In some countries, the relevant information is exchanged between insurers to identify potential frauds. Insurers are transparent about this and operate in compliance with data protection and privacy requirements. Such exchanges of information among insurers (in varying forms) exist in Croatia, Estonia, Finland, Germany, Ireland, Malta, the Netherlands, Norway, Portugal, Slovenia, Spain, Sweden and the UK, and are currently being considered in Cyprus (Insurance Europe, 2013).
- **Cross-Border Cooperation:** Nordic countries meet regularly to discuss trends, issues and common challenges, since trends in one country have been seen to spread to neighboring countries (Insurance Europe, 2013).
- **Formalized groups** to investigate insurance fraud. In France, insurers set up a national body in 1989 – (Agence pour la lutte contre la fraude à l'assurance, ALFA) to investigate suspicious claims. ALFA promotes counter-fraud activities, by training and certification of fraud investigators, advice on how to handle fraudulent cases that target several insurers at a time, and advice on managing relationships with law enforcement agencies. In Sweden, insurance undertakings have special investigation units charged with detecting insurance fraud and make police reports of detected or suspected frauds In UK, the Insurance Fraud Bureau (IFB) focuses on detecting and preventing organized and cross-industry insurance fraud. The IFB co-ordinates the industry response to the identification of criminal fraud networks and works closely with the police and other law enforcement agencies. It encourages people to report suspected or known frauds anonymously through an insurance cheat-line. The impact of the IFB has been hugely positive since its launch in July 2006, with numerous arrests and tens of millions of pounds of savings for insurers and ultimately their customers.

- **Co-operation with law enforcement agencies:** In several countries like Croatia, Denmark, Estonia, Germany, Ireland, the Netherlands, Portugal, Spain, Sweden and the UK, Insurers have increased their co-operation with law enforcement agencies (Insurance Europe, 2013).
- **Technology:** In order to detect frauds Insurers increasingly use technology like electronic devices to detect the authenticity of documents submitted for claims and social media and other websites to authenticate information supplied.
- **Training:** In various countries like UK, Germany, Finland and Denmark, Insurance staff and police are trained to raise awareness of fraud, to show how to detect it and to highlight the new and ever-changing methods used by fraudsters.

11.4.4   EXISTING ALLIANCES

- Several alliances already exists which cover different aspect of fraud prevention mechanisms. New York Alliance Against Insurance Fraud (NYAAIF) was founded in 1999 and works towards public awareness programs on insurance fraud. It was found that a consistent and hard-hitting message to consumers has proven to be an effective and efficient method for increasing awareness about fraud, reducing the public's tolerance for this crime and encouraging citizens to report fraud.
- Argos is an investigative organization of a group of French insurance companies which researches, identifies and recovers vehicles and other property reported stolen by owners. In 2008, Argos found or identified 10,163 vehicles (10,445 in, 2007). This decrease was in the context of fewer vehicle thefts. The main role of Argos is the identification of stolen vehicles, for which it is compensated by insurers. It is developing links with vehicle pounds and cooperation with other European countries.
- Hellenic Association of Insurance Companies (HAIC) – equivalent of CEA for Greece) which is responsible for the relations between the private insurance companies and the state is working on the development of an anti-fraud data system and other fraud-prevention mechanisms.

11.4.5   FRAUD MANAGEMENT SYSTEM

Fraud Management system and standard is thus the most sought solution for insurers. The system will provide a means of exchanging information between insurers for the data required for tackling fraud. It also integrates it with certified network of private investigators. It should integrate the system with police to get information from stolen vehicles. Such system should identify the claims incurred by a particular vehicle on the basis of its loss history. Loss adjusters can introduce suitable measures if any abnormalities are detected.

It should give its members access to claims data in order to enable the detection of multiple and fraudulent motor insurance claims. It should also be able to identify suspicious situation using data mining techniques. It should look for matches against known fraudsters and for fraud indicators.

11.5   MATHEMATICAL MODEL

We will develop a mathematical model to analyze the decision that will be made by firms for the level of participation in the alliance.

We define "participants" as the firms, which actively participate in the formalization of standards and adopters as the firms, which do not participate in the formalization but will join the alliance by adopting their system as per the standards, once it is available.

11.5.1   BUILDING BLOCKS

11.5.1.1   Value of Standard

Let's say α is the per unit value of the alliance total output T. α will depend upon the firms characteristics, like size, locations, exposure to fraud risk, etc. Higher the exposure to risk, higher the benefit is from preventing the fraud.

Let α_a be the threshold adopter value, the value below which the firm will not become part of the alliance. Let α_p be the threshold participant's value, the value below which the firm will not participate in the formalization of standard. Let α_m be the maximum value for any participant.

11.5.1.2   Participation Benefit

There will be benefit associated with the participation of firm in the formalization of the standard. The benefit comes from the advanced knowledge of the direction of the standard. Let this benefit be denoted by B. B is function of effort e and α. More the effort and the value the firm sees in alliance, positive will be the benefit. However, $B(e, \alpha) = 0$ in following condition - when e=0 that is, there is no effort is being put by the participant or $\alpha = 0$ that is, firm sees no value in the output.

Firms need to balance between the benefit and cost of formalization of standard.

11.5.1.3   Participation Cost

There will be cost associated with the participation of firm in formalization of the standard. This will be in the form of membership fees and early adoption that is, if the standard is not mature. There is also a risk of the failure of the alliance output, due to technological change (Kexin Zhao et al., 2006). Let this cost and risk be denoted by C. We assume that the C be uniform for all the firms.

11.5.1.4   Adoption Cost

When the standard is ready, there will be cost associated with the adoption of the system.

The adoption cost is the cost of modification of the existing system (application, hardware, etc.) for adoption of the standard and the cost of learning and cost of enhancing the processes (Hall, 2003).

The adoption cost is a function of the complexity, trialability and observability (Hall, 2003).

While the complexity directly impacts the cost, the trialability and observability facilitates the adoption. Trialability is the degree to which the benefit of standard may be experienced. More is the trialability; less is the uncertainty and faster is the adoption. Observability is the degree to which the result of the standards is visible. The easier it is to observe, the faster it will be adopted (Rogers, 1995). Let's assume the adoption cost to be A.

11.5.1.5   Network Effect

More the number of members of this standard, higher will be the quality of the data (Yoo and Choudhary, 2002; Mukhopadhyay, 2003). Thus if a person is already convicted for fraud in any company, this data being available to the companies, the fraud can be avoided. Higher the number of companies, faster will be the fraud detection. Also, more the data is available of the risks, better will be the aggregation of risks and fraud data.

Let η be the network effect. Assuming that α is uniformly distributed between 0 and α_m, the network benefit will be $= \eta(\alpha_m - \alpha_a)$.

11.5.2   UTILITY FUNCTION

Thus the firms utility function

$\pi =$ Per unit value of Output * Total Output + Network Benefit – Adoption cost + Participation Benefit – Participation cost

- $\pi = \alpha \times T + \eta(\alpha_m - \alpha_a) - A + B(e, \alpha) - C$

We will now work on getting optimal value for adopters α_a, below which the profitability will be 0.

Thus, $\pi_a = \alpha_a \times T_a + \eta(\alpha_m - \alpha_a) - A + B(e, \alpha) - C$
Since for adopters, $B(e, \alpha) = C = 0$

- $\pi_a = \alpha_a \times T_a + \eta(\alpha_m + \alpha_a) - A$

11.5.2.1   Adoption Condition

Consider the case when the output of alliance effort T is zero, that is, the standard is available without any effort from any of the members.

- $\pi_a = \alpha_a \times T_a + \eta(\alpha_m + \alpha_a) - A$
- $\pi_a = \alpha_a \times 0 + \eta(\alpha_m + \alpha_a) - A$
- $\pi_a = \eta(\alpha_m + \alpha_a) - A$

Thus, if $A < \eta \times \alpha_m$ it is a no brainer for any firm to decide on the adoption as the network benefit is more than the adoption cost.

Considering the case where $A > \eta \times \alpha_m$, the utility function for border-line case is

- $\alpha_a \times T_a + \eta(\alpha_m - \alpha_a) - A = 0$
- $\alpha_a = (A - \eta \times \alpha_m)/(T_a - \eta) = 0$
- $\alpha_m - \alpha_a = \alpha_m - (A - \eta \times \alpha_m)/(T_a - \eta)$

[Subtracting from α_m on both the sides]

- $\alpha_m - \alpha_a = $ adopters network $= [\alpha_m(T_a - \eta) - (A - \eta \times \alpha_m)]/(T_a - \eta)$
- $\alpha_m - \alpha_a = [\alpha_m \times T_a - \alpha_m \times \eta - A - \eta \times \alpha_m]/(T_a - \eta)$
- $\alpha_m - \alpha_a = [\alpha_m \times T_a - A]/(T_a - \eta)$
- $\alpha_m - \alpha_a = [\alpha_m \times T_a - A+]/(T_a - \eta)$

11.5.2.2   Adoption Network Variation with Standard Value

Let's analyze the adopter's network variation with respect to the value of the standard.

- $\partial(\alpha_m - \alpha_a)/\partial T_a = \partial([\alpha_m \times T_a - A]/(T_a - \eta))/\partial T_m$
- $\partial(\alpha_m - \alpha_a)/\partial T_a = \{(T_a - \eta)\ \partial([\alpha_m \times T_a - A])\partial T_m - [\alpha_m \times T_a - A] \times$
  $\partial(T_a - \eta)/\partial T_a\}/(T_a - \eta)^2$
- $\partial(\alpha_m - \alpha_a)/\partial T_a = (T_a - \eta) \times \alpha_m - [\alpha_m \times T_a - A] \times 1/(T_a - \eta)^2$
- $\partial(\alpha_m - \alpha_a)/\partial T_a = (\alpha_m \times T_a - \eta \times \alpha_m - \alpha_m \times T_a + A]/(T_a - \eta)^2$
- $\partial(\alpha_m - \alpha_a)/\partial T_a = (-\eta \times \alpha_m - + A]/(T_a - \eta)^2$
- $\partial(\alpha_m - \alpha_a)/\partial T_a = (A - \eta \times \alpha_m]/(T_a - \eta)^2$

Since we are considering $A > \eta \times \alpha_m$ the above equation results in a positive value that is, > 0.

Thus, the number of adopters is positively related to total output of the standard being formalized.

11.5.2.3   Adoption Network Variation with Network Strength

Analyzing the adopter's network variation with respect to the strength of network that is, η.

- $\partial(\alpha_m - \alpha_a)/\partial\eta = \partial([\alpha_m \times T_a - A]/(T_a - \eta))/\partial\eta$
- $\partial(\alpha_m - \alpha_a)/\partial\eta = \{(T_a - \eta)\ \partial([\alpha_m \times T_a - A])\partial\eta - [\alpha_m \times T_a - A] \times$
  $\partial(T_a - \eta)/\partial\eta\}/(T_a - \eta)^2$

- $\partial(\alpha_m - \alpha_a)/\partial\eta = \{0 - [\alpha_m \times T_a - A] \times \partial(T_a - \eta)/\partial\eta\}/(T_a - \eta)^2$
- $\partial(\alpha_m - \alpha_a)/\partial\eta = - [\alpha_m \times T_a - \eta] \times (-1)/(T_a - \eta)^2$
- $\partial(\alpha_m - \alpha_a)/\partial\eta = [\alpha_m \times T_a - A]/(T_a - \eta)^2$

If $T_a > A/\alpha_m$ then the above value will be positive. The adopter's network will be positively related to the strength of network effect.

Thus, higher the network benefit, larger the value an additional user brings to other adopters as well as itself.

11.5.2.4 Adoption Network Variation with Adoption Cost

Analyzing the adopter's network variation with respect to the adoption cost that is, A.

- $\partial(\alpha_m - \alpha_a)/\partial A = \partial([\alpha_m \times T_a - A]/(T_a - \eta))/\partial A$
- $\partial(\alpha_m - \alpha_a)/\partial A = \{(T_a - \eta) \, \partial([\alpha_m \times T_a - A]\partial A - ([\alpha_m \times T_a - A] \times \\ \partial(T_a - \eta)/\partial A\}/(T_a - \eta)^2$
- $\partial(\alpha_m - \alpha_a)/\partial A = \{(T_a - \eta)(-1) - [\alpha_m \times T_a - A] \times \partial(T_a - \eta)/\partial A\}/(T_a - \eta)^2$
- $\partial(\alpha_m - \alpha_a)/\partial A = \{(T_a - \eta)(-1) - 0\}/(T_a - \eta)^2$
- $\partial(\alpha_m - \alpha_a)/\partial A = \{(T_a - \eta)(-1)\}/(T_a - \eta)^2$
- $\partial(\alpha_m - \alpha_a)/\partial A = \{(-1)\}/(T_a - \eta)^2$

Thus the number of adopters is negatively related to the adoption cost.

11.5.3 PARTICIPATION THRESHOLD

The utility function of participants and adopters are as below:

- $\pi_p = \alpha \times T + \eta(\alpha_m - \alpha_a) - A + B(e, \alpha) - C$

Since for adopters, $B(e, \alpha) - C = 0$

- $\pi_a = \alpha \times T + \eta(\alpha_m - \alpha_a) - A$

The marginal participant prefers weakly active participation to adoption.

- $\alpha_a \times T_a + \eta(\alpha_m - \alpha_a) - A + B(e, \alpha) - C \geq \alpha_p \times T + \eta(\alpha_m - \alpha_a) - A$

Also, the participant will participate if it has positive utility i.e.

- $\alpha_p \times T_a + \eta(\alpha_m - \alpha_a) - A + B(e, \alpha) - C \geq 0$
- $B(e, \alpha) - C \geq 0$

Thus for all firms with $\alpha(\alpha_p \leq \alpha \leq \alpha_m)$ the firms will participate than being an adopter.

If $\alpha_p > \alpha_m$ that is, the threshold value of participation is greater than α_m, then no firm will have incentive to participate.

Thus participation of firm is not dependent on the total output T, or network effect η or A (adoption cost). It is purely a function of Benefit B(e, α) and cost associated with the participation.

11.5.4 ADOPTION THRESHOLD

If α_a is threshold for adoption and α_p is threshold for participation and α_m is the maximum value of α, then by Participation threshold theory, all firms with $\alpha(\alpha_p \leq \alpha \leq \alpha_m)$ will participate than being an adopter. All firms with $\alpha \leq \alpha_a$ will not adopt as they don't find any value in being part of the alliance.

So only condition in which the adopters exist is $\alpha_a \leq \alpha_p$, in which case the proportion of adopters is $\alpha_p - \alpha_a$.

We now prove that the firms adopting the standard will have its profit as increasing function of α. And that this increase is higher for participants than adopters.

The utility function for participant is equal to

- $\pi_p = \alpha \times T + \eta(\alpha_m - \alpha_a) - A + B(e, \alpha) - C$

- $\dfrac{\partial \pi_p}{\partial \alpha} = T + \partial B(e, \alpha) / \partial$

The utility function for adopter is equal to

- $\pi_a = \alpha \times T + \eta(\alpha_m - \alpha_a) - A$

- $\dfrac{\partial \pi_a}{\partial \alpha} = T > \dfrac{\partial \pi_p}{\partial \alpha}$

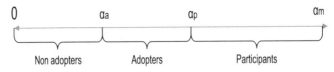

FIGURE 11.1 Adoption threshold limits.

11.6   DISCUSSION

The above mathematical model can be described in the form of below model.

Adoption Network Variation with standard value:

If all the other values are constant, the adopter's utility increases with the value of the standard being formalized. Thus the number of adopters will increase with the value of the standard.

Adoption Network Variation with network strength:

If all the other values are constant, the adopter's utility increases with the strength of network. Thus higher the network benefit, larger the value an additional user brings to other adopters as well as itself.

Adoption Network Variation with adoption cost:

If all the other values are constant, the utility of adopter decreases with the increase in adoption cost. Thus the number of adopters is negatively related to the adoption cost.

Participation threshold:

The participation of firm is purely a function of Benefit and cost associated with the participation. If the firm doesn't find the participation benefit to be higher than the participation cost, it would prefer to wait and watch

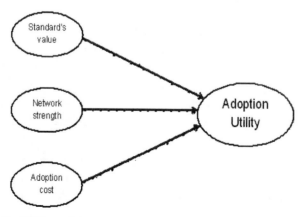

FIGURE 11.2   Utility model.

and adopt once the standard is ready. Thus the participation is not dependent on the standard value, network effect and adoption cost.

Adoption threshold:
Since the participation benefit is a function of value of standard, the expected utility of the standard for participant increase at higher rate than for adopters.

11.7   CONCLUSION

This chapter started with an overview of industrial revolution brought by the advances in Information and technology. With this revolution, the landscape of alliances has changed significantly. The aspects of alliance, which are relevant for handling insurance fraud are covered in details in the subsequent subsections.

The benefits of standardization and examples of standardization with specific focus to financial institutions are covered. Network economics is explained with special focus on the externalities.

A detailed statistics on the insurance fraud in various countries is provided showing impact of recession on the fraud.

In addition to theoretical contribution, a model is presented to show that the contribution of members in alliance is critical to the sustainability and success of standard consortia.

The outcome of the analysis is that to enhance firm's adoption of the standard, the condition $A < \eta \times \alpha_m$ should be established. This condition can be achieved by minimizing the adoption cost, maximizing the network effect and maximizing the utility of the alliance output.

To reduce the adoption cost (A) the alliance can involve the software developing firms to enhance their systems to incorporate the new standard. This will ensure that from the technological perspective an optimized solution is arrived which limits the cost of adoption. The uncertainty faced by the potential adopters can be minimized by the trialability and observability of the standard. The figures of the fraud detection using the standard can be published and the standards should be made available on trial for potential adopters to realize the potential of alliance benefits.

The network effect (η) cannot only be increased by involving the insurance companies, but also involving government departments, investigative agencies and consumers themselves. The government involvement in standard development can be through framing fraud prevention policies and integrating police departments, which can work closely with insurance firms in checking the suspicious claims and organizing road safety campaigns. Police and investigative agencies can exchange information on stolen vehicles with the insurance companies. The police officers and investigative agencies can be trained on technical insurance matters to enable them identify fraud behavior. Awareness programs for consumers about the implication of fraud and providing mechanism for them to report fraud can be the best way to get first-hand information about the potential fraud. All parties working together in developing and implementing the standard can increase the network effect dramatically.

The value of the standard (α_m) can be increased by utilizing appropriate business intelligence, data mining and predictive analytics thus identifying fraudulent claims before they are subject to insurers' standard claims process. Better training to staff handling claims about the usage of standard will lead to better awareness and increased productivity. Thus making training as a part of package can increase the value of the standard.

To enhance firm's participation in the alliance, it is important to increase firm's awareness of the benefits brought by the standard. Higher the value from the standard, harder the firms will work to make it success. Some ways of enhancing firm's participation is by improving the participation benefit (B) of its members, for example, voting rights in defining standard formalization, providing exclusive resources to members only (Kexin Zhao et al., 2006; Rogers, 1995). Other way is to reduce the participation cost (C), which can be achieved by using Internet technologies for communication between members, for example, mailing list, online forums, etc.

The limitation of the study is that the developing nations, BRIC Nations (Brazil, China, India and China), which are the four most important economies in the world, referred to as emerging markets, are excluded. There is very limited consolidated data available for these nations. An effort is required to consolidate data available from various sources. This should uncover the cultural attitude towards risks and fraud.

ACKNOWLEDGEMENTS

The authors acknowledge with gratitude the support provided by SSP (www.ssp-uk.com) for carrying out this research.

KEYWORDS

- **alliance**
- **collaborative development**
- **fraud**
- **insurance**
- **network externalities**
- **technology diffusion**

REFERENCES

1. American National Standards Institute (ANSI) http://www.ansi.org/ Accessed on 11th October 2014.
2. ANIA, Italian insurance association http://www.ania.it/export/sites/default/it/pubblicazioni/rapporti-annuali/Italian-Insurance-Statistical-appendix/Italian-Insurance-in-2012–2013.pdf, Accessed on 10th October 2014.
3. ANIA, Italian insurance association, http://www.ania.it Accessed on 11th October 2014.
4. ARGOS http://www.gieargos.org/PortailARGOS/Default.aspx Accessed on 20th October 2014.
5. Arizona Department of Insurance http://www.azinsurance.gov/ Accessed on 11th October 2014.
6. Artis, M., Ayuso, M., Guillen, M., 'Modeling different types of automobile insurance fraud behavior in the Spanish market,' Insurance: Mathematics and Economics, 1999, 24, 67–81.
7. Association of British Insurers, Insurance cheats feel the heat, 30-May, 2014.
8. Association of British Insurers, Research report on General insurance claims fraud, July 2009.
9. Axelrod, R., The evolution of cooperation. New York: Basic Books 1984.
10. Badaracco, J. L., Jr., The knowledge link – How firms compete through strategic alliances. Boston: Harvard Business School Press, 1991.
11. BDO Stoy Hayward, 'Fraud Track 6: On the Ropes,' June 2009.

12. Beamish, P. W. Multinational joint ventures in developing countries. London: Routledge, 1988.
13. Berg, S. V. "Standardization Issues in telecommunication: Competition, Cooperation and Coercion?" mimeo University of Florida, revised September 24, 1984a.
14. Berg, S. V. "Technological Externalities and a theory of technical compatibility standards" mimeo University of Florida, revised November 7, 1984b.
15. Besen S. M., Johnson, L. L. "Compatibility Standards, Competition, and Innovation in the Broadcasting Industry." The RAND Corporation. R-3453-NSF, November 1986.
16. Bodendorf, F., Reinheimer, S. 'Offer Evaluation in an Electronic Air Cargo Market.' Proceedings of the Fifth European Conference on Information Systems (ECIS'97), 868–881. 1997.
17. British Retail Consortium, BRC Retail Crime Survey 2013.
18. Buonanno, P., Montolio, D., 'International Review of Law and Economics, 2008, 28, 89–97.
19. CEA Statistics No.38 The European Motor Insurance Market, February 2010.
20. Child, J., Management and organization. New York: Halstead Press, 1974.
21. CIFAS, 'Fraud Trends and Recession Go Hand in Hand,' first quarter, 2009.
22. Clarke, M., 'The control of insurance fraud. A comparative view.' The British Journal of Criminology, 1990, 30:1, 1–23.
23. D'Aveni, R. A., Hypercompetition. Free Press, New York, 1994.
24. Dierickx, I., Cool, K., Asset stock accumulation and the sustainability if competitive advantage. Management Science, 1989, 35(12), 1504–1511.
25. Doz, Y. L., Hamel, G., Alliance advantage: The act of creating value through partnering. Cambridge, Mass.: Harvard Business School Press, 1998.
26. EY, EY Asia-Pacific insurance outlook, Continuous evolution, 2014.
27. Farrell, J., and Saloner, G. Coordination through committees and markets. RAND Journal of Economics, 1988, 19, 235–252.
28. Farrell, J., Saloner, G., Standardization, Compatibility, and Innovation, Rand Journal of Economics, Vol. 16, No 1, Spring 1985.
29. Financial Services Authority, 'Financial Risk Outlook 2009.'
30. Fraud Advisory Panel, 'Fraud Facts,' 2, January, 2009.
31. Gomes-Casseres, B., The alliance revolution. Cambridge, Mass.: Harvard University Press; 1996.
32. Hall, B., Innovation and Diffusion, Forthcoming in Fagerberg, J., D. Mowery, and R. R. Nelson (eds.), Handbook of Innovation, Oxford University Press, October 2003.
33. Hamel, G., Doz, Y. L., Prahalad, C. K., Collaborate with your competitors and win. Harvard Business Review, 1989, 67, 133–139.
34. Hammond, Janice H.; Kelly, Maura G., Quick response in the apparel industry, Harvard Business School, 1990, Note N9-690-038.
35. Hansen, M., Madnick, S., Siegel, M., Process Aggregation Using Web Services, MIT Sloan School of Management, February 2002.
36. Harrigan, K. R., Strategies for joint ventures. Lexington: Lexington Books, 1985.
37. Hellenic Association of Insurance Companies http://greekinsurancemarket.co.uk/ accessed on 20-Oct-2014.
38. Hobday, M., The limits of Silicon Valley: A critique of network theory. Technology Analysis and Strategic Management, 6, 231–243.

39. Hoyt, R., 'The Effect of Insurance Fraud on the Economic System,' Journal of Insurance Regulation, 1988, 304–315.
40. IFB, Insurance Fraud Bureau, http://www.insurancefraudbureau.org/ Accessed on 11th October 2014.
41. Insurance Bureau of Canada, 'Auto Fraud in the Fast Lane,' 2006.
42. Insurance Council of Australia, 'Insurance Fraud in Australia,' September 1994.
43. Insurance Europe The impact of insurance fraud, Insurance Europe, 2013.
44. Insurance Research Council, cited by the Coalition Against Insurance Fraud in 'Auto Insurance,' 2008.
45. ISVAP, Italian insurance regulatory authority, www.isvap.it Accessed on 11th October 2014.
46. Italian insurance association, ANIA http://www.ania.it Accessed on 11th October 2014.
47. Jarillo, J. C., On strategic networks. Strategic Management Journal, 1988, 9, 31–41.
48. Jarillo, J. C., Comments on 'Transaction costs and networks.' Strategic Management Journal, 1990, 11, 497–499.
49. Jarillo, J. C., Stevenson, H. H., Co-operative Strategies -The payoffs and the pitfalls. Long range planning, 1991, 24(1), 64–70.
50. Joseph Farrell, Garth Saloner, Coordination through committees and markets, RAND Journal of Economics Vol. 19. No. 2, Summer 1988.
51. Katz, M., and Shapiro, C., Network Externalities, Competition, and Compatibility. The American Economic Review, 75, 3, June 1985.
52. Kexin Zhao, Mu Xia, Michael J. Shaw, Journal of Management Information Systems, 2006.
53. Kogut, B., Zander, U., Knowledge of the firm, combinative capabilities, and the replication of technology. Organization Science, 1992, 3(3), 383–397.
54. KPMG, 'Fighting Fraud,' 2009, 27, Spring.
55. Lei, D., Slocum, J. W., Global strategy, competence-building and strategic alliances. California Management Review, 1992, 80–97.
56. Madhok, A., Tallman, S. B., Resources, transactions and rents: Managing value through interfirm collaborative relationships. Organizational Science, 1998, 9(3), 326–339.
57. Malone, T. W., Crowston, K. 'The Interdisciplinary Study of Coordination.' ACM Computing Surveys, no. 1, vol. 26, 87–119, 1994.
58. Malone, T. W., Rockart, J. F. 'Computers, Networks and the Corporation.' Scientific American, no. 3, vol. 265, 92–99, 1991.
59. Moore, J. F. The death of competition: Leadership and strategy in the age of business ecosystems. New York: Harper Business, 1996.
60. Mowery, D. C., Oxley, J. E., Silverman, B. S., Strategic alliances and interfirm knowledge transfer. Strategic Management Journal, 1996, 17(Winter Special Issue), 77–91.
61. New York Alliance Against Insurance Fraud (NYAAIF) http://www.fraudny.com accessed on 20-Oct-2014.
62. Oliver, C., Determinants of inter-organizational relationships: integration and future directions. Academy of Management Review, 1990, 15(2), 241–265.

63. Osborne, D., 'Crime and the UK Economy,' 1995, cited in Scorcu, A., Cellini, R., 'Economic Activity and Crime in the Long Run: An Empirical Investigation on Aggregate Data from Italy, 1951–1994,' International Review of Law and Economics, 1996, 18, 279–292.

64. Pfeffer, J., Salancik, G. R. The external control of organizations. New York: Harper & Row, 1978.

65. Pisano, G. P., The governance of innovation: Vertical integration and collaborative arrangements in the biotechnology industry. Research Policy, 1991, 20, 237–249.

66. Pisano, G. P., Mang, P. Y., Collaborative product development and the market for know-how: Strategies and structures in the biotechnology industry. In R. A. Burgelman, R. S. Rosenbloom (Eds.), Research on Technological Innovation, Management and Policy (pp. 109–136). Greenwich, Conneticut: JAI Press., 1993.

67. PricewaterhouseCoopers, 'Fraud in a Downturn: A review of how fraud and other integrity risks will affect business in 2009,' February, 2009.

68. Rogers E. M., Diffusion of Innovations, Fourth Edition, The Free Press, 1995.

69. Royal and SunAlliance, 'Britons see Insurance Fraud as more acceptable during recession,' February, 2009.

70. Scorcu, A., Cellini, R., 'Economic Activity and Crime in the Long Run: An Empirical Investigation on Aggregate Data,' International Review of Law and Economics, 1996, 18, 279–292.

71. Shapiro, C., Varian, H. R., Information rules – a strategic guide to the network economy: Harvard Business School Press, 1998.

72. Teece, D. J., Profiting from technological innovation: Implications for integration, collaboration, licensing and public policy. Research Policy, 1986, 15, 285–305.

73. Tsushima, M., 'Economic Structure and Crime: The Case of Japan,' Journal of Socio-Economics, 2002, 25(4), 497–515.

74. Whetten, D. A., Interorganizational relations: A review of the field. Journal of Higher Education, 1981, 52, 1–28.

75. World Wide Web Consortium (W3C), XML Essentials http://www.w3.org/standards/xml/core Accessed 20th October 2014.

76. Yoo, B., Choudhary, V., Mukhopadhyay, T. A model of neutral B2B intermediaries. Journal of Management Information Systems, 19, 3 (Winter 2002–2003), 43–68.

PART 5:

E-VOTING

CHAPTER 12

ASSESSING THE FACTORS THAT COULD IMPACT THE ADOPTION OF E-VOTING TECHNOLOGIES WITH A SOUTH AFRICAN CONTEXT

MOURINE ACHIENG[1] and EPHIAS RUHODE[2]

[1]P.O. Box 652, Cape Town 8000, RSA Cape Peninsula University of Technology, Cape Town, South Africa, Tel: +27769070044, E-mail: sachiengm@gmail.com

[2]P.O. Box 652, Cape Town 8000, RSA Cape Peninsula University of Technology, Cape Town, South Africa, Tel: +27-21-460-3284; E-mail: RuhodeE@cput.ac.za; Ruhode@gmail.com

CONTENTS

12.1   ABSTRACT

This article examines the factors that could potentially influence the adoption of e-voting technologies within the South African context. The study explored the challenges associated with the current manual paper voting electoral process and how these challenges are related to the factors that could influence the adoption of e-voting technologies. The approach taken was inductive based interpretivism paradigm and technology adoption models. Data was collected via questionnaires and semi-structured interviews. Diffusion of Innovation constructs relative advantage, compatibility, and complexity was found to be potential factors. Other factors included availability of infrastructure and resources, awareness, trust in the innovation and digital divide. These results are significant to the Electoral management body that need to consider these factors before introducing e-voting technologies in South Africa.

12.2   BACKGROUND AND PURPOSE

South Africa has conducted five national elections since it became a democratic state back in 1994. This year's (2014) elections marked the

20th anniversary of democracy. The Independent electoral commission (IEC) of South Africa, who are in charge of organizing and running the elections, has been faced with various challenges running these elections over the past years. One of the challenges the ICE has faced over the years is to improve voter turnout especially amongst the youths. Since this years' (2014) elections was the first for the "born frees" (those voters born after the first elections were held in 1994 post the apartheid era), these voters were not only born into a free South Africa, but also came into a society that is increasingly using technology in all aspects of their lives. As such they have been exposed to the use of technology much earlier into their lives compared to older voters. One would imagine that the IEC would to be aiming at making the voting process more appealing to this group of voters.

With a youth group that is accustomed to using technology in their everyday begs the question whether the South African voting population is ready for an e-voting system, Or whether the lack of ICT infrastructure and resources especially in rural communities and informal settlements could hinder the adoption of such a system. There are many issues that need to be firstly addressed, such as infrastructure un-readiness, older generation of voter's reluctance to accept technology, and the lack of standard and framework prior to implementation.

The extent to which South Africa is ready for an e-voting system has not be extensively explored therefore this paper argues on the need for thorough study on adoption of e-voting technologies and also looks at the challenges still lingering from the country's previous elections and the issues that might arise with the implementation of an e-voting system in South Africa. The study was undertaken with the purpose of exploring how the adoption of e-voting technologies would diffuse within South African context with the focus being Cape Town and also explore the factors that could influence the adoption of e-voting from the perspective of the citizens and the Independent Electoral Commission.

Every democracy requires an electoral process that is free and fair; it is in this political field can the use of technology be measured. This research defines democracy as a form of governance in which the power is vested in and exercised by all eligible citizens to participate equally through their elected representatives. Gianluca (2006:5) points out that e-governance is generally referred to as the use of ICTs to harness changes, looking not only at the increasing use of ICTs as a technological tool for delivering services online and improving the efficiency of administrations, but as a new model for opening up government services to citizens, thereby increasing transparency and participation and making governments more responsive and centered upon its citizen's needs.

Macintosh (2004) defined e-democracy as "the use of information and communication technologies to engage citizens, support the democratic decision-making processes and strengthen representative democracy." While e-democracy is the use of technology for strengthening the mechanisms of democratic decision making, e-voting and e-participation are components of e-democracy that focus on the means for achieving this (*ibid*). E-participation according to Islam (2008) is the use of modern ICT supported platform to facilitate participation in democracy and governance. E-participation requires those who govern to pay attention to political opinions of those they govern; it is not just delivery of public services online.

Qadah and Taha (2007) defined the term electronic voting as the "use of computers or computerized equipment to cast votes in an election." The authors continue to emphasize that "e-voting aims at increasing participation, lowering costs of running elections and improving the accuracy of the results." According to the ACE Electoral knowledge Network (2010), countries such as the USA, Brazil, and India have successfully implemented e-voting to address various challenges associated with the manual paper based electoral process. It is in light of this that the study explored the challenges and prospects of adopting an e-voting system and how South Africa could leverage on the opportunities it presents. It should also be noted that the study does not look into specific electronic voting systems, but rather the adoption and diffusion of electronic voting technologies in general.

12.3 LITERATURE REVIEW

Much has been written about the adoption of electronic voting technologies by countries around the world. Whereas there are countries such as Brazil and India that have embraced the idea of introducing technology into the voting process, others have doubts about the security issues and other risks that would come with the introduction of such a technology, therefore opting not adopt the technology or abandon it completely (ACE Electoral knowledge Network, 2010). Literature has shown that there are those countries that had initially adopted e-voting system but later on stopped the because of the issues that surrounding the technology (Thakur, 2013).

South Africa took to the poll this year to mark their 20th anniversary of democracy. This year's election saw the so called "born frees" (those who were born after the first elections were held in, 1994) voters cast their ballots for the first time. These voters were not only born into a free South Africa, but also came into a society that is increasingly using technology in all aspects of their lives. As such they have been exposed to the use of technology much earlier into their lives compared to the older generation of voters.

Most of these "born frees" are accustomed to the convenience of technology, they might find long queues at the polling stations and the use of ballot papers less appealing. The born free would prefer electronic applications like the once they are accustomed to. With a society that is accustomed to using technology in their everyday begs the question whether South Africa is ready for an e-electoral process. Smith and Macintosh (2004) emphasized that a modern e-enabled system of democratic governance seems to require some sort of modernization of the electoral process, whether through e-counting methods or an e-voting system. According to Macintosh (2004), a powerful symbol of a democracy is the participation of citizens in the free and fair elections of representatives to govern them. Macintosh continues to mention that voting is seen as the act that currently defines the relationships between citizens, governments and democracy (2004). As such e-voting takes on an influential symbolic role in e-democracy.

In March of 2013, the Independent Electoral Commission held a conference to investigate if the South African electoral process needed to be bettered using a technology (Seminar on Electronic Voting and Counting Technologies, 2013). The IEC has always been known as a technologically savvy organization and it has won numerous awards as a result (IEC, 2009). The idea of exploring the possibility of using e-voting did not emanate from within the IEC; in 2009 during his acceptance speech of the election results the president suggested that perhaps the electoral process could use an electronic voting system.

The adoption of e-voting technologies is not only happening in the western countries but also in other democracies in the African continent. Most of these countries are trying to reduce the use of papers in the voting process but e-voting does include a bit of paper produced for verifiability purposes. The e-voting system does not just produce any kind of paper; but the kind that is machine-readable. After marking their choice, voters insert the ballot into the machine, which reads the choice of the voter, tallies the result and issue them immediately. With this kind of system there would be no manual counting of ballot papers, but they are verifiable should the results be disputed.

The worldwide experience of implementing e-voting is mixed with respect to adoption, non-adoption or adoption followed by abandonment (Goldsmith, 2011). Brazil and India are the best examples of countries that have successfully adopted e-voting technologies. In Brazil, phones and ATMs features have been adopted which gives the voting machines a touch of familiarity especially the illiterate voters. One of the faults of the e-voting system is its lack of transparency. Because the e-voting process does not involve a ballot paper, there is no trail to verify the results. Voters have to trust the machines and those who handle them. In other words, e-voting is not entirely seamless it assures instant results but lacks transparency.

According to Thakur (2013), the adoption of e-voting technologies is not a "sudden or immediate switchover technology; rather the vision is one of a phased move to multi-channel elections in which voters re offered an increasing range of means to register their vote." Thakur noted that countries that have successfully adopted e-voting did so through phased transparent trials, pilots, followed by full implementations (2013).

The context in which e-voting technologies would be adopted is equally important in determining whether a country should adopt an e-voting system. Just as literature has revealed there are countries that have found the e-voting system unsuitable and reverted to manual voting and counting. Countries that have reverted to manual voting include The Netherlands, Germany, and Ireland. The reasons for going back to manual voting vary from country to country Goldsmith (2011). The Netherlands reverted back to manual voting because of the inconvenience of constant malfunctioning of machines. Germany could not take the lack transparency. The German constitution prescribes that counting of votes should be public. Because e-counting is not visible to the eye, the courts declared it unconstitutional (*ibid*).

The adoption of e-voting technologies has a lot to do with the need of the technology for the particular country. After a few inconvenience, some countries can easily ditch e-voting. Trust and the size of the electorate are important in the decision to whether or not to adopt e-voting technologies. For successful consideration and implementation of electronic voting and counting technologies transparency and openness are very essential. Change can be unsettling and it is crucial that stakeholders trust the electoral process. Thad Hall argued in a presentation at the EVOTE 2012 Conference that "it is not the technology that is used that matters, but the way in which the technology is implemented that ultimately determines the success of an election technology project."

The study cannot discuss at the benefits of e-voting technologies without looking the risks and challenges associated with its adoption as well. According to Gupta (2011), the biggest challenge of deploying e-voting system is not technology but change management. Gupta says that change management is important not only in terms of cultural change, but also in terms of changing operations and processes that this automated e-voting system may introduce. Gupta (2011) divides the challenges into two broad categories; technology challenges and organizational/human factor challenges. Cetinkaya and Centinkaya (2007) stated that "from a technological view accepting e-elections is not a problem as a significant part of the population are using technology in different aspects of their daily life. Therefore, the challenge comes from the high sensitivity of e-voting systems towards some subjects such as security, privacy and trusting … suppliers of e-voting systems."

The technological developments in South Africa have opened up the possibility of adopting e-voting and counting technologies and this clearly provides some opportunities and challenges. Svensson and Leenes (2003) argued that on one hand, the electronic voting technology may help make voting more cost effective and convenient for voters, which may in turn increase voter turnout. However, on the other hand, e-voting may introduce new risks that may affect the electoral values such as secrecy of the vote and placing of voting as an observable institution in modern democracies (Svensson and Leenes, 2003).

12.4 TECHNOLOGY ACCEPTANCE MODELS

The study made use of theoretical analytical framework to understand factors of technology adoption. The theoretical analytical framework therefore provides a lens through which this research phenomenon was viewed. Burton-Jone and Hubona (2005) in their study pointed out that the prediction of the adoption and use of information technology has been a key interest for many scholars since the start of research in information systems.

Adoption theories according to Straub (2009), examine an individual and the choices they make to accept or reject a particular innovation. The author goes on to define adoption theory as a "micro-perspective on change, focusing not on the whole but rather the pieces that make up the whole." Straub (2009) further discusses that in contrast, diffusion theories describe how an innovation spreads through a population. Diffusion theories may consider factors like time and social pressures to explain the process of how a population adopts or adapts to or rejects a particular innovation. Thus diffusion theory takes a macro-perspective on the spread of an innovation across time (*ibid*).

The study did recognize the existence of several other technology adoption theories in the Information Systems field, however the study focused on just a few of the theories used frequently in technology adoption field such as TAM, UTUAT, TRA and DoI.

12.4.1 THEORY OF REASONED ACTION (TRA)

The Theory of Reasoned Action was formulated by Fishbein and Ajzen (1980) and posits that behavioral intentions are influenced by attitudes and subjective norms. The behavioral intentions ultimately lead to the actual behavior. A limitation of TRA perhaps is its inability to account for non-controllable variables affecting behavior. From a theoretical point of view, the TRA is intuitive, parsimonious, and insightful in its ability to explain behavior (Bagozzi, 1982). The TRA assumes that individuals are usually rational and will consider the implications of their actions prior to deciding whether to perform a given behavior (Ajzen & Fishbein, 1980). According to Foxall (2007), The TRA deals with predictions rather than outcome of behaviors. In the TRA, behavior is determined by behavioral intentions, thus limiting the predictability of the model to situations in which intentions and behavior are highly correlated (Yousafzai et al., 2010).

12.4.2 TECHNOLOGY ACCEPTANCE MODEL (TAM)

Davis' (1989) TAM is widely used to study user acceptance of technology, it is based on the theory of reason action (TRA) which states that beliefs influence intentions and intentions influences one's action. According to TAM, perceived usefulness (PU) and Perceived ease of use (PEOU) influences one's attitude toward system usage, which influences one's behavioral intention to use a system, which in turn determines actual system usage. Mathieson et al. (2001) did an extensive comparison between TPB and TAM and found that in most cases TAM is easier to apply when predicting IS usage. Davis (1989) developed TAM to predict individual adoption and use of new Information Technologies. Furthermore, TAM was developed after the introduction of information systems into organizations therefore it would be beneficial for researches looking at technology adoption in organizations.

TAM like its other predecessors has major limitations and according to Li, Nasco, and Clark (2007), TAM theorized that an individual's behavioral intentions to adopt a particular piece of technology are determined by the person's attitude towards the use of the technology. TAM was developed to understand employee acceptance of new technology and most research using the model has focused on cognition rather than affect. The emphasis on cognition might be appropriate for organizational context where adoption is mandated and users have little choice regarding the decision. But it is an insufficient explanation for consumer context in which potential; users are free to adopt or reject new technology based on how they feel as well as how they think. TAM has its dominant focus on business-to-business (B2B) research in the working environments (Asare et al., 2011:194).

Other criticisms of TAM are that it fails to explain how adoption or usage can be improved using the variables and that apart from the perceived ease of use and perceived usefulness; TAM fails to incorporate other factors, which have an influence on technology acceptance (Venkatesh et al., 2003; Al-Qeisi, 2009). TAM has difficulties in explaining the gap between predicted adoption and actual adoption. In other words the link between attitude and behavior and between intention and behavior are not convincingly dictated or the research results have been contradictory (Juntumaa, 2011:7–9).

12.4.3 UNIFIED THEORY OF ACCEPTANCE AND USE OF TECHNOLOGY (UTAUT)

Venkatesh et al. (2003) launched a merged information technology model; namely the Unified theory of acceptance and use of technology (UTAUT) due to the technology adoption research diversity and complexity, and the failure of TAM to consider organizations' external factors that influence adoption (Well et al., 2010:814). The original UTAUT framework focused on the mandatory use of technology in a work environment, but it has also been utilized in voluntary settings (Chiu et al., 2010:449). The purpose of UTAUT-model therefore is to understand system and to provide constructs which are meant to be independent of any particular theoretical perspective (Venkatesh et al., 2003:447). Furthermore, the model posits that adoption intention has significant positive influence on technology usage in every research settings (Venkatesh et al., 2003:456). In other words, UTAUT is inherently a general adoption theory, which is not context-dependent.

After discussing these three theories, the study adopts DoI theory taking into consideration only three of its attributes (relative advantage, compatibility and complexity). The study chose this theory because DoI focuses predominantly on the pre-adoption phase not the actual use of the innovation or new applications (Ozdemir et al., 2008:216). The three constructs used have the most consistent significance relationship to innovation adoption. The reason is that because e-voting is yet to be adopted in South Africa, the other two constructs; triability and observability cannot be measured. Relative advantage was relevant for the context of this research as the study aimed at determining the participants (voters and the Electoral Management Body) perception of the relative advantage/benefits of e-voting technologies compared to the current paper-based electoral process.

According to Taylor and Todd (1995a), compatibility is an important construct that can positively influence adoption. They give an example stating that "if the use of an innovation violates a cultural or social norm it is less likely to be adopted (Taylor and Todd, 1995a: 141)." For this study, the researcher sought to explore the participants' perception of e-voting technologies being compatible to their way of lives. The third and last construct to be examined in this study was complexity, which is equivalent to TAM's construct perceived ease of use (Agarwal and Karhanna, 1998:3; Moore and Benbast, 1999; Carter and Bélanger, 2005). This study tries to measure participant's perception of the complexity of e-voting technologies.

12.4.4 DIFFUSION OF INNOVATION (DOI)

According to Rogers (2003), DoI theory seeks to explain the process and factors that influence the adoption of new innovations. The DoI theory also according to Bhattacherjee (2012), explains how innovations are adopted within a population of potential adopters and the innovation decision process, factors determining the rate of adoption and categories of adopters. Rogers (2003:5) defines diffusion as "the process by which an innovation is communicated through certain channels over time among the members of a social society." The author further differentiates the adoption process from the diffusion process in that the diffusion process occurs within society, as a group process; whereas, the adoption process pertains to an individual. Rogers (2003:12) defines the adoption process as "the mental process through which an individual passes from first hearing an innovation to finally adopting the innovation."

Brown (1999) pointed out that the purpose of DoI theory is to "provide individuals from disciplines interested in the diffusion of an innovation with a conceptual paradigm for understanding the process of diffusion and social change." Diffusion of Innovation theory provides well developed concepts and a large body of empirical results applicable to the study of technology evaluation, adoption and implementation, as well as tools, both quantitative and qualitative, for assessing the likely rate of diffusion of technology, and identifies numerous factors that facilitate or hinder technology adoption and implementation (Fichman, 1992). These factors include the innovation-decision process, attribution of the innovation and innovators' characteristics.

The innovation-decision process is the process through which an individual (or other decision-making unit) passes through different phases starting from first knowledge about the innovation to formulating an attitude towards it, to a decision regarding adoption or rejection, to implementation of the new idea, and to confirmation of this decision (Rogers, 2003). This process consists of five phases; knowledge, persuasion, decision, implementation and confirmation as illustrated in Figure 12.1.

In the Knowledge phase, an individual is exposed to an innovation. An attitude about the innovation, which can be favorable or unfavorable is formed in the Persuasion phase. The individual engages in activities that lead to a decision to adopt or reject the innovation in the decision phase. The innovation is put to use in the implementation phase. Even after deciding to

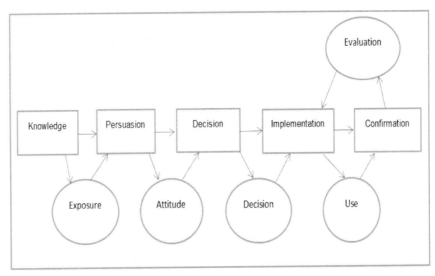

FIGURE 12.1   Phases in the innovation–decision process (Source: Rogers, 2003).

adopt, an individual may in the confirmation phase evaluate the decision to continue or discontinue use of the innovation.

Within the context of this study, during the Knowledge *phase*, voters would be exposed to an innovation (e-voting technologies); they would then form an attitude about the innovation in the *persuasion phase,* which could be favorable or unfavorable. The voters and IEC would then engage in activities that lead to a decision to adopt or reject the innovation in the *decision phase.* The innovation would then be put to use in the *implementation phase.* Even after deciding to adopt, voters/IEC may in the confirmation phase evaluate the decision to continue or discontinue use of the innovation. Rogers (2003:15), points out that an individuals' perception of the attributes of an innovation and not the attributes as classified objectively by experts or change agents, affect the rate of adoption.

12.5   RESEARCH METHODOLOGY

The study is an exploratory qualitative study with an interpretivist approach. Interpretivism is based on the observation that there are major differences between the natural world and the social world. Interpretivist researchers thus attempt to understand a phenomenon through accessing the meanings assigned to them by participants (Bhattacherjee, 2012). The aim of interpretivism is to understand the individual experiences of those being studied, how they think and feel and how they act/re-act in their habitual contexts.

The study's objectives was to explore how the adoption of e-voting would diffuse within South African context and also explore the factors that could influence the adoption of e-voting from the perspective of the citizens and the Independent Electoral Commission. To select the appropriate sample from the population was a challenging task for a number of reasons. Firstly, there is a high level of variation within the South African society, including a variation in age, culture, education, and relation to technology and secondly electronic voting technology is a new or nonexistence concept for South Africa citizens. The approach that this research used for selecting the sample was based on the notion that when a new technology is introduced into a society, it is not expected to be adopted and used immediately by the entire population. On the contrary, the process of adoption passes through stages, starting with a small group of people who later after successfully adopting the technology encourages others to take up the new technology.

Rogers (2003) also explains that an innovation will diffuse through population overtime, and the rate of adoption will vary between those who adopt early, referred to as "innovators" and "early adopter" and those who adopt the innovation much later, referred to as "laggards." The remaining two segments, "early majority" and "late majority," account for the majority of users who adopt innovation overtime. Figure 12.2 is a representation of a diffusion curve, which is a graphical representation of cumulative frequency of individual adoptions. It illustrates how diffusion overtime is composed of individual making adoption decisions.

Rogers (2003) also explains that an innovation will diffuse through population overtime, and the rate of adoption will vary between those who adopt early, referred to as *"innovators"* and *"early adopter"* and those who adopt the innovation much later, referred to as *"laggards."* The remaining two segments, *"early majority"* and *"late majority,"* account for the majority of users who adopt innovation overtime. Figure 12.2 is the diffusion curve, which is a graphical representation of cumulative frequency of individual adoptions. It illustrates how diffusion overtime is composed of individual making adoption decisions.

Based on this view, the research selected two sampling units within the Cape Town area. The first sample unit was the voters who were expected to be more aware of technology and either eager to use technology or

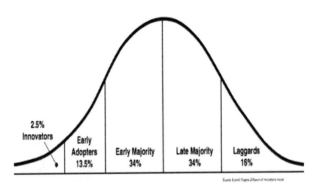

FIGURE 12.2   The bell shaped curve (Source: Rogers, 1995).

were currently using a technology. The second sampling unit was the Independent Electoral Commission officials who are in charge of running the elections in South Africa. These two sampling units were assessed for their views on e-voting, their view on South Africa's readiness to adopt an e-voting system and how much they knew about e-voting technologies.

Jansen (2010:3) mentions that surveys in a qualitative approach does not aim at establishing occurrences in a specific situation but aims at the diversity of a specific topic within a given area of study. The study used an online survey questionnaire and in-depth semi-structured interview as the primary source of data. The questionnaire, which was designed inline with guidelines for questionnaire design recommended by Babbie and Mouton, contained both open-ended and closed-ended questions and evolved around participants' view of e-voting technologies in general in comparison with the manual paper based voting systems currently used (Babbie and Mouton, 2008). The questions asked were clear and simple to avoid double meanings. The questionnaire was then pre-tested with a small group of people before it was administered to the study population. This was done to see if the respondents were able to understand the questions and also to identify which questions the respondents were reluctant to answer (Ham, 2007). Those involved in the pre-test were no longer eligible for inclusion in the final survey sample.

Appropriate revision of the questions was made and the final draft of the questionnaire was administered for the study. A link to the survey was then sent to participants via email. This study utilized purposive sampling, which is a type of nonprobability sampling to obtain participants for this study. Purposive sampling involved the researcher making a conscious decision about which individuals would best provide the desired information required for this study (De Vaus, 2002; Burns and Grove, 2007). There is also a snowballing effect taking place as the participants invited others to the survey.

A total of 400 participants agreed to participate in the survey; owing to time constraints a larger sample size could not be obtained. An in-depth semi-structured interview was conducted with officials from the IEC who oversee the running of elections in South Africa. The semi-structured interview included questions about the electoral process, challenges with the current electoral process and how they tackle those challenges, the

IEC's knowledge of electronic voting and counting technologies, South Africa's readiness for an e-voting or counting system and what factors the IEC thought could hinder the adoption of any of this systems.

This study used thematic data analysis for analyzing the data; Braun and Clarke (2006) define thematic analysis as a qualitative analytic method for "identifying, analyzing and reporting patterns (themes) within data." It minimally organizes and describes your data set in rich detail. However, frequently it goes further than this and interprets various aspects of the research topic." In general, thematic analysis involves the searching across a data set to find repeated patterns of meaning (Braun and Clarke, 2006). Furthermore, according to Fereday and Muir-Cochrane (2006), thematic analysis is a form of pattern recognition within the data, where emerging themes become the categories for analysis (Fereday and Muir-Cochrane, 2006). After going through several literatures on analytical techniques, the study chose to use thematic analysis because of some of the advantages it presents; "can carefully summarize key features of a large body of data and offer a thick description of the data set and can highlight similarities and difference across the data set (Braun and Clarke, 2006)."

The DoI framework was used as a theoretical lens for analysis to validate the analysis process. The themes that emerged from the data were categorized based on the theoretical constructs, which are relative advantage, compatibility, and complexity.

12.6   RESULTS/FINDINGS

The study revealed that there is a great deal of evidence that South African voters still associate the current manual voting system with various challenges. Literature analysis reveal some of these challenges which include lack of infrastructure and resources to aid the running of the elections especially in the informal settlements as one of the major challenges the IEC is faced with. These findings are supported by the findings from the interview with the IEC who confirm that they often forced to improvise resources and infrastructure during elections in these communities, in contrast they do not have the same experience in urban areas where infrastructure are more developed. These finding could perhaps be associated with the vast inequalities that still exist amongst communities in South Africa and especially in the Western Cape Province specifically in Cape Town where this study was conducted. The study is by no means saying that a challenge such as this could be solved by the adoption of e-voting technologies but perhaps such a technology would open up possibilities of the South African government introducing ICT infrastructures in the informal settlements (Figure 12.3).

Another challenge that emerged from the findings was the immense task of counting votes after the casting of votes. With the pressure of releasing results as soon as the elections are cast the IEC are faced with the task of counting manually the votes cast from all the 9 provinces with accuracy. Counting of paper ballots after the elections is usually a tedious job especially when voting goes beyond scheduled times. A delay in releasing election results is often more associated with election violence and always leads to the questioning of the credibility of the results. This is a challenge that perhaps could be easily mitigated by introducing the e-counting technology, which would provide results fast and accurately.

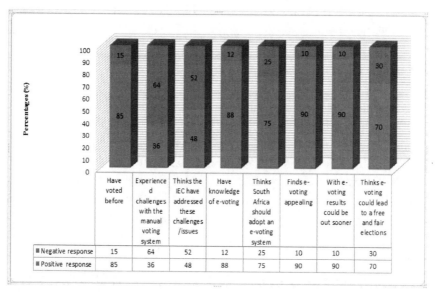

	Have voted before	Experienced challenges with the manual voting system	Thinks the IEC have addressed these challenges /issues	Have knowledge of e-voting	Thinks South Africa should adopt an e-voting system	Finds e-voting appealing	With e-voting results could be out sooner	Thinks e-voting could lead to a free and fair elections
■ Negative response	15	64	52	12	25	10	10	30
■ Positive response	85	36	48	88	75	90	90	70

FIGURE 12.3 Percentage of responses comparing e-voting and manual paper based voting.

From the voters point of view some of the challenges they associate the current paper based voting process is the long queues at polling station during elections. However, it is worth mentioning that not all voters are bothered by the long queues, this is a problem that is more associated with young voters who are seem reluctant to stand in queues for long periods of time. Most young voters surveyed saw this as one of the deterring factors to go to polling station on election days to cast their votes.

The findings of the study also revealed some of the factors that could influence the adoption of e-voting technologies within the South African context. Understanding these factors could assist the decision makers (in this case the Electoral Management Body) predict and manage under what conditions to make decisions about adopting e-voting technologies. Armed with this information, the Electoral management Body can then makes informed decisions and can the leverage on the opportunities that e-voting technologies presents.

Table 12.1 represents major challenges that emerged from the study.

Since the study used the DoI as a lens of analysis the study used three constructs (relative advantage, complexity, and compatibility) form the framework. Apart from these three constructs the study also reveals other factors including; trust in the technology, availability of infrastructure and resources, digital divide, and awareness.

12.6.1   TRUST IN THE TECHNOLOGY

The findings show that trust in the technology is a likely factor that could influence the adoption of electronic voting. The participants of the study thought security and privacy issues were factors that might prevent them from trusting and therefore adopting electronic voting technologies. Based on their knowledge or experience of other electronic systems that have been affected by security and privacy issue, these participants thought that if e-voting were not secure enough, their voting right could be under threat and their voting information altered or misused by hackers.

12.6.2   AVAILABILITY OF INFRASTRUCTURE AND RESOURCES

The findings point out that the availability of ICT-enabled infrastructure especially in the informal settlements would have a positive influence in

TABLE 12.1 Challenges of the Current Manual Voting System

1. Lack of infrastructure and resources	This is a challenge for the IEC especially within the informal settlements. Often the IEC are forced to improvise resources and infrastructure, which makes it very difficult to run elections in these communities.
2. Immense task of counting of votes	Counting of paper ballots after the elections is usually a tedious job especially when voting goes beyond scheduled times. A delay in releasing election results is often more associated with election violence and always leads to the questioning of the credibility of the results.
3. Illiteracy	Understanding of the ballot papers can be a challenge especially amongst elderly citizens some who are illiterate; these voters usually end up requiring assistance, which sometimes leads to them being coerced into changing their votes.
4. Spoilt ballot papers	This challenge stems from the illiteracy, which sometimes leads to spoilt ballot papers
5. Voter intimidation	Having party representatives at voting stations on voting days could lead to voter intimidation.

the adoption of e-voting technologies. Equal availability of these ICT-enabled infrastructures in both the rural and urban communities increases the intentions of adopting an e-voting system. The availability of ICT-enabled infrastructure makes it easier for the ICE to implement an e-voting system. The findings reveal that the IEC perceive that increased resources would be needed by providing additional skilled staff that would have the knowhow of operating e-voting technologies, and the provision of funds to assist in acquiring these technologies.

12.6.3 DIGITAL DIVIDE

The digital divide is quite evident between the various communities in Cape Town this would make some portion of the citizens to enjoy the conveniences of e-voting technology. The citizens from the rural areas and informal settlements would be excluded from e-voting for the lack of operational capacity or other economic reasons. A large and diverse literature has highlighted the significance of closing the digital divide gap if the adoption technologies are to be successful. Digital divide is the gap between those who use and have access to digital technologies and those who do not (Reddick et al., 2000; Oostveen and den Besselaar, 2004; Bozinis, 2007). Reddick et al. (2000) stresses that the previous definition is not complete, as non-users are not homogenous.

12.6.4   AWARENESS

The adoption of technology can be seen as a process that begins with awareness of the technology and progresses through a series of phases that ends with an appropriate and effective usage. According to Rogers (2003) the innovation-decision process is the process through which an individual (or other decision-making unit) passes from first knowledge about the innovation to formulating an attitude towards it, to a decision regarding adoption or rejection, to implementation of the new idea, and to confirmation of this decision. Figure 12.1 shows the five phases of innovation-decision process; in the Knowledge phase, an individual is exposed to an innovation, a person (or other decision-making unit) first becomes aware of the technology. The study concludes that awareness of technology before its adoption is crucial and may increase the perceived usefulness of the technology, thus contributing to the adoption of the new technology.

The findings of the study revealed 88% of the survey voters were aware of electronic voting technologies in other countries Figure 12.1. The findings also reveal that the IEC are aware of e-voting technologies. The awareness of e-voting in South Africa may increase the perception of relative advantage, compatibility and ease of use amongst the voters thus increasing the chances of them adopting e-voting. The more IEC are aware of e-voting the more they can make amicable decision when implementing an e-voting system should they decide to. This could be either by collaborating with countries that have experience with e-voting.

Table 12.2 provides a summary of the above-discussed factors that could influence adoption of e-voting technologies within the South African context.

TABLE 12.2   Factors that Could Influence the Adoption of e-Voting Technologies in South Africa

DoI Constructs:	
1. Perceived Relative Advantage of e-voting technologies	The findings reveal that the favoritisms of e-voting technologies over paper-based system is based on the perception by participants that e-voting would be convenient in time and costs savings, reduction of human error, increase transparency and accountability, etc.
2. Perceived ease of use (Complexity)	The findings reveal that the IEC are the view that the people under resourced communities might not have the necessary skills to be able to use an e-voting system as they have not been exposed such computerized systems.
3. Compatibility	The findings reveal that those voters who have used other online applications systems like Internet banking and e-filling of taxes have a perception that an e-voting system would not be any different from these applications and are more likely to adopt it because it already fits into their livelihoods.
Other factors include:	
1. Trust in the technology	The findings suggest that trust in an e-voting technology is a likely influence in the adoption process. Trust in the security of the system influences the voters' opinion about e-voting technologies considerably.
2. Availability of infrastructure and resources	The development of basic infrastructure to capture the advantages of the new technologies and communications tool is of vital importance for the implementation of e-voting technologies.
3. Digital divide	The digital divide is quite evident between the various communities in Cape Town this would make some portion of the citizens to enjoy the conveniences of e-voting technology.
4. Awareness	Technology adoption can also be seen as a process that begins with awareness of the technology and progresses through a series of steps that end inappropriate and effective usage.

12.7   DISCUSSIONS

This research has identified seven factors that are measures for characteristics that would influence the adoption of e-voting technology within the South African concept. Three of this factors stem from the Diffusion of innovation theory used in this study as a theoretical lens of analysis; relative advantage (less time at polling stations, reduce the cost of the physical ballot paper and other overheads, ease distribution of electoral materials and reduce electoral delays), according to Carter and Belanger, 2004; Lean et al., 2009, the more e-government services/e-services are perceived to have a relative advantage then more citizens and other stakeholders are likely to adopt and use them.

Gimpel and Schuknecht (2003) found that accessibility makes a significant difference in voter turn. E-voting technology provides a relative advantage in the convenience of casting a vote under less geographic and temporal restrictions. Complexity (ease of use and understanding of e-voting technology) Carter and Belanger (2005) and Phang et al., 2005 in their studies found that ease of use is a significant determinant of the intentions of people with limited skills in the use of computerized application to adopt the innovation. Compatibility (how similar was the e-voting with other innovations available in South Africa for it to be considered compatible with the lifestyle of the voters). Research suggests that citizens who are considered e-savvy and use the Internet regularly to communicate and complete transactions are likely to view e-voting as consistent with the way they like to interact with other entities.

The other four are new factors that emerged from the data collected.

12.7.1 TRUST IN THE TECHNOLOGY

The study's review of literature on trust in technologies revealed that trust is an important factor in adoption of e-government services (Welch et al., 2005; Belanger and Hiller, 2006; Colesca and Dobrica, 2008; Carter and Weerakkody, 2008; Lean et al., 2009). Literature also shows that a voter's perception of security of the electoral process is equally important to the actual security itself. Since procedural security is evident and understandable to voters, it has a comparative advantage when it comes to developing and supporting the social acceptance for the new e-processes (Xenakis and Macintosh, 2005).

12.7.2 AWARENESS

Technology adoption can also be seen as a process that begins with awareness of the technology and progresses through a series of steps that end in appropriate and effective usage. Awareness: - potential users learn enough about the technology and its benefits to decide whether they want to investigate further.

12.7.3   AVAILABILITY OF INFRASTRUCTURE AND RESOURCES

The development of basic infrastructure to capture the advantages of the new technologies and communications tool is of vital importance for the implementation of e-voting technologies. However, an ICT infrastructure does not consist simply of telecommunications and computer equipment. E-readiness and ICT literacy are also necessary in order for people to be able to use and see the benefits of e-governance initiatives like e-voting technologies. Having the education, freedom and desire to access information is critical to e-governance efficacy. Presumably, the higher the level of human development, the more likely citizens will be inclined to accept and use e-governance services like e-voting technologies.

12.7.4   DIGITAL DIVIDE

DiMaggio and Hargittai (2001) expanded the definition of the "digital divide" to "digital inequality" between people who are already online and those who are not. This inequality comes from the inequality in technical means, inequality in skills and inequality in the availability of social network. These inequalities affect the quality of knowledge the user receives in the IT world. The digital inequality rather evident in Cape Town where the study takes place most citizens from the informal settlements may not have the necessary skills to use the ICT applications.

12.8   CONCLUSION AND RECOMMENDATIONS

South Africa over the last two decades has seen a growth in the technological development, which has opened up the possibilities of the adoption of e-voting technologies. As much as this provides various opportunities and benefits it also creates an opening for risks and challenges as well. E-voting may introduce new risks and affect the electoral values such as secrecy of the vote and placing of voting as an observable institution in modern democracies.

The study concluded that for the adoption of e-voting technologies to be successful in South Africa, relative advantage, compatibility and complexity (ease of use) of e-voting technologies should be taken into consideration. How the citizens view e-voting based on this three constructs is very crucial for its success. There are also other factors that have emerged from the literature and the findings that together with these constructs play as important part in the adoption of e-voting in the future.

Literature on other countries with experience in e-voting has shown that the introduction of technology in elections is a challenging project that requires careful deliberation and planning. Literature has also shown that e-voting can greatly reduce direct human control and influence in the electoral process. This provides an opportunity for solving some electoral problems, but also introduces a whole range of new concerns as result of this e-voting usually attracts a lot of criticism and opposition. The recommendation of the study is that the important principles of traditional democratic elections should remain even with the introduction of such a technology if the trust of the voters is to be maintained.

For the Electoral Management Body (IEC) to get citizens to adopt and use an e-voting system, they must genuinely view e-voting as useful to the voters. The system must be efficient and should meet the specific needs of the intended users who are the voters. The IEC also must conduct widespread and attractive awareness campaigns targeting those citizens in rural areas or informal settlements, appropriately informing them about the real benefits (including lower costs, time saving, reducing long queues at polling stations, etc.) if the adoption of e-voting is to be effective. The IEC should also take into consideration the cultural diversity of South Africa, in terms of the literacy level and Internet experience. Moreover, knowledge, resources and support should be provided to the various communities such as, providing computers and Internet access at the community level in public places, especially in areas that have lack of ICT enabled infrastructure.

This research has revealed that although e-voting has many potential benefits over the manual voting system. There should be careful deliberation by decision makers (IEC) before making decision on the adoption of an e-voting system. The IEC must take into consideration all factors that could influence voters both positively and negatively into consideration. This study is of the view that for South Africa to be able to leverage on the opportunities that e-voting technologies present, all factors that may influence the adoption of e-voting revealed in study should be addressed. Otherwise the full potential of e-voting may never be achieved. The study provides empirical support and validates the findings of previous research in the technology adoption field.

KEYWORDS

- **diffusion of innovation**
- **e-voting technologies**
- **interpretivism**

REFERENCES

1. ACE Electoral Knowledge Network 2010 http://aceproject.org/ace-en/focus/e-voting/countries [accessed August 2013].
2. Agarwal, R., Karahanna, E. 1998. "On the Multi-dimensional Nature of Compatibility Beliefs in Technology Acceptance" presented at the *Annual Diffusion of Innovation Group in Information Technology (DIGIT), Helskinki, Finland.*
3. Ajzen, I., Fishbein, M. 1980. *Understanding Attitude and Predicting Social Behavior* Englewood Cliffs: Prentice-Hall.
4. Al-Qeisi, K. 2009. "Analyzing the Use of UTAUT Model in Explaining an Online Behavior: Internet Banking Adoption," *Doctoral dissertation*, Department of Marketing and Branding, Brunel University.
5. Anthony, K. Asare, Thomas, G. Brashear Alejandro, Elad Granot, Vishal Kashyap, 2011. "The role of channel orientation in B2B technology adoption," *Journal of Business and Industrial Marketing*, 26(3), 193–201.
6. Babbie, E., Mouton, J. 2001. *The Practice of Social Research*. Cape Town: Oxford University Press South Africa.
7. Bagozzi, Richard. 1982. A Field Investigation of Causal Relations among Cognitions, Affect, Intentions, and Behavior, Journal of Marketing Research, 19, 562–584.
8. Belanger, F., Hiller, J., 2006. A framework for e-government: privacy implications. Business Process Management Journal 12 (1), 48–60.
9. Bhattacherjee, A., 2012. Social Science Research: principles, methods, and practices. Available at: http://scholarcommons.usf.edu/cgi/viewcontent.cgi?article=1002&context =oa_textbooks[]accessed May 2013].
10. Braun, V., Clarke, V. 2006. Using thematic analysis in psychology. *Qualitative Research in Psychology*, 3, 77–101.
11. Brown, K. M. 1999. Diffusion of Innovations. http://hsc.usf.edu/~kmbrown/Diffusion_of_Innovations_Overview.htm[accessed October 2013].

12. Burns, N., Grove, S. 2007. Understanding nursing research: Building an evidence-based practice. 4. St. Louis, MO: Elsevier; pp. 60–96.

13. Burton-Jones, A., Hubona, G. S. 2005 Individual differences and usage Behavior: Revisiting a Technology Acceptance Model Assumption. In *ACM SIGMIS Database*, 17, 1825–1834.

14. Carter, L., Bélanger, F., 2005. The utilization of e-government services: citizen trust, innovation and acceptance factors. *Information Systems Journal*, 15(1), 5–25. Available at: http://onlinelibrary.wiley.com/doi/10.1111/j.1365-2575.2005.00183.x/full [Accessed April, 2013].

15. Carter, L., Belanger, F., 2004. The influence of perceived characteristics of innovating on e-government adoption. *Electronic Journal of E-government*, 2(1),.11–20. Available at: http://www.ejeg.com/issue/download.html?idArticle=18 [accessed April, 2013].

16. Carter, L., Weerakkody, V., 2008. E-government adoption: A cultural comparison. *Information Systems Frontiers*, 10(4), 473–482. Available at: http://www.springer-link.com/index/10.1007/s10796-008-9103-6 [Accessed April, 2013].

17. Cetinkaya, O., Centinkaya, D. 2007. *"Towards Secure E-Elections in Turkey: Requirements and Principles" International Workshop on Dependability and Security in e-Government (ARES'07), Vienna, Austria, pp. 903–907, 10–13 April 2007. http://www.ceng.metu.edu.tr/~corhan/Papers/desegov07.pdf.* [Accessed October, 2013].

18. Chiu, Yen-Ting Helena, Fang, Shih-Chieh, Tseng, Chuan-Chuan, 2010. Early versus potential adopters: Exploring the antecedents of use intentions in the context of retail service innovations. *International Journal of Retail and Distribution Management.* Vol. 38, No. 6, pp. 443.

19. Colesca, S., Dobrica, L. 2008. Adoption and use of e-government services: the case of Romania, Journal of Applied Research and Technology, Universidad Nacional Autonoma Mexico, 6(3), 204–217.

20. Davis, F. D. 1986. "A Technology Acceptance Model for Empirically Testing New End-User Information Systems: Theory and Results," *Doctoral dissertation, Sloan School of Management, Massachusetts Institute of Technology.*

21. deVaus, D. A. 2002. Surveys in Social Research, 5th edition, Allen & Unwin, Crows Nest, Australia, pp. 379.

22. DiMaggio, P., Hargittai, E., Neuman, W. R., Robinson, J. P. 2001. Social implications of the Internet. Annual Review of Sociology, 27(1), 307–336.

23. Fereday, J., Muir-Cochrane, E., 2006. Demonstrating rigor using thematic analysis: A hybrid approach of inductive and deductive coding and theme development. International *Journal of Qualitative Methods, 5(1)* http://www.ualberta.ca/~iiqm/backissues/5_1/pdf/fereday.pdf [Accessed October, 2013].

24. Fichman, R. G., 1992. Information technology diffusion: a review of empirical research. *Proceedings of the International Conference on Information Systems*, p. 195.

25. Foxall, G. R., 2007. Behaviorism Gordon, R. Foxall., 55, 1–55.

26. Gianluca, M. 2006. E-Governance in Africa, from Theory to Action: A Practical-Oriented Research and Case Studies on ICTs for Local Governance. In *Proceedings of the 2006 International Conference on Digital Government Research*; ACM: New York, NY, USA; 2006; Volume 151: 209–218.

27. Goldsmith, B., 2011. *Electronic Voting and Counting Technologies*, International Foundation for Electoral Systems (IFES).

28. Gupta, V. 2011. E-voting Move to Intelligent Suffrage. *SETLabs Briefings,* 9(2), 3–9.

29. Ham, C. 2007. Green Labeling: Investigation into the marketing of FSC certified timber along the domestic timber value chain in South Africa. MBA thesis work, Stellenbosch University South Africa.

30. IEC. 2009a. Voting questions and answers. http://bit.ly/PFodqu [Accessed June 2013].

31. Islam, M. S., 2008. Towards a sustainable e-Participation implementation model. European Journal of ePractice, 5(10).

32. Jansen, Harrie, 2010. The Logic of Qualitative Survey Research and its Position in the Field of Social Research Methods [63 paragraphs]. Forum Qualitative Sozialforschung. Forum: Qualitative Social Research, 11(2), Art. 11, http://nbn-resolving.de/urn:nbn:de:0114-fqs1002110.[Accessed April, 2013].

33. Juntumaa, M. 2011. Putting consumers' IT adoption in context: failed link between attitudes and behavior. Aalto university publications series. Doctoral Dissertation 14/2011.

34. Lean, O. K., Zailani, S., Fernando, Y. 2009. "Factors influencing intention to use e-government services among citizens in Malaysia," International Journal of Information Management, 29(6), 458–475.

35. Macintosh, A., 2004. Characterizing e-participation in policy-making. *System Sciences, 2004. Proceedings of the 37th*, 00(C), pp. 1–10. Available at: http://ieeexplore.ieee.org/xpls/abs_all.jsp?arnumber=1265300 [Accessed June, 2013].

36. Mathieson, K., Peacock, E., Chin, W. W. 2001. "Extending the Technology Acceptance Model: The Influence of Perceived User Resources," The DATA BASE for Advances in Information Systems, 32(3), 86–112.

37. Moore, G., Benbasat, C. I. 1991. Development of instruments to measure the perceptions of adopting an information technology. Inform. Syst. Res 2, 192–222.

38. Ozdemir, S., Trott, P., Hoecht, A. 2008. Segmenting Internet banking adopter and non-adopters in the Turkish retail banking sector. *International Journal of Bank,* 26(4).

39. Qadah, G., Taha, R., 2007. Electronic voting systems: Requirements, design, and implementation. Computer Standards Interfaces, 29(3), 376–386. Available at: http://linkinghub.elsevier.com/retrieve/pii/S0920548906000754. [Accessed August, 2013].

40. Rogers, E. M. 2003. Diffusion of innovation (5th ed.). New York: Free Press.

41. Ruane, J. M. 2005. *Essentials of Research Methods: A Guide to Social Science Research.* Malden: Blackwell Publishing.

42. Straub, E. T., 2009. Understanding Technology Adoption: Theory and Future Directions for Informal Learning. *Review of Educational Research,* 79(2), 625–649. Available at: http://rer.sagepub.com/cgi/doi/10.3102/0034654308325896 [Accessed June, 2013].

43. Svensson, J. Leenes, R., 2003. E-voting in Europe: Divergent democratic practice. *Information Polity,* 8(1), 3–15.

44. Taylor, S., Todd, P. 1995a. Decomposition and crossover effects in the theory of planned behavior: A study of consumer adoption intentions. *International Journal of Research in Marketing,* 12(2), 137–155.

45. Thakur, S. 2013. Overview of e-voting cross national experience. *Seminar on Electronic Voting and Counting Technologies, Cape Town.* March 2013.

46. Venkatesh, V., Morris, M. G., Davis, G. B., Davis, F. D. 2003. User acceptance of information technology: toward a unified view. *MIS Quarterly* 27(3), 425–478.

47. Welch, E. W., Hinnant, C. C., Moon, M. J. 2005. Linking Citizen Satisfaction with E-government and Trust in Government. *Journal of Public Administration Research and Theory,* 15(3), 371–391.

48. Wells, John, Campbell, Damon, Valacich, Joseph and Featherman, Mauricio. 2010. The Effect of Perceived Novelty on the Adoption of Information Technology Innovations: A Risk/Reward Perspective. Decision Science, 41(4), 813–843.

49. Xenakis, A., Macintosh, A., 2005. Procedural security and social acceptance in e-voting. In, *2005. HICSS'05. Proceedings of the* 1–9. Available at: http://ieeexplore. ieee.org/xpls/abs_all.jsp?arnumber=1385476 [Accessed June, 2013].

50. Yousafzai, S. M., Foxall, G. R., Pallister, J. G. 2010. "Explaining Internet banking behavior: theory of reasoned action, theory of planned behavior, or technology acceptance model?" *Journal of Applied Social Psychology*, 40 (5), 1172–202.

INDEX